Nematology Research in China Vol.9

中国线虫学研究

（第九卷）

彭德良 等 主编

中国农业科学技术出版社

图书在版编目(CIP)数据

中国线虫学研究．第九卷 /彭德良等主编．--北京：中国农业科学技术出版社，2023.6

ISBN 978-7-5116-6307-8

Ⅰ.①中…　Ⅱ.①彭…　Ⅲ.①线虫动物-研究-中国　Ⅳ.①Q959.17

中国国家版本馆 CIP 数据核字(2023)第 107698 号

责任编辑　姚　欢
责任校对　王　彦
责任印制　姜义伟　王思文

出 版 者　中国农业科学技术出版社
北京市中关村南大街 12 号　邮编：100081
电　　话　(010) 82106631(编辑室)　(010) 82109702(发行部)
(010) 82109709(读者服务部)
网　　址　https://castp.caas.cn
经 销 者　各地新华书店
印 刷 者　北京科信印刷有限公司
开　　本　185 mm×260 mm　1/16
印　　张　16.25
字　　数　350 千字
版　　次　2023 年 6 月第 1 版　2023 年 6 月第 1 次印刷
定　　价　80.00 元

《中国线虫学研究（第九卷）》

编辑会

前　言

我国植物线虫科技工作者克服三年新冠疫情的影响，坚持面向世界农业科技前沿、面向国家重大需求、面向现代农业建设主战场、面向人民生命健康的要求，坚持深入生产实际，从生产中发现重大线虫病害问题，着眼于解决我国农业生产中的线虫病害发生危害日益严重的现实问题，在农作物重大线虫领域的基础研究和应用研究两个方面均取得很大的进展，研究深度和广度都有进一步的提升，受到国内外同行的重视和关注，在植物线虫致病分子机制、线虫早期诊断与检测、生物防治、种质资源抗性、综合治理技术与策略等方面取得长足进步，对发展我国植物线虫学科和人才队伍建设具有重要推动作用。

《中国线虫学研究（第九卷）》共收集了112篇研究论文、摘要简报和综述，内容涉及农林植物线虫发生分布、诊断与检测监测技术、生物学、分子生物学、生物防治、化学防治与综合治理，以及昆虫病原线虫的应用基础研究等线虫学研究的各个方面，反映了近三年来我国线虫学工作者在相关领域的基础理论、应用技术以及病害综合治理方面的最新研究成果。

本书出版得到了中国农业科学院植物保护研究所、植物病虫害综合治理全国重点实验室、湖南农业大学大学、佛山市盈辉作物科学有限公司、广东真格生物科技有限公司、拜耳作物科学（中国）有限公司、日本石原产业株式会社等单位的资助与支持，中国农业科学技术出版社对本书的出版给予了大力帮助。在此，我们一并表示衷心的感谢！

在编辑文稿时，本着文责自负的原则，按照论文规范性要求进行收录整理，对个别文句进行了修订。由于时间仓促，疏漏和不足之处难免，敬请作者和读者批评指正。

编　者

2023年6月

目　录

安徽省首次发现肾形肾状线虫危害瓜蒌 …………………………………………………… 朱　衎等（1）
复配剂 WV32、WV33 及缓释颗粒剂 FW203 防治植物根结线虫病研究 ……………………… 张怡萱等（12）
福建茶树根际 1 种针线虫的种类鉴定 ……………………………………………………… 丁怡倩等（16）
山东省马铃薯根结线虫病病原鉴定和侵染动态观察 ………………………………………… 代明明等（21）
食用菌菌渣防治番茄根结线虫病研究初探 …………………………………………………… 覃丽萍等（32）
福建省水稻潜根线虫的种类鉴定及种群结构 ………………………………………………… 黄泓晶等（34）
A New Record of *Laimaphelenchus spiflatus* Gu *et al.*, 2020 (Rhabditida: Aphelenchoididae) from Shanxi Province, with Proposal of *L. liaoningensis* Song *et al.*, 2020 as a Junior Synonym …………………………………………………………………… Wang Liyi 等（40）
Advances in Morphological and Molecular Identification of PWN ……………………… Gu Jianfeng 等（48）
Cloning and Characterization of microRNA396 and Its Targets in *Cucumis metuliferus* Resistant to *Meloidogyne incognita* …………………………………………………………… Ye Deyou 等（54）
Development and Application of Recombinase Polymerase Amplification assay for rapid and visual Detection of *Pratylenchus coffeae* ………………………………………………… Wu Caiyun 等（55）
Effects of long-term Consecutive Monoculture of Yam on the Fungal Community and Function in the Rhizospheric Soil …………………………………………………………………… Yao Jian 等（56）
GmLecRKs-GmCDL1-GmMPK3/6 通路在大豆孢囊线虫抗性中的功能研究 ……………………… 张　磊等（57）
Host-induced Silencing of a Nematode Chitin Synthase Gene Decreases Abundance of Rhizosphere Fungal Community While Enhancing *Heterodera glycines* Resistance of Soybean ………… Tian Shuan 等（58）
Identification of key MicroRNAs in *Cucumis metuliferus* under *Meloidogyne incognita* Stress … Ye Deyou 等（60）
Isolation and Characterization of Streptomycetes Strains JXGZ01 with Nematicidal Activity Against Root-knot Nematode, *Meloidogyne incognita* ………………………………… Xu Xueliang 等（61）
Population Dynamics of *Meloidogyne graminicola* in Soil in Different Types of Rice Agroecosystems in Hunan Province, China ………………………………………… Yang Zhuhong 等（62）
Effect of Different Initial Population Densities of *Meloidogyne graminicola* on Growth and Yield of Upland Rice cv. Hanyou73 ……………………………………………………… Yang Zhuhong 等（64）
Recombinase Polymerase Amplification Coupled with CRISPR-Cas 12a Technology for Rapid and Highly Sensitive Detection of *Heterodera avenae* and *Heterodera filipjevi* …………… Shao Hudie 等（65）
Resistance to *Heterodera filipjevi* in Wheat: An Emphasis on Classical and Modern Management Approaches …………………………………………………………… Neveen Atta Elhamouly 等（66）
Thirty Years of Plant Parasitic Nematode Research in China ……………… Lizzete Dayana Romero Moya 等（67）
miRNA 在植物与线虫互作及细菌诱导植物免疫中的研究进展 ………………………………… 杨　帆等（68）
Phased 的 T2T 基因组揭示异源多倍体线虫起源模式为未减数配子与单倍体配子的杂交 …… 代大东等（72）
Volutella ciliate Q7 对马铃薯腐烂茎线虫的作用 ……………………………………………… 马　娟等（73）
百岁兰曲霉对水稻干尖线虫的作用机制研究 ………………………………………………… 贾建平等（74）
3 株不同来源淡紫紫孢菌对植物线虫的活性 ………………………………………………… 吴　艳等（75）
7 种药剂对粗茎秦艽根结线虫病的防治效果 ………………………………………………… 李云霞等（76）

孢囊线虫效应蛋白 Hg11576 靶向大豆 GmHIR1 抑制寄主免疫的分子机制研究 …………… 姚 珂等（77）
马铃薯主栽品种及育种资源抗金线虫 *H*1 基因分子鉴定 ……………………………… 江 如等（78）
8 种植物挥发物对马铃薯金线虫卵孵化的抑制效果 …………………………………… 姚汉央等（79）
马铃薯金线虫群体遗传分化及毒性特征分析 …………………………………………… 江 如等（80）
马铃薯金线虫（*Globodera rostochiensis*）生物学特征研究 ……………………………… 邓春菊等（81）
马铃薯品种资源对马铃薯金线虫耐病性的筛选 ………………………………………… 易 军等（82）
马铃薯不同品种对马铃薯孢囊线虫的抗性鉴定 ………………………………………… 兰世超等（83）
杀线剂田间防控马铃薯金线虫（*Globodera rostochiensis*）效果 ……………………… 黄立强等（84）
腐烂茎线虫 ISSR-PCR 反应体系的建立与优化 ………………………………………… 韩 变等（86）
腐烂茎线虫类毒液过敏原蛋白基因 *DdVAP*2 功能研究 ………………………………… 常 青等（88）
马铃薯腐烂茎线虫脂肪酸与视黄醇结合蛋白家族基因的全基因组水平鉴定及其功能
分析 ……………………………………………………………………………………… 王 喆等（89）
马铃薯不同品种对腐烂茎线虫的室内抗病性评价 ……………………………………… 霍宏丽等（91）
大豆孢囊线虫对酸碱盐化学信号响应的分子调控机制 ………………………………… 姜 野等（92）
大豆孢囊线虫新种群 X12 致病基因分析 ………………………………………………… 都文振等（94）
大豆孢囊线虫（*Heterodera glycines*）HgUIM1 的功能研究 ………………………… 张刘萍等（95）
禾谷孢囊线虫效应子 Ha17370 靶向植物 CBSX ………………………………………… 张笑寒等（97）
杂交小麦品种对小麦孢囊线虫的抗性鉴定 ……………………………………………… 于敬文等（98）
南方根结线虫分泌粒蛋白基因 *MiSCG5L* 功能研究 …………………………………… 叶梦迪等（99）
南方根结线虫转录因子 Mi_03370 影响其发育及侵染能力的功能研究 ……………… 廖宇澄等（100）
象耳豆根结线虫 MeMSP1 效应子与植物谷胱甘肽转移酶家族蛋白（GSTs）互作帮助
线虫寄生 ………………………………………………………………………………… 陈永攀等（101）
象耳豆根结线虫克服 Mi-1 抗性基因相关效应子的筛选 ……………………………… 曾媛玲等（102）
爪哇根结线虫危害白及 ………………………………………………………………… 杨艳梅等（103）
西班牙根结线虫侵染危害柚子根部的首次鉴定报道 ………………………………… 王 丽等（105）
陕西省丹参根结线虫种类鉴定及侵染能力测定 ………………………………………… 潘 嵩等（106）
国内首次发现南方根结线虫侵染黄花菜 ………………………………………………… 魏佩瑶等（108）
广西水稻产区稻菜轮作田水稻根结线虫病发生情况调查及鉴定 ……………………… 黄金玲等（109）
首次在安徽省发现玉米短体线虫侵染危害玉米 ………………………………………… 王 硕等（110）
广西火龙果根结线虫病的病原鉴定 ……………………………………………………… 伍朝荣等（111）
河南省白术根结线虫病的病原鉴定 ……………………………………………………… 许相奎等（112）
甘肃省 3 种中药材根结线虫病病原鉴定 ………………………………………………… 石明明等（113）
基于构建腐烂茎线虫全基因组转录调控网络揭示植物线虫动态调控模型及龄期特异调控
模块 ……………………………………………………………………………………… 丛子文等（115）
基于线粒体 *COI* 序列分析中国北方地区腐烂茎线虫群体遗传结构 ………………… 李云卿等（117）
接种松材线虫后红松转录组分析 ………………………………………………………… 曹业凡等（118）
烟草过表达根结线虫异分支酸水解酶基因抑制 SA 介导的免疫 ……………………… 方辰杰等（120）
土壤样本中菲利普孢囊线虫 TaqMan 实时荧光 PCR 快速检测及定量分析 ………… 竖晋卓等（121）
拟禾本科根结线虫 qPCR 检测研究 ……………………………………………………… 朱 衔等（122）
水稻干尖线虫对不同水稻品种的趋性比较 ……………………………………………… 宛 宁等（124）
水稻抗拟禾本科根结线虫基因筛选和鉴定 ……………………………………………… 王东伟等（125）
水稻类钙调素蛋白 OsCML 在调控抗拟禾本科根结线虫中的功能研究 ……………… 魏 英等（126）

水稻生物钟基因 *CCA1* 在拟禾本科根结线虫与寄主互作中的功能研究 …………………… 朱诗斐等（128）
甜菜孢囊线虫调控寄主基因可变剪切的效应蛋白筛选及作用机制初探 …………………… 张梦涵等（129）
番茄 *LeMYB330* 基因过表达植株广谱抗病性分子机理研究 …………………… 康志强等（130）
番茄候选感病基因 *T106* 的全长克隆和功能验证 …………………… 闫曦蕊等（131）
负调控因子 OsWD40-193 与 OseEF1A1 的相互作用抑制水稻对尖细潜根线虫（*Hirschmaniella muccronata*）的抗性 …………………… 单崇蕾等（132）
钾离子转运蛋白 OsHAK 在水稻抗拟禾本科根结线虫中的功能研究 …………………… 张家芊等（134）
不同密度玉米孢囊线虫对玉米产量损失的研究 …………………… 李荣超等（135）
玉米孢囊线虫孵化特性及不同品种玉米对其抗性研究 …………………… 王　媛等（136）
乙烯信号通路参与大豆孢囊线虫抗性基因 *GmAAT* 表达调控机制研究 …………………… 何　龙等（137）
植物寄生线虫取食管调控取食分子量大小机制初探 …………………… 赵　薇等（138）
丁香酚抑杀南方根结线虫三种生测方法比较研究 …………………… 陈宗雄等（139）
不同品种根系分泌物对马铃薯金线虫卵孵化的影响 …………………… 余曦玥等（140）
和硕县加工辣椒根结线虫病病原鉴定及其防治药剂筛选 …………………… 曹　铭等（141）
寄主挥发物 BHT 对小卷蛾斯氏线虫觅食策略的影响 …………………… 唐凡希等（143）
健康与根结线虫病柑橘根际土壤微生物群落对比分析 …………………… 杨姗姗等（144）
咖啡短体线虫不同种群对模式植物本氏烟的寄生性和致病性测定 …………………… 牛文龙等（145）
喹唑啉类化合物作为新型杀线虫骨架的发现 …………………… 王　盛等（146）
利用抗坏血酸过氧化物酶揭示生防放线菌 XFS-4 对大豆孢囊线虫的抗性研究 …………… 项　鹏（147）
Study on Endophytic Fungi Identification and Nematicidal Activity of *Chaetomium ascotrichoides* 1-24-2 from Five-year-old Pine Wood Nematode-resistant Masson Pine（*Pinus massoniana*） …………………… Zheng Lijun 等（148）
莓实假单胞 Sneb1990 通过多种途径增强番茄抗南方根结线虫 …………………… 王　帅等（150）
南方根结线虫对不同用途玉米苗期侵染观察 …………………… 高海英等（151）
内蒙古地区腐烂茎线虫的适生区预测 …………………… 杨　帆等（153）
山苍子精油对南方根结线虫的防效评价 …………………… 王　丽等（154）
蔬菜根结线虫高效生物熏蒸植物筛选 …………………… 李秀花等（155）
烟酰胺酶抑制剂在大豆对孢囊线虫的抗性中的作用机制及应用研究 …………………… 陈璐莹等（157）
野艾蒿根系孢囊线虫种类鉴定及其孵化特性 …………………… 陈京环等（158）
野生豆孢囊线虫侵染对不同抗性大豆品种生理生化指标的影响研究 …………………… 郑刘春等（160）
西甜瓜根际微生态分析及其根结线虫生防菌株筛选 …………………… 豆浓笑等（161）
淡紫紫孢菌 PLHN 与甲维盐减量复配防治象耳豆根结线虫的作用机理 …………………… 杨紫薇等（162）
党参根结线虫病优势生防菌长枝木霉 TL16 作用方式探究 …………………… 张　洁等（164）
一种提高大豆气生根转化效率的技术及其在大豆孢囊线虫抗性分析中的应用研究 ………… 王　靓等（166）
一株巨大芽孢杆菌对拟禾本科根结线虫防治效果研究 …………………… 叶　姗等（167）
种子处理对东北地区水稻干尖线虫病防治作用研究 …………………… 杨　芳等（168）
一个松材线虫病疫点拔除的成功案例 …………………… 胡先奇等（169）
昆虫病原线虫外代谢物对觅食行为的调控研究 …………………… 张　忱等（172）
昆虫病原线虫分泌蛋白对寄主激活的响应及致病功能分析 …………………… 常豆豆等（173）
新型1,2,4-噁二唑类杀线虫剂的设计与合成 …………………… 张　琪等（175）
一种新病害——空心菜根腐病的病原鉴定 …………………… 秦　玲等（176）
新型不饱和酮类化合物的杀腐烂茎线虫活性、构效关系和作用机制 …………………… 周搏航等（177）

大蒜腐烂茎线虫对大蒜的损害阈值 …… 成泽珺等（178）
淡紫紫孢菌活性孢子储存技术研究 …… 张雯欣等（179）
拟禾本科根结线虫效应蛋白 MgCBP1 在线虫寄生中的作用 …… 黄春晖等（180）
OsBet v1 蛋白通过木聚糖酶抑制蛋白介导水稻对拟禾本科根结线虫的抗性 …… 王　婧等（181）
香芹酚防治象耳豆根结线虫及缓解噻唑膦药害的作用研究 …… 龙昌文等（182）
Challenges of Research in Plant Nematology in Colombia …… Lizzete Dayana Romero Moya 等（183）
生物源农药防治根结线虫病研究进展 …… 包玲凤等（185）
矛线目线虫的多样性研究概述 …… 李红梅等（194）

中国线虫学研究（第九卷）Nematology Research in China（Vol. 9）：1-11

安徽省首次发现肾形肾状线虫危害瓜蒌*

朱　衎**，郑思远，严　丽，方　圆，吴慧平，潘月敏，鞠玉亮***
（安徽农业大学植物保护学院，植物病虫害生物学与绿色防控安徽普通高校重点实验室，合肥　230036）

摘　要：肾形肾状线虫是作物根部半内寄生植物线虫，被列为危害最严重的十大植物线虫之一。本文对安徽省新发生瓜蒌肾形线虫病的病原进行鉴定并对其生活史进行初步研究。利用形态学和分子生物学将病原线虫鉴定为肾形肾状线虫 *Rotylenchulus reniformis* Linford & Oliveira，1940，系统发育分析表明其属于 A 型。肾形肾状线虫在瓜蒌根部完成生活史需要 24d，每条雌虫平均产卵 64 粒，繁殖系数 $R=11.02\pm0.62$。本文系首次报道肾形肾状线虫在安徽省发生危害，瓜蒌为肾形肾状线虫的新寄主。本研究为进一步掌握肾形肾状线虫的分布区域及潜在寄主，为该病害的综合防控提供了理论依据。

关键词：瓜蒌；肾形肾状线虫；形态学特征；系统发育；生活史

First Report of *Rotylenchulus reniformis* Infecting *Trichosanthes kirilowii* in Anhui Province, China*

Zhu Kan**, Zheng Siyuan, Yan Li, Fang Yuan, Wu Huiping, Pan Yuemin, Ju Yuliang***
(*School of Plant Protection, Anhui Agricultural University, Key Laboratory of Biology and Sustainable Management of Plant Disease and Pests of Anhui Higher Education Institutes, Anhui Agricultural University, Hefei*　230036, *China*)

Abstract: *Rotylenchulus reniformis* is a semi-endoparasite in crop roots, which is known as one of the top ten most serious plant nematodes. In this paper, the pathogen of *Trichosanthes reniformis* disease newly occurred in Anhui province was identified and its life history was preliminarily studied. The pathogen was identified to be *R. reniformis* Linford & Oliveira, 1940 with morphology and molecular biology, and the phylogenetic analyses showed that it belonged to type A. The life cycle of *R. reniformis* were 24 days in the root of *Trichosanthes kirilowii*. Each female lays 64 eggs on average, and the reproduction coefficient is $R=11.02\pm0.62$. This is the first report on the occurrence of *R. reniformis* in Anhui province, and it is the first time to confirm that *T. kirilowii* is a new host of *R. reniformis*. This study provides a theoretical basis for further understanding the distribution area and potential host of *R. reniformis*, and for the comprehensive prevention and control of the disease.

Key words: *Trichosanthes kirilowii*; *Rotylenchulus reniformis*; Morphological characteristics; Phylogenetic analysis; Life cycle

---

* 基金项目：安徽省重点研究与开发计划（202204c06020028）；国家自然科学基金（31801714）
** 第一作者：朱衎，硕士研究生，从事植物线虫研究。E-mail：1253239065@ qq. com
*** 通信作者：鞠玉亮，副教授，从事植物线虫研究。E-mail：juyull@ 163. com

瓜蒌 *Trichosanthes kirilowii* Maxim. 属葫芦科栝楼属多年生草质藤本植物，其果实、种皮、种子、根均可入药，兼具药用、食用、保健功能，经济价值较高[1]。目前，瓜蒌是当前农业结构调整中重要的特色经济作物之一，在安徽、山东、江苏等地均有大规模种植。植物线虫病害是严重制约瓜蒌生产的主要土传病害，且随着瓜蒌种植年限的增加危害愈发严重[2]。肾形肾状线虫 *Rotylenchulus reniformis* Linford & Oliveira，1940 是作物根部半内寄生植物线虫，通常在热带和亚热带地区发生危害，与根结线虫等并称为危害最严重的世界十大植物线虫[3]。肾形肾状线虫寄主范围广泛，可危害豆科、葫芦科、茄科、十字花科、旋花科、菊科、禾本科、天南星科、大戟科、锦葵科、棕榈科、苋科、番荔枝科、木棉科、罂粟科等 300 多种植物[4]。肾形肾状线虫生活史较特别，在卵内发育成的 2 龄幼虫进入土壤，在不取食的状态下经 3 次蜕皮发育为未成熟雌虫（蠕虫形雌虫）和雄虫，蠕虫形雌虫虫体前部 1/3 处侵入植物根内取食并建立类似孢囊线虫的取食位点，根外部虫体逐渐膨大，经 7～9d 最终发育为成熟的肾形雌虫[5]。肾形肾状线虫侵染根系，导致根系坏死、萎缩、须根减少，植物地上部分表现矮小、黄化等症状，可导致作物减产 40%～60%[3]。

目前，肾形肾状线虫已在我国广东、上海、湖北、海南、四川、广西、福建、山东和浙江等地的蔬菜或园林植物上发生危害[6]。吴慧平等[7]在开展安徽省盆景花卉根际植物线虫类群调查工作中，在铜陵市南天竹 *Nandina domestica* 根际土壤中分离到肾形线虫属线虫，未鉴定到具体种，而后尚未发现肾形肾状线虫在安徽省农作物上发生危害。笔者于 2020 年和 2021 年对安徽省部分瓜蒌产区植物线虫病害进行调查，发现除南方根结线虫 *Meloidogyne incognita* 危害外，在部分瓜蒌产区发现肾形线虫，且危害严重。基于此，本研究在前期调查的基础上，利用形态学和分子生物学手段对安徽省瓜蒌肾形线虫病的病原进行种类鉴定，并对其遗传进化进行分析，利用室内接种的方法对其生活史进行初步研究。

1 材料与方法

1.1 田间调查与样品采集

分别于 2020 年和 2021 年对安徽省部分瓜蒌产区肾形线虫病害进行调查。采用田间五点取样法，并有针对性地选择长势较弱的瓜蒌，剪去瓜蒌植株地上部分，挖取瓜蒌根系，同时取根际 15cm 深的土壤 1kg，将瓜蒌根系和根际土壤带回实验室进一步进行线虫的分离鉴定。

1.2 线虫分离及形态学鉴定

将瓜蒌根际土壤充分混匀，取 100mL 土壤，利用贝曼漏斗法分离肾形线虫各龄期幼虫、蠕虫形雌虫及雄虫，并统计虫口密度。肾形线虫的形态学鉴定主要基于蠕虫形雌虫及雄虫的形态，临时玻片制作及形态测计参考张燕等[8]的方法并稍作改动。

1.3 分子生物学鉴定

单条雌虫 DNA 提取参考邓艳凤等[6]的方法并稍作改动：体视显微镜下在瓜蒌根部挑取单条雌虫，置于载玻片上用 ddH_2O 清洗 2 次，置于含 16μL ddH_2O 的 PCR 管中；液氮处理 5min，65℃水浴 5min，重复 2 次；加入 4μL 5×Lysis Buffer，65℃处理 1h，95℃处理 10min，-20℃保存备用。

利用特异性引物 Rr-F3/Rr-B3 对本研究所采集 7 份样品进行分子生物学鉴定，阳性对照为肾形肾状线虫由南京农业大学植物线虫实验室惠赠，阴性对照分别为南方根结线虫

M. incognita、北方根结线虫 *M. hapla*、爪哇根结线虫 *M. javanica*、拟禾本科根结线虫 *M. graminicola* 和 ddH_2O[9]。PCR 反应体系：DNA 模板 2μL，10μmol/L 引物 Rr-F3/Rr-B3 各 0.5μL，12.5μL 2×Taq PCR MasterMix，ddH_2O 补足至 25μL。PCR 反应程序：94℃预变性 3min；94℃变性 30s，58℃退火 30s，35 个循环，72℃延伸 30s；72℃延伸 10min。扩增产物在 2%琼脂糖凝胶上电泳检测。

1.4 系统发育分析

选择肾形肾状线虫 ITS、18S、28S、mtDNA-CoI 基因引物 18S/26S[10]、18S-A/18S-S3[11]、D2A/D3B[12]、JB3/JB5[13]进行 PCR 扩增。PCR 产物电泳检测后，将条带亮且与目标基因大小一致的产物送至南京擎科生物科技有限公司测序。测序结果在 NCBI 数据库中进行 BLAST 比对，并下载 NCBI 数据库中的相关线虫序列。利用 MEGA11 的邻接法（Neighbor-joining method）构建系统进化发育树，采用 Boostrap 值分析分支聚类的可靠性。

1.5 室内接种试验

体视显微镜下在瓜蒌根部挑取肾形肾状线虫卵块，并于 26℃培养箱中孵化，3d 后收集 2 龄幼虫并配制 2 000 条/mL 的线虫液，用于室内接种试验。取瓜蒌种子机械破壳后于温水中催芽，5d 后播种于含灭菌土壤（黏土和沙土按 2∶1 混合）的花盆中，置于 27℃温室中培养。瓜蒌苗培养 10d 后每盆接种 2 000 条/mL 的线虫液 2.5mL，即 5 000 条肾形肾状线虫 2 龄幼虫，接种后继续置于温室中培养，定期浇水管理。线虫接种 30d 后，统计根部卵块数量、每个卵块含卵量、根部肾形肾状线虫量、土壤中肾形肾状线虫数量，计算繁殖系数 $R=P_f/P_i$（R 为繁殖系数，P_f为最后线虫数量，P_i为初始接种线虫数量）。此外，瓜蒌接种肾形肾状线虫后，每 2d 取 1 次样，每次取 5 棵瓜蒌，用清水轻轻洗去根部泥沙，用次氯酸钠-酸性品红染色[14]，观察线虫侵染及发育情况。

2 结果与分析

2.1 瓜蒌肾形肾状线虫病危害症状

通过对安徽省部分瓜蒌产区肾形线虫病调查发现，该病害主要在巢湖市瓜蒌种植区发生。部分田块发生严重，肾状属线虫数量高达 1 050 条/100mL 土壤，导致瓜蒌大面积枯萎死亡，严重威胁瓜蒌产量和品质（图 1a）。感染肾状属线虫的瓜蒌地上部生长缓慢、植株矮小，且在瓜蒌生长后期呈明显的早衰现象，高温、干旱条件下瓜蒌易枯死。肾状属线虫侵染瓜蒌根系，老根表现畸形、肿大，严重时出现坏死症状（图 1b）；新生根须根减少，根部呈现明显成熟雌虫和胶质卵块，根系局部出现坏死症状（图 1c-e）。

表 1 瓜蒌肾形线虫病调查

采集地	土壤类型	采集时间/年-月	土壤中线虫数量/[条/（100mL）]
巢湖市中庙镇小李村 1	黏质土	2021-07	140
巢湖市中庙镇小李村 2	黏质土	2021-07	150
巢湖市中庙镇小李村 3	黏质土	2021-07	1 008

（续表）

采集地	土壤类型	采集时间/年-月	土壤中线虫数量/［条/（100mL）］
巢湖市柘皋镇驷马村	黏质土	2021-09	41
巢湖市柘皋镇大路袁村 1	黏质土	2021-09	1 050
巢湖市柘皋镇大路袁村 2	黏质土	2021-09	428
巢湖市柘皋镇大路袁村 3	黏质土	2021-09	18
潜山市梅城镇河湾村	沙质土	2020-09	—
潜山市梅城镇平桥村	沙质土	2020-09	—
潜山市黄铺镇望虎村	沙质土	2020-05	—
临泉县杨桥镇甄庄村	黏质土	2020-06	—
临泉县杨桥镇郭沟村	黏质土	2020-06	—

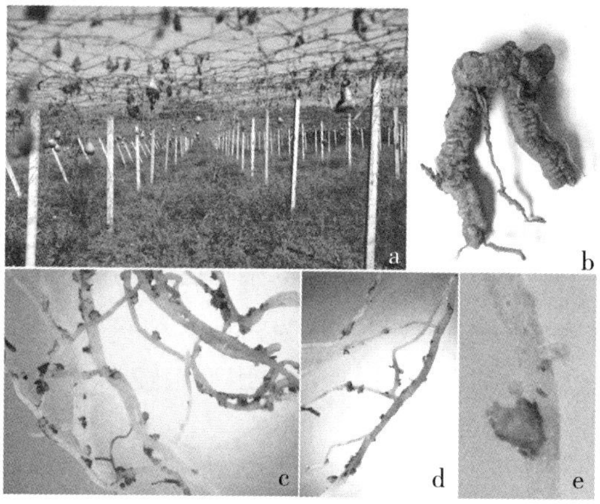

a. 地上部症状；b-d. 根部症状；e. 雌虫及胶质卵块

图 1　肾状属线虫危害瓜蒌症状

2.2　形态学鉴定

未成熟雌虫：蠕虫形，加热杀死后虫体呈“C”形，唇区隆起、圆锥形、无缢缩；头架、口针强壮，口针基部球圆形；食道腺长，覆盖肠端，口针基部球距食道腺开口距离与口针长度约相等；中食道球卵圆形，瓣门清楚，排泄孔位于食道峡基部；双生殖管对生，卵巢不成熟；尾长圆锥形，具透明尾端。

成熟雌虫：虫体腹弯，膨大呈肾形；颈区弯曲不规则；阴门突起受精囊圆形至不规则形，通常有精子；肛门后虫体呈半球形，具细长尾尖突；产卵于胶质卵囊内。

雄虫：食道退化，中食道球不发达；口针纤弱；交合刺纤长，向腹面弯曲呈弓状，交合伞不明显。

基于形态学鉴定，本研究所述肾状属线虫主要形态测量值与海南省已报道肾形肾状线虫群体 RRLC1[15] 基本吻合，确定引起瓜蒌肾形线虫病的病原为肾形肾状线虫 *Rotylenchulus reniformis* Linford & Oliveira，1940（图 2、表 2）。

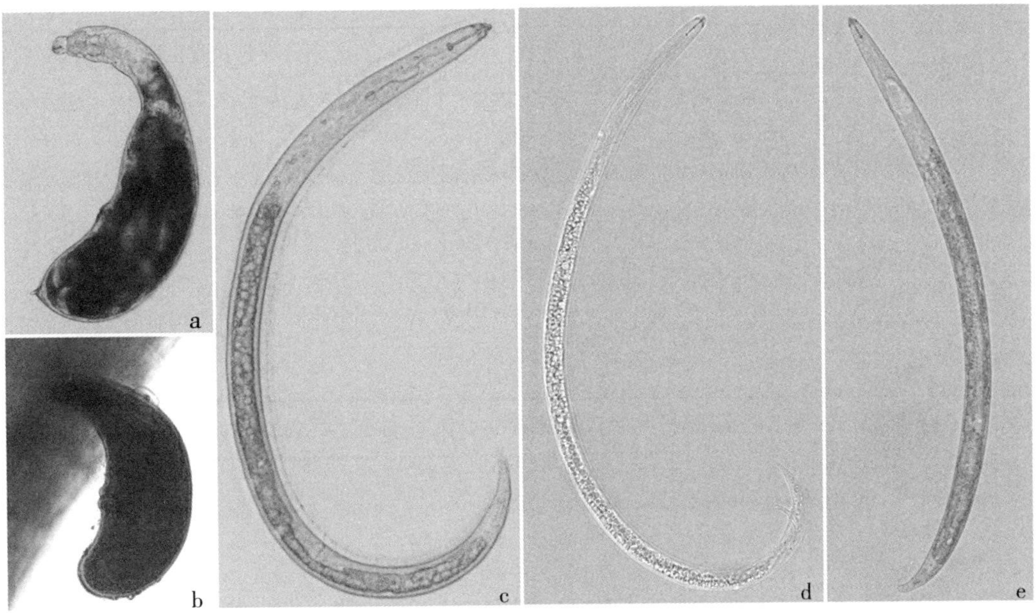

a，b. 雌虫；c. 未成熟雌虫；d. 雄虫；e. 2 龄幼虫

图 2　肾形肾状线虫形态

表 2　肾形肾状线虫形态特征测量值

形态参数	本研究所用群体		海南肾形肾状线虫群体 RRLC1[15]	
	未成熟雌虫	雄虫	未成熟雌虫	雄虫
n	20	20	20	20
L/μm	390. 2±25. 2 (336. 0~424. 3)	421. 9±32. 3 (368. 7~489. 6)	390. 9±33. 4 (325. 6~456. 7)	438. 4±44. 4 (361. 8~488. 7)
a	23. 3±1. 4 (20. 3~25. 1)	26. 7±3. 0 (23. 4~33. 8)	24. 0±1. 6 (19. 3~27. 8)	25. 8±1. 9 (20. 8~28. 3)
b	4. 6±0. 2 (4. 3~4. 9)	5. 3±0. 3 (5. 0~5. 9)	—	—
c	14. 5±1. 4 (12. 4~16. 6)	16. 9±1. 6 (13. 8~19. 7)	13. 7±0. 9 (12. 9~18. 0)	15. 1±1. 5 (14. 3~20. 7)
c′	2. 8±0. 4 (2. 3~3. 5)	3. 2±0. 5 (2. 7~4. 1)	—	—
V	72. 1±1. 3 (70. 7~74. 0)	—	75. 9±6. 8 (67. 5~79. 1)	—
口针长/μm	17. 2±1. 1 (15. 4~18. 6)	13. 5±0. 8 (12. 0~14. 807)	14. 9±1. 2 (12. 6~16. 9)	13. 8±1. 2 (11. 1~15. 3)
DGO/μm	16. 1±2. 3 (12. 3~19. 5)	—	15. 0±1. 7 (12. 4~17. 2)	—
最大体宽/μm	16. 8±0. 8 (14. 8~17. 6)	15. 9±1. 0 (14. 5~17. 3)	16. 3±0. 9 (14. 7~17. 4)	17. 0±1. 1 (15. 1~18. 9)

（续表）

形态参数	本研究所用群体		海南肾形肾状线虫群体 RRLC1[15]	
	未成熟雌虫	雄虫	未成熟雌虫	雄虫
肛门处体宽/μm	9.7±0.8 （8.0~10.7）	7.9±1.2 （5.2~9.2）	—	—
尾长/μm	27.1±3.0 （23.4~33.4）	25.0±2.3 （21.8~29.5）	28.7±2.9 （25.3~34.1）	29.1±1.8 （26.3~32.4）
透明尾/μm	6.9±0.7 （5.7~7.6）	7.5±1.5 （6.0~10.0）	7.3±1.5 （4.2~9.4）	7.9±1.5 （5.3~10.2）
交合刺长/μm	—	19.9±1.9 （16.2~22.2）	—	19.1±1.4 （16.7~21.1）

注：*n*—样本数；L—体长；a—体长与最大体宽的比值；b—体长与头端至食道与肠连接处长度的比值；c—体长与尾长的比值；c′—尾长与肛门处体宽的比值；DGO—口针基部球末端至背食道腺开口的距离；V—阴门至头部的长度与体长的比值×100。

2.3 特异性引物验证

以单条雌虫 DNA 为模板，利用已报道的 rDNA-D2D3 区特异性引物 Rr-F3/Rr-B3[9] 进行 PCR 扩增。结果发现，采自安徽省巢湖市中庙镇和柘皋镇瓜蒌田的 7 份样品均出现 190bp 阳性扩增条带，与阳性对照一致，而对照组无扩增条带，进一步确定其为肾形肾状线虫（图 3）。

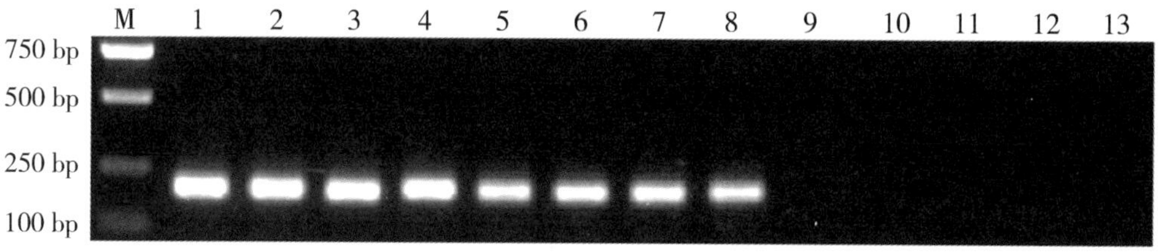

1-7. 待测样品；8. 肾形肾状线虫；9. 南方根结线虫；10. 北方根结线虫；11. 爪哇根结线虫；12. 拟禾本科根结线虫；13. ddH_2O；M. DL2000 Marker

图 3 肾形肾状线虫 PCR 扩增产物

2.4 系统发育分析

将本研究所测肾形肾状线虫 ITS、28S、COI 基因序列在 NCBI 数据库中进行 BLAST 比对，ITS 序列种内相似度为 88.05%~99.66%，28S 种内相似度为 94.95%~100%，COI 种内相似度为 99.18%~100%，可见肾形肾状线虫 ITS 区遗传变异较大，与已报道结果一致。基于 ITS、28S、COI 的系统发育分析，本研究肾形肾状线虫 ITS 和 28S 均与 A 型（type A）群体遗传距离较近，而与 B 型（type B）群体遗传距离较远，表明本研究肾形肾状线虫属于 A 型。此外，本研究肾形肾状线虫 ITS 与孤雌生殖群体（AY335190）遗传距离较远，而与两性生殖群体（AY335191、FJ374686）遗传距离较近。

2.5 肾形肾状线虫生活史

肾形肾状线虫接种瓜蒌后，每 2d 取瓜蒌根系用次氯酸钠-酸性品红染色，观察线虫发

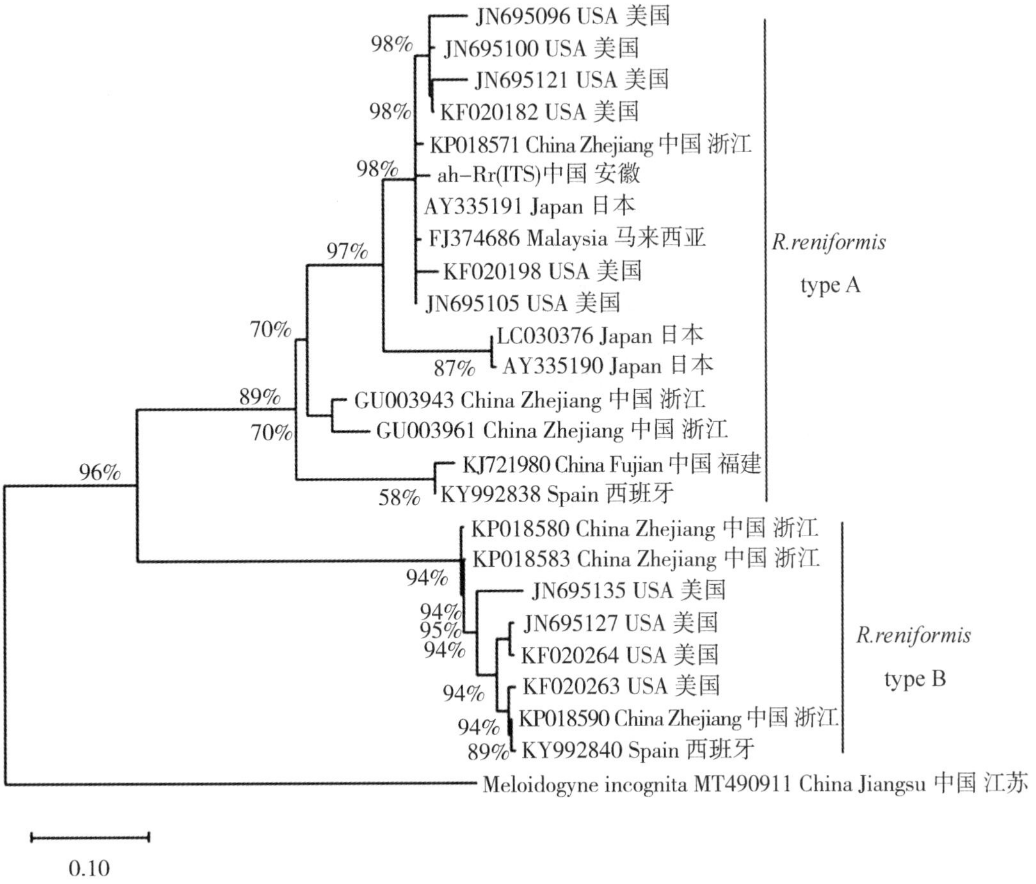

图 4　基于 ITS 对肾形肾状线虫群体系统发育分析

育情况。结果发现，接种 2 龄幼虫 6d 后可在瓜蒌根部发现未成熟雌虫侵染根系（图 7a），接种 8d 后未成熟雌虫在根外的虫体部分逐渐膨大（图 7b-e），接种 16d 后在根部可见成熟雌虫（图 7f-g），接种 20d 时成熟雌虫周围可见胶质卵囊（图 7h-i），每个卵囊平均含 64 个卵，接种 24d 后卵囊内可见 2 龄幼虫游出。该结果表明，在 27℃室内条件下，肾形肾状线虫寄生瓜蒌根部的生活史周期为 24d。此外，将 5 000 条肾形肾状线虫 2 龄幼虫接种瓜蒌 30d 后，统计根部卵块数量、每个卵块含卵量、根部肾形肾状线虫量、土壤中肾形肾状线虫数量，计算繁殖系数 $R=11.12\pm0.62$（表 3）。通过室内接种试验，首次证实瓜蒌为肾形肾状线虫的新寄主。

表 3　肾形肾状线虫接种瓜蒌的繁殖系数

处理	根部卵块量/个	每卵块含卵量/个	根部肾形肾状线虫量/条	土壤中肾形肾状线虫量/条	线虫总量/pf	繁殖系数
处理一	54	63	6 091	45 500	54 993	11. 00
处理二	56	58	6 295	49 160	58 703	11. 74
处理三	50	69	6 455	43 200	53 105	10. 62

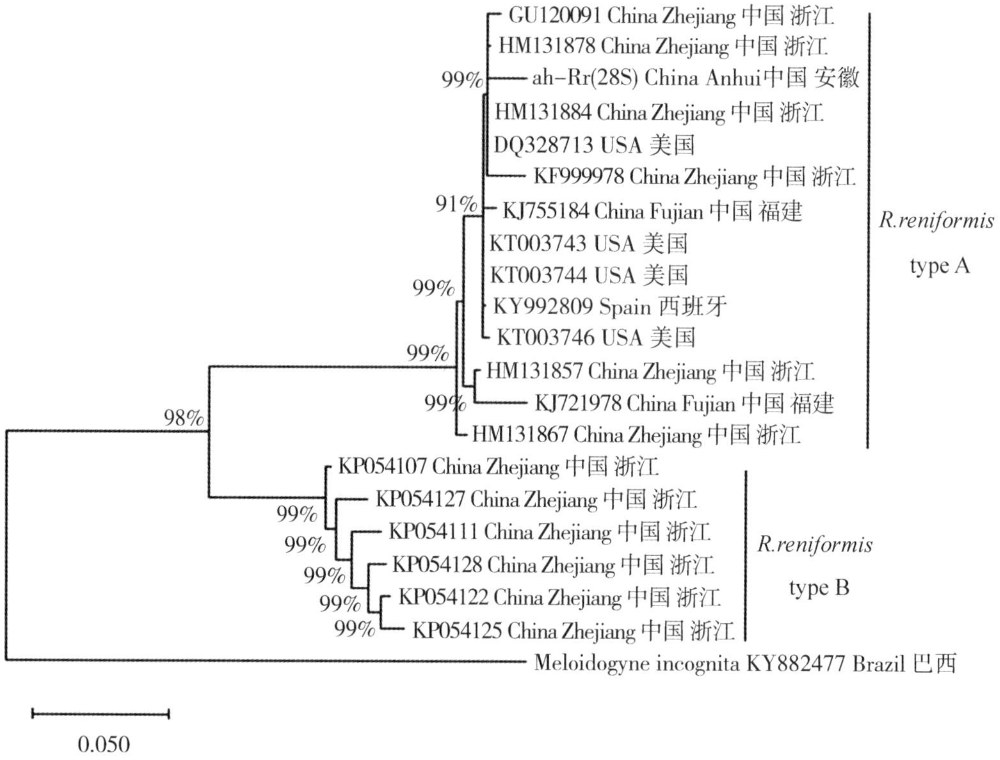

图 5　基于 28S 对肾形肾状线虫群体系统发育分析

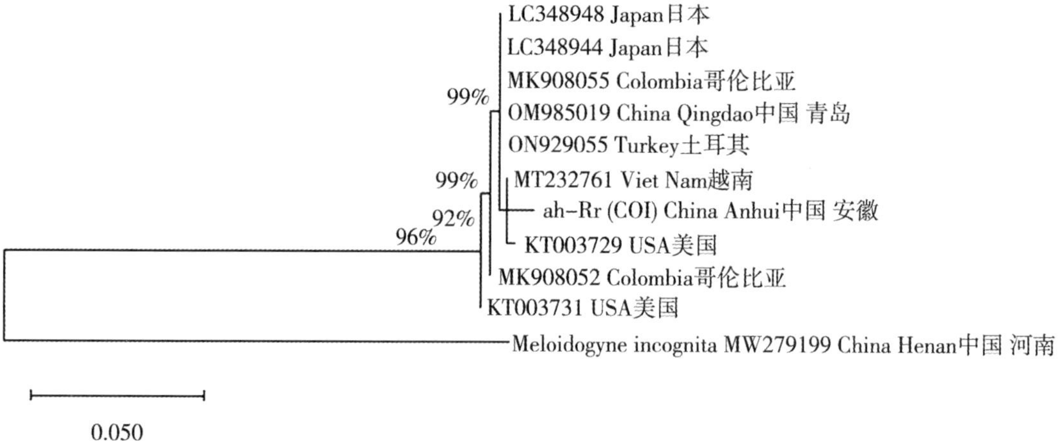

图 6　基于 COI 对肾形肾状线虫群体系统发育分析

3　讨论

植物线虫是引起我国农作物病害的重要病原物之一，对大田作物造成的经济损失为 10%~20%，严重时可达 30%~50%，部分地区可超 80%。随着全球气候变化、种植制度改

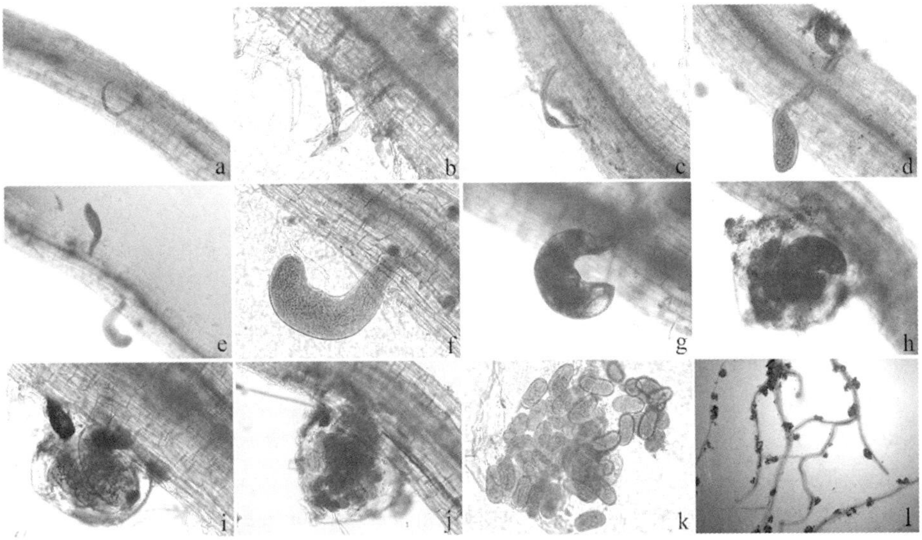

a. 未成熟雌虫侵染根系；b-e. 未成熟雌虫由蠕虫形发育为肾形；
f-g. 成熟雌虫；h-i. 产卵状态雌虫；j-l. 卵囊

图 7 肾形肾状线虫在瓜蒌根部的发育动态

革、高值农业迅速发展等因素，植物线虫病害发生趋势愈发严重，将成为我国第二大类植物病害[16]。肾形肾状线虫作为一种危害严重的植物线虫，近年来在蔬菜、果树、园林植物上相继被发现和报道。肾形肾状线虫形态学鉴定主要基于未成熟雌虫、成熟雌虫及雄虫的形态特征。本研究中的肾形肾状线虫主要形态测量值与国内已报道的肾形肾状线虫群体基本吻合，个别特征略有差别[8, 15]。不同形态测量值的差异可能与肾形肾状线虫所处温湿度、土壤环境、寄主种类等因素相关。肾形肾状线虫 rDNA-ITS 区在不同群体甚至同一群体内遗传变异较大，Vovlas 等[17]和 Palomares-Rius 等[18]利用 ITS、28S 对肾形肾状线虫进行系统发育分析，对 A 型（type A）和 B 型（tybe B）进行有效区分，并将肾形肾状线虫与肾状属其他线虫进行有效区分。基于 *ITS*、*28S*、*COI* 基因进行的遗传进化分析表明，本研究所采集肾形肾状线虫属于 A 型，且与两性生殖肾形肾状线虫（AY335191、FJ374686）遗传距离较近，无 A 型和 B 型同时发生的现象。

肾状属线虫生活史较特别，2 龄幼虫在土壤中不取食的状态下经过 3 次蜕皮发育成未成熟雌虫，未成熟雌虫为取食虫态。幼虫蜕皮过程通常会保留前一龄期的角质层，且每次蜕皮后虫体较前一龄期短小[5]。笔者对分离自土壤的肾形肾状线虫镜检时，发现土壤中存在各龄期幼虫，且大量线虫处于蜕皮阶段，前一龄期的角质层覆盖虫体的现象较普遍。对肾形肾状线虫在瓜蒌根上的生活史研究发现，瓜蒌根系上卵囊较密集，但每个卵囊含卵量较少，约为 64 粒/卵囊，单个雌虫产卵量明显少于根结线虫的产卵量。前人研究表明，肾形肾状线虫在缺乏寄主或干燥条件下，线虫可在土壤中保持某一蠕虫阶段而不生长，其生活史可在两年内完成[5, 19]。田间调查时也发现，在部分黏质土且略板结的土质较差的田块，土壤中肾形肾状线虫数量超过 1 000 条/100mL 土壤，表明肾形肾状线虫对逆境的适应能力相对较强。此外，5 000 条 2 龄幼虫接种瓜蒌 30d 后，繁殖系数为 $R=10.33\pm0.61$，高于肾形肾状线虫在大豆根部的繁殖能力[20]，本文推测瓜蒌是肾形肾状线虫的合适寄主。

综上所述，本文首次报道肾形肾状线虫在安徽省发生危害，对其形态特征、遗传进化、生活史等进行初步描述和研究，且首次报道瓜蒌为肾形肾状线虫的新寄主。本研究为进一步掌握肾形肾状线虫的分布区域及潜在寄主，为采取有效的防治措施提供理论依据。

参考文献

[1] 柴欣，朱霖，戚爱棣，等．栝楼属植物化学成分研究进展［J］. 辽宁中医药大学学报，2013，15（1）：66-70.

[2] 高丙利．植物线虫综合治理概论［M］. 北京：中国农业科学技术出版社，2021：109-111.

[3] JONES J T，HAEGEMAN A，DANCHIN E G J，*et al.* Top 10 plant-parasitic nematodes in molecular plant pathology［J］. Molecular Plant Pathology，2013，14（9）：946-961.

[4] ROBINSON F，INSERRA R N，CASWELL-CHEN E P，*et al.* *Rotylenchulus* species：identification，distribution，host ranges，and crop plant resistance［J］. Nematropica，1997，27（2）：127-180.

[5] BIRD A F. Growth and moulting in nematodes：changes in the dimensions and morphology of *Rotylenchulus reniformis* from start to finish of moulting［J］. International Journal for Parasitology，1983，13：201-206.

[6] 邓艳凤，徐红兵，田忠玲，等．肾形肾状线虫个体间 rDNA 内转录间隔区的变异［J］. 浙江大学学报，2015，41（3）：252-260.

[7] 吴慧平，陶卫平，杨荣铮．安徽省盆景花卉线虫分布初报［J］. 植物检疫，1998，12（2）：81-82.

[8] 张燕，MATAFEO A，石红利，等．肾形肾状线虫（*Rotylenchulus reniformis*）浙江杭州群体形态学和分子特性及寄主范围［J］. 植物病理学报，2011，41（1）：37-43.

[9] 孙丹丹．广东省园林植物肾形肾状线虫分子特征及分子检测技术的研究［D］. 广州：华南农业大学，2018.

[10] VRAIN T，WAKARCHUK D，LAPLANTELEVESQUE A，*et al.* Intraspecific rDNA restriction fragment length polymorphism in the *Xiphinema americanum* group［J］. Fundamental and Applied Nematology，1992，15（6）：563-573.

[11] PHILLIPS M，FERRAZ L，ROBBINS R，*et al.* Confirmation of the synonymy of *Paratrichodorus christiei*（Allen，1957）Siddiqi，1974 with P-minor（Colbran，1956）Siddiqi，1974（Nematoda：Triplonchida）based on sequence data obtained for the ribosomal DNA 18S gene［J］. Nematology，2004，6（1）：145-151.

[12] HOLTERMAN M，WURFF A V D，ELSEN S V D，*et al.* Phylum-wide analysis of SSU rDNA reveals deep phylogenetic relationships among nematodes and accelerated evolution toward crown clades［J］. Molecular Biology & Evolution，2006，23（9）：1792-1800.

[13] SOFIE D，JAN V，ANNELIEN R，*et al.* Exploring the use of cytochrome oxidase c subunit 1（COI）for DNA barcoding of free-living marine nematodes［J/OL］. PLoS ONE，5（10）：e13716. DOI：10. 1371/journal. pone. 0013716.

[14] BYRD D W，KIRPKATRICK T，BARKER K R. An improved technique for clearing and straining plant tissues for the detection of nematodes［J］. Journal of Nematology，1983，15（1）：142-143.

[15] 林小漫，芮凯，符美英，等．海南岛柑橘肾形线虫病的发生及病原鉴定［J］. 基因组学与应用生物学，2020，39（11）：5100-5105.

[16] 彭德良．植物线虫病害：我国粮食安全面临的重大挑战［J］. 生物技术通报，2021，37（7）：1-2.

[17] VOVLAS N，TIEDT L R，CASTILLO P，*et al.* Morphological and molecular characterisation of one new and several known species of the reniform nematode，*Rotylenchulus* Linford & Oliveira，1940（Hoplolaimidae：Rotylenchulinae），and a phylogeny of the genus［J］. Nematology，2016，18（1）：67-107.

[18] PALOMARES-RIUS J E，CANTALAPIEDRA-NAVARRETE C，ARCHIDONA-YUSTE A，*et al.* Prevalence and molecular diversity of reniform nematodes of the genus *Rotylenchulus*（Nematoda：Rotylenchulinae）in the Mediterranean Basin［J］. European Journal of Plant Pathology，2018，150（21）：439-455.

[19] HEALD C M，INSERRA R N. Effect of temperature on infection and survival of *Rotylenculus reniformis*［J］. Journal of Nematology，1988，20（3）：356-361.

[20] 张燕，石红利，郭恺，等．肾形肾状线虫的生活史及繁殖能力［J］. 浙江大学学报，2011，37（1）：7-12.

中国线虫学研究（第九卷）Nematology Research in China（Vol. 9）：12-15

复配剂 WV32、WV33 及缓释颗粒剂 FW203 防治植物根结线虫病研究*

张怡萱**，许杏滢，左　婷，邱燕婷，文艳华***
（华南农业大学植物保护学院植物线虫研究室，广州　510642）

Studies on the Control Effect of Plant Root-knot Nematode Diseases with WV32，WV33 and FW203*

Zhang Yixuan**，Xu Xingying，Zuo Ting，Qiu Yanting，Wen Yanhua***
（*Lab of Plant Nematology*，*College of Plant Protection*，
South China Agricultural University，*Guangzhou*　510642，*China*）

摘　要：根结线虫（*Meloidogyne* spp.）是造成世界农业生产严重损失的一类植物寄生线虫，寄主广泛，尤其严重危害茄科、葫芦科植物。传统化学杀线剂对环境以及人体健康存在威胁，也容易产生抗药性，植物源及天然成分杀线剂成为新的研究热点。本文以本实验室前期从植物中分离鉴定出的杀线活性物质，及筛选出的天然活性成分为有效成分配制成 WV32、WV33、FW203 三种制剂，采用盆栽拌药处理，人工接种根结线虫的方法，研究了两种复配剂及缓释颗粒剂对辣椒及番茄根结线虫病的防治效果，为进一步的大田防治应用提供依据。

关键词：天然成分；根结线虫；防效

1　材料与方法

1.1　材料

1.1.1　供试植物材料

辣椒（*Capsicum annuum*）：帝王 336 线椒，山东寿光市春秋种业有限公司。

番茄（*Solanum lycopersicum*）：新金丰 1 号，广州长合种子有限公司。

1.1.2　供试线虫

南方根结线虫（*Meloidogyne incognita*）华南农业大学植物线虫研究室大棚盆栽保存。

1.1.3　供试药剂

本实验室以植物源成分及天然成分为有效成分的复配剂 WV32、WV33 及缓释颗粒剂 FW203。

---

* 基金项目：公益性行业（农业）科研专项经费项目（201503114）
** 第一作者：张怡萱，硕士研究生，主要从事植物杀线剂研究
*** 通信作者：文艳华，副教授，从事植物线虫学研究。E-mail：yhwen@ scau. edu. cn

2 方法

2.1 供试线虫的培养收集

南方根结线虫卵粒的收集：从本实验室大棚盆栽接种了南方根结线虫的植株病根上，摘取成熟卵囊，以 1% 次氯酸钠浸泡并猛烈震荡 4min，然后用清水冲洗多遍，用 500 目网筛收集分散的卵粒备用。

根结线虫二龄幼虫的收集：从病株根结上挑取新鲜卵囊，放入垫有一层维达纸巾的筛网上，置于加有灭菌水的 6cm 培养皿里，在 25℃培养箱孵化 3~5d。待孵化大量二龄幼虫时，收集备用。

2.2 供试植物育苗方法

育苗所用土：翠筠靓土营养土。

辣椒育苗方法：取适量种子，放置于垫有两层圆形滤纸的 12cm 培养皿中，滤纸用清水湿润，置于 26℃恒温培养箱中，露白后穴施于育苗盆中，每格一颗种子。

番茄育苗方法：取适量种子放置于垫有两层圆形滤纸的 12cm 培养皿中，滤纸用清水湿润，置于 26℃恒温培养箱中，露白后穴施于育苗盆中，每格一颗种子。

2.3 盆栽试验方法

试验在华南农业大学农学院网室内进行。取上述灭菌后的营养土与细沙按体积比 2∶1 混匀。取长势相同、两叶一心期的供试植物苗，每盆移栽 1 棵。

辣椒接南方根结线虫盆栽试验：2022 年 9 月育苗，辣椒育苗期 40d。每盆拌药剂用量设以下处理：WV321、WV322、WV323 由低到高 3 个浓度，以不施药接虫为接虫对照 CK，共 4 个处理，随机排列。WV32 组每个处理 10 个重复，量取 10L 的混合土与 10 盆量的药剂充分混合，将混匀的营养土装于直径 16cm×高 12cm 的花盆内，每盆 1L。移苗后 35d 接种南方根结线虫二龄幼虫（J2）和卵粒混合悬浮液，每株接（560 J2+1 050 卵粒）/mL 的悬浮液 1mL，接种 44d 后扣盆调查，清洗根部，记录地上部分鲜重、地下部分鲜重、根结数、病级和每株植物根部卵粒数，若根部根结较少，则计数根结数，计算根结防效（公式 1）。

$$\text{根结防效（\%）}=\frac{\text{对照根结数}-\text{处理根结数}}{\text{对照根结数}}\times 100 \tag{1}$$

番茄接南方根结线虫盆栽试验：2022 年 10 月育苗，番茄育苗期 16d。每盆拌药剂用量设以下处理：FW203 高、中、低剂量；WV331、WV332、WV333 由低到高 3 个浓度，以不施药接虫为接虫对照 CK，每个处理重复 10 次，随机排列。量取 10 L 的混合土与 10 盆量的药剂充分混合，将混匀的土装于直径 16cm×高 12cm 的花盆内，每盆装 1L 土，移苗后 3d 接种南方根结线虫二龄幼虫（J2）和卵粒混合悬浮液，每株接（360 J2+1 110 卵粒）/mL 的悬浮液 1mL。接种后 46d 每个处理组取 4 盆扣盆调查。

3 结果与分析

3.1 复配剂 WV32 对辣椒南方根结线虫病的防治效果

接种 44d 结果表明，WV32 对辣椒根结线虫病有较好的防治效果。在 WV321、WV322、WV323 处理中，防效分别为 83.1%、82.2%、85.7%，三者防效差异不大。在各剂量处理中，地上部鲜重及根重与接虫 CK 对比有明显增加，WV321、WV322、WV323 的地上部鲜重分别为 65. 8g、78. 4g、66. 5g，与接虫 CK 的地上部鲜重 9. 2g 对比有明显增加。综合地上部鲜重与根结防效，WV322 为最适浓度（表 1、图 1）。

表 1　复配剂 WV32 防治辣椒根结线虫病盆栽试验（2022 年 12 月 16 日接虫，44d 后扣盆）

处理	地上部鲜重/g	根重/g	每克根结数/个	根结防效/%
WV321	65. 8±16. 6b	9. 5±2. 4a	7. 8±2. 8b	83. 1±5. 9a
WV322	78. 4±5. 8a	9. 4±2. 0a	8. 2±1. 9b	82. 2±3. 9a
WV323	66. 5±9. 5b	10. 1±1. 5a	6. 8±2. 5b	85. 7±5. 5a
CK	9. 2±2. 6c	3. 4±0. 6b	46. 3±24. 9a	

注：表中数据为 10 次重复±SE，经 DMRT 检验，同列数据具相同字母的表示差异不显著（P=0. 05）。

A. 地上部植株生长情况；B. 地下部植株生长情况

图 1　复配剂 WV32 防治辣椒根结线虫病盆栽试验（2022 年 12 月 16 日接虫，44d 后扣盆）

3. 2　复配剂 WV33 及缓释颗粒剂 FW203 对番茄根结线虫病防治效果

接种 46d 结果表明，WV331、WV332、WV333 处理防效分别为 91. 36%、91. 73%、83. 96%；每克根的卵粒数分别为 558 粒、513 粒、1 138 粒，明显少于接虫对照组 CK 的 11 178 粒；在 WV331、WV332 及 WV333 中，各处理组的地上部鲜重与根重与接虫对照 CK 对比均明显增加；综合地上部鲜重与根结防效，WV332 为最适浓度（表 2、图 2）。

接种 46d 结果表明，缓释颗粒剂 FW203 低、中、高 3 种剂量处理，防效分别为 67. 33%、80. 34%、88. 83%，防治效果随处理剂量的增加而增加；FW203 高、中、低剂量处理的每克根卵粒数分别为 2 104 个、2 506 个、964 个，明显低于未施药对照组的

11 178 粒；地上部鲜重与根重均明显高于未施药接虫对照 CK，综合地上部鲜重与根结防效，中浓度为最适浓度。

表 2　WV33 及 FW203 对番茄根结线虫病盆栽防治效果（接种 46d 调查）

处理	地上部鲜重/g	根重/g	每克根结数/个	根结防效/%	每克卵粒数/粒
CK	6.92±0.51d	1.57±0.12c	118±13a		11 178
WV331	19.49±3.07bc	4.57±0.43ab	10±2d	91.36±1.83a	558
WV332	21.53±1.21abc	5.30±0.24a	10±2d	91.73±1.43a	513
WV333	19.27±0.70bc	3.44±0.18b	19±2cd	83.96±1.93abc	1 138
FW203 低	23.57±2.47ab	4.04±0.24ab	39±6b	67.33±5.38d	2 104
FW203 中	26.62±1.44a	5.05±0.10a	23±5bcd	80.34±3.79bc	2 506
FW203 高	16.41±3.46c	3.73±0.94b	13±5cd	88.83±4.52ab	964

注：表中数据格式为 x±S.E，x 代表 4 次重复平均值，S.E 为标准误；同列数据后标有相同字母表示在 5%差异不显著。

图 2　复配剂 WV33 处理番茄地上部植株生长情况（接种 46d）

注：A、B、C、D、E、F、G、H 分别代表 WV331、WV332、WV333、FW203 低、FW203 中、FW203 高、CK2、CK1。

参考文献（略）

中国线虫学研究（第九卷）Nematology Research in China（Vol. 9）：16-20

福建茶树根际 1 种针线虫的种类鉴定

丁怡倩[1]*，李　硕[1]，王　婷[1]，章淑玲[2]，刘国坤[1]**
（[1]福建农林大学生物农药与化学生物学教育部重点实验室，福州　350002；
[2]福建农业职业技术学院，福州　350119）

摘　要：笔者 2022—2023 年对福建省茶树根际寄生线虫进行调查时，发现了一种针线虫分离频率高，达 53. 1%。该线虫在长势衰退的茶树根部每 100g 土平均可分离出 1 020 条。线虫经形态观察与测量，与美丽针线虫（*Paratylenchus lepidus*）种的描述一致。对 rDNA-ITS 区及 28S D2-D3 区序列扩增，利用最大似然法和贝叶斯法构建了系统发育树分析表明该线虫序列与美丽针线虫种群序列处于同一分支且支持率高，进一步明确为美丽针线虫。

关键词：茶树；针线虫；形态学；分子生物学

Identification of a Species of *Paratylenchus* in the Rhizosphere of Tea Plants in Fujian Province

Ding Yiqian[1]*，Li Shuo[1]，Wang Ting[1]，Zhang Shulin[2]，Liu Guokun[1]**
（[1] *Key Laboratory of Biopesticide and Chemical Biology*，*Ministry of Education*，*Fujian Agriculture and Forestry University*　350002，*Fuzhou*，*China*；[2] *Fujian Vocational College of Agricultural*，*Fuzhou*　350119，*China*）

Abstract：Investigation of rhizosphere parasitic nematodes of tea in Fujian Province has been done in 2022-2023，*Paratylenchus* sp. was found which frequency is 53. 1%. the average 1 020 nematodes were isolated from per 100g soil in tea plantations with declining growth. The morphological characteristics of the nematode were consistent with the description of *Paratylenchus lepidus* species. The rDNA-ITS and 28S D2-D3 regions of the nematode has been amplificated respectively，phylogenetic trees were constructed using Maximum Likelihood and Bayesian methods，phylogenetic analysis showed that the nematode sequence was group with the sequence of *P. lepidus* available from the database respectively，which further confirmed as *P. lepidus*.

Key words：Tea tree；Nematode；Morphology；Molecular biology

植物寄生线虫是茶树上的重要病原物，其中主要包括短体线虫（*Pratylenchus* spp.）、根结线虫（*Meloidogyne* spp.）、拟鞘线虫（*Hemicriconemoides* spp.）、螺旋线虫（*Helicotylenchus* spp.）、针线虫（*Paratylenchus* spp.）[1,2]等。针线虫（*Paratylenchus* spp.）为根系迁移性外寄

* 第一作者：丁怡倩，硕士研究生，研究方向为植物线虫病害及线虫病害防治。E-mail：1648754971@qq. com
** 通信作者：刘国坤，教授，研究方向为植物线虫病害及线虫病害防治。E-mail：liuguok@ 126. com

生物，主要通过口针侵染寄主植物的根表皮细胞或根毛，口针长的类群危害皮层薄壁细胞。针线虫在茶树上的危害报道相对较少，主要有福建省茶园报道弯曲针线虫（*P. curvitatus*）和突出针线虫（*P. projectus*），种群大时造成茶树侧根萌发数量少，细短，黄化，无再分枝[3,4]。美丽针线虫（*P. lepidus*）在安徽省茶园报道，种群数量大[5]。笔者 2022—2023 年在对福建茶树线虫进行调查时发现种群数量大的 1 种针线虫，现将该线虫进行了鉴定。

1 材料与方法

1.1 样本采集及线虫保存

线虫样本采集自福建省福州市、南平市、宁德市多个茶园 5~30cm 茶树根际土壤层。称取 100g 土壤样本及少量根组织（剪碎）采用浅盘法分离 24h。

1.2 形态鉴定

线虫缓慢加热杀死后制成临时玻片[6]，在 Nikon 显微镜（Eclipse Ni-U 931609，日本）下进行观测，用与之配套的相机（DS-Ril）拍照，用测量软件和 de Man 公式[7]对各形态特征进行测量和数据处理。

1.3 分子生物学鉴定

1.3.1 DNA 的提取与扩增

采用单条线虫进行 DNA 提取，提取方法采用蛋白酶 K 法[8]。采用 25μL 扩增体系：7.5μL ddH_2O，12.5μL Ex Premix TaqTM（TAKARA，大连宝生物公司），上下游引物各 1μL，DNA 模板 3μL。

28S D2-D3 区扩增引物对为 D2A/D3B（5′-ACAAGTACCGTGAGGGAAAGTTG-3′；5′-TCGGAAGGAACCAGCTACTA-3′）[9]，扩增反应条件：95℃，3min；95℃，30s，54℃，50s，72℃，1min，30 个循环；72℃，10min，4℃，保存。

rDNA-ITS 区扩增引物对为 V567/26S（5′-TTGATTACGTCCCTGCCCTTT-3′；5′-TTTCACTCGCCGTTACTAAGG-3′）[10]，扩增反应条件：94℃，4min；94℃，30s，55.6℃，50s，72℃，1min，30 个循环；72℃，10min，4℃，保存。

将上述 PCR 产物经 1%琼脂糖凝胶电泳检测确定为目的条带后，PCR 产物送至上海生工生物公司进行双向测序。

1.3.2 系统发育树的构建

将目标序列上传至 BLAST 比对，选取序列相似度最高的类群作为建树序列，利用 G-INS-I 方法进行序列的多重比对，去除无效位点，转换序列格式，运用程序 MrBayes on XSEDE，采用 GTR+I+G 模式构建贝叶斯树[11]；运用程序 RAxML-HPC2 on XSEDE，采用 GTRCAT 模式构建极大似然树[12]。

2 结果与分析

2.1 种群调查

在 98 个衰退茶树根际土壤样品中，有 52 个样本中分离到针线虫，分离频率为 53.1%，在长势衰退的茶树根部针线虫的数量大，每 100g 土平均可达 1 020 条。

2.2 形态特征鉴定

雌虫：虫体经温热杀死处理后，虫体向腹面弯曲呈开放性 C 型（图 1A）。头部略圆锥形，头骨架骨质化弱，末端稍平截，口针纤细。环形食道，中食道球瓣门明显。峡部细窄。后食道球梨形，食道与肠平接。排泄孔位于后食道球中部偏上的位置。侧区有 4 条侧线，间距相似。成熟的雌虫阴门处不膨胀，阴门位于虫体的 81.5%左右处，具有明显的阴门侧膜（图 1C）。尾部锥形，弯向腹面，尾渐细，尾末端圆或钝指状，背面近端处有缺刻，尾环渐细密。

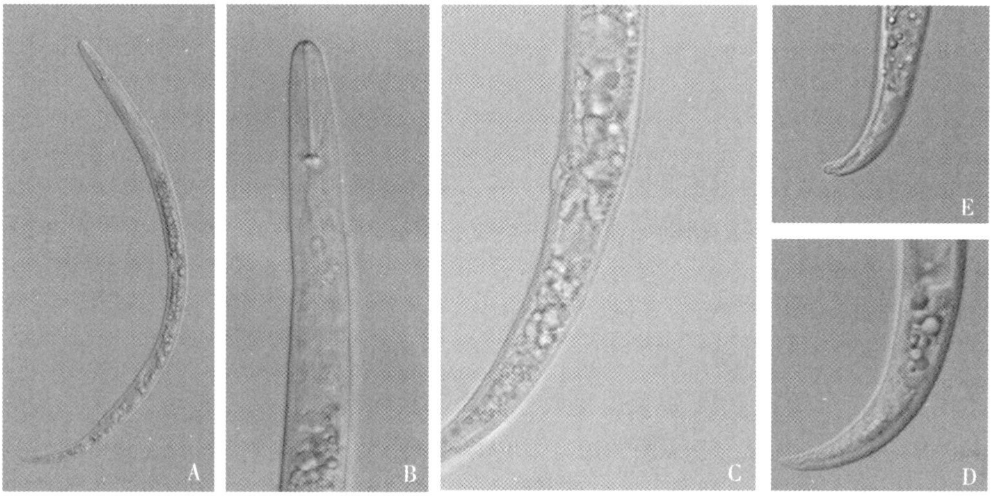

A. 整体；B. 体前部；C. 阴门盖；D、E. 尾部

（标尺：A = 25μm，B、D = 10，C、E = 20μm）

图 1　美丽针线虫的形态特征

雄虫：未发现。

其形态特征（表 1）符合美丽针线虫（*P. lepidus*）原始描述种的分类特征，主要分类特点为口针长 23~25μm，排泄孔位于后食道球中部偏上的位置，有阴门盖。

表 1　美丽针线虫福建种群测量值　　　测量单位：μm

形态指标	测定种群
n	15
L	339.0±35.6（287.6~412.5）
a	23.8±2.6（18.3~27.6）
b	5.7±0.5（5.0~6.8）
c	11.7±1.3（9.3~13.7）
c′	3.3±0.3（2.8~4.0）
V	81.5±1.4（77.0~83.7）
V′	26.6±2.7（21.8~30.7）
Stylet length	24.50±0.87（23.09~25.76）
Excretory pore（Ep）	76.1±5.8（69.0~88.6）
Pharynx（Ph）	59.4±4.6（52.2~72.1）

（续表）

形态指标	测定种群
Head to vulva	276. 6±29. 4（234. 7~339. 0）
Max body diam	14. 5±2. 5（11. 5~20. 0）
Tail length	29. 39±4. 80（22. 0~37. 7）

2.3 分子鉴定

28S D2-D3 区 2 条序列扩增为 772bp、771bp，GenBank 获序列号分别为 OQ924042、OQ924043，与 *P. lepidus*（MW716308）一致性为 99. 10%、99. 87%，构建的系统发育树（图 2A）表明其均与 *P. lepidus* 处在同一分支且支持率高。

ITS 区 2 条序列扩增为 993bp、1 034bp，GenBank 获序列号分别为 OQ924133、OQ924134，与 *P. lepidus*（MK886695）一致性为 98. 39%、98. 64%，构建的系统发育树（图 2B）表明其均与 *P. lepidus* 处在同一分支且支持率高。

通过形态特征鉴定与分子生物学鉴定，该线虫为美丽针线虫（*P. lepidus*）。

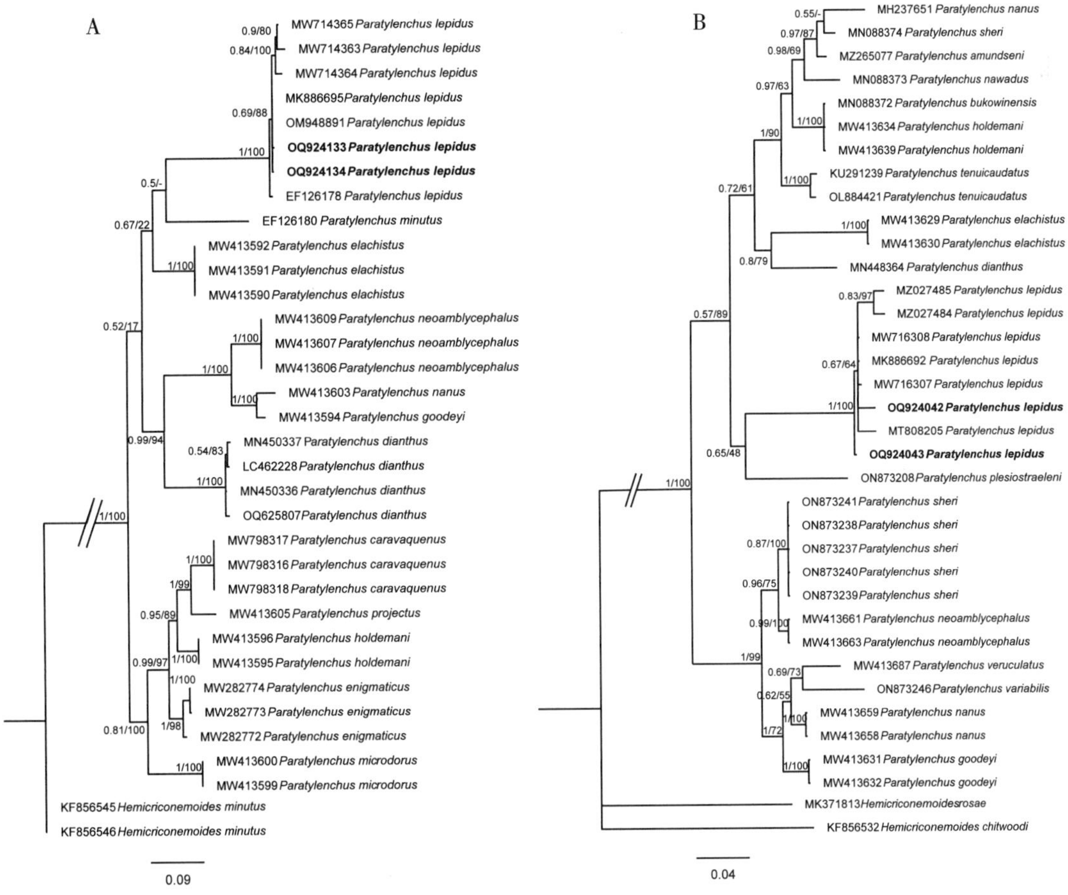

A. 28S 区系统发育树；B. ITS 区系统发育树

图 2　基于美丽针线虫的 28S D2-D3 区和 ITS 区的系统发育树

3 讨论

本次在茶叶根部线虫种类调查过程，发现一个针线虫优势种，通过形态与分子鉴定确定为美丽针线虫。针线虫是一类重要的植物外寄生线虫，其寄主范围广泛，可造成严重的危害，例如，侵染危害生菜、樱桃李、胡萝卜、烟草和玫瑰，导致产量和品质下降[13]。针线虫是茶树上发生普遍的植物寄生性线虫，在一些茶园的茶树根围线虫区系中针线虫量占92%以上，弯曲针线虫能削弱茶树根系生长和导致减产[3]。美丽针线虫首次报道于斯里兰卡茶树根际[14]，在我国台湾和安徽省茶园有报道，寄生茶树根部，影响茶树生长、茶叶品质[5]。本研究为首次在福建在本次茶树根际中发现的美丽针线虫，分离频率高达 77.8%，在茶园中分布广泛，特别是在弱势植株中种群数量大，但是否因美丽针线虫的危害造成，尚需进一步研究。

参考文献

[1] SPIEGEL Y. Plant Parasitic Nematodes in subtropical and tropical agriculture：2nd Edition. [J]. European Journal of Plant Pathology，2005，113（4）：439.

[2] STANTON J. Book Review-Plant Parasitic Nematodes in subtropical and tropical agriculture. [J]. Australasian plant pathology，1990，19（4）：146-147.

[3] 张绍升，翁自明，马俊英，等．针线虫对茶树的致病作用及其发生规律 [J]. 中国茶叶，1991（2）：32-34.

[4] 章霜红，陈成金，叶新民，等．山茶花根部针线虫的鉴定 [J]. 植物保护，1997（2）：16-18.

[5] 马慧勤．安徽省茶园杂草与茶树根际线虫的生态关系 [D]. 合肥：安徽农业大学，2018.

[6] 张绍升．植物线虫病害诊断与治理 [M]. 福州：福建科学技术出版社，1999.

[7] 谢辉．植物线虫分类学 [M]. 北京：高等教育出版社，2000.

[8] 王江岭，张建成，顾建锋．单条线虫 DNA 提取方法 [J]. 植物检疫，2011（2）：32-35.

[9] DE LEY P，DE LEY I T，MORRIS K，*et al.* An integrated approach to fast and informative morphological vouchering of nematodes for applications in molecular barcoding [J]. Philosophical transactions of the Royal Society of London. Series B，Biological sciences，2005，360（1462）：1945-1958.

[10] SUBBOTIN S A，MAAFI Z T，MOENS M. Molecular identification of cyst-forming nematodes（Heteroderidae）from Iran and a phylogeny based on ITS-rDNA sequences [J]. Nematology，2003，5（1）：99-111.

[11] RONQUISF F，TESLENKO M，MARK P V D，*et al.* Mrbayes 3. 2：Efficient bayesian phylogenetic inference and model choice across a large model space [J]. Systematic Biology，2012，61（3）：539-542.

[12] STAMATAKIS A，HOOVER P，ROUGEMONT J. A rapid bootstrap algorithm for the RAxML web servers [J]. Systematic Biology，2008，57（5）：758-771.

[13] 苗文韬．针线虫属 4 种群的综合鉴定与描述 [D]. 杭州：浙江大学，2021.

[14] Raski D J. Revision of the Genus *Paratylenchus* Micoletzky，1922 and Description of New Species. Part Ⅱ of Three Parts [J]. Journal of nematology，1975，7（3）：274-295.

中国线虫学研究（第九卷）Nematology Research in China（Vol. 9）：21-31

山东省马铃薯根结线虫病病原鉴定和侵染动态观察*

代明明[1,2**]，梁　晨[1,2]，宋雯雯[1,2]，史倩倩[1,2]，赵洪海[1,2***]
（[1]青岛农业大学植物医学学院/山东省植物病虫害绿色防控工程研究中心，青岛　266109；
[2]东营青农大盐碱地高效农业技术产业研究院，广饶　257347）

摘　要：根结线虫（*Meloidogyne* spp.）为根系专性内寄生线虫，基本所有维管束植物都受其侵染危害，是威胁全球农业生产的十大病原线虫之一。笔者课题组前期于山东省潍坊市马铃薯田中发现根结线虫已对马铃薯造成损害。本研究为明确山东省部分马铃薯产区根结线虫的种类及侵染动态，从山东省潍坊市和平度市采集罹病样品进行线虫分离鉴定，采用形态学、rDNA-ITS 区序列分析和特异性引物扩增相结合的方法，对危害马铃薯的根结线虫进行种类鉴定。结果如下：形态学结果表明，马铃薯根结线虫与南方根结线虫（*Meloidogyne incognita*）的形态特征一致；基于 rDNA-ITS 区序列进行系统发育分析表明，马铃薯根结线虫种群与 GenBank 中报道的南方根结线虫种群以 99%的支持率聚集在同一支；特异性引物检测结果也证实马铃薯根结线虫为南方根结线虫（*M. incognita*）。通过在马铃薯生育期内取样调查，对马铃薯根结线虫的生长发育动态进行监测发现，南方根结线虫（*M. incognita*）在马铃薯一个生长周期中发生两代，以土壤中 2 龄幼虫（J2）初侵染马铃薯根系，于根部定殖、生长并产卵，40~45d 后卵孵化产生的第二代 J2 则于结薯期至成熟期危害根系和薯块，导致马铃薯表面粗糙，商品率下降。因此，生产上结合马铃薯喜冷凉、不耐高温的习性，可适当提前种植或选种早熟品种，实现时间上躲避或减轻第二代南方根结线虫 J2 对马铃薯块茎的危害。

关键词：马铃薯；根结线虫；种类鉴定；侵染动态

Species Identification and Infection Dynamics Observation of Root-knot Nematode on Potato in Shandong Province*

Dai Mingming[1,2**], Liang Chen[1,2], Song Wenwen[1,2], Shi Qianqian[1,2], Zhao Honghai[1,2***]
([1] *College of Plant Health and Medicine, Qingdao Agricultural University/Shandong Green Control of Plant Diseases and Pests Engineering Research Center, Qingdao*　266109, *China*;
[2] *Academy of Dongying Efficient Agricultural Technology and Industry on Saline and Alkaline Land in Collaboration with Qingdao Agricultural University, Guangrao*　257347, *China*)

Abstract: The root-knot nematode is a root-specific endoparasitic nematode that infests basically all vascular plants, and is one of the top ten pathogenic nematodes threatening agricultural production worldwide. The root knot nematode

---

* 基金项目：黄三角国家农高区省级科技创新发展专项资金项目（2022SZX23）；山东省重点研发计划（重大科技创新工程）项目专题（2022CXGC020710-6）
** 第一作者：代明明，硕士研究生，从事植物线虫学研究。E-mail：1061743919@ qq. com
*** 通信作者：赵洪海，教授，从事植物线虫学研究。E-mail：hhzhao@ qau. edu. cn

has been found to cause damage to potatoes in the potato fields of Weifang, Shandong Province. In this study, in order to identify the species and infestation dynamics of root-knot nematodes in some potato production areas in Shandong Province, samples were collected from Weifang and Pingdu for nematode isolation and identification, and a combination of morphology, rDNA-ITS region sequence analysis and specific primer amplification was used to identify the species of root-knot nematodes affecting potatoes. The results were as follows: morphological results showed that the potato root-knot nematode was consistent with the southern root-knot nematode (*Meloidogyne incognita*); phylogenetic results based on rDNA-ITS region sequences showed that the potato root-knot nematode population clustered with the southern root-knot nematode population reported in GenBank with 99%; The specific primer assay results also confirmed that it was the southern root-knot nematode. The growth and developmental dynamics of *M. incognita* were monitored by sampling during the potato reproductive period. The result was that two generations of *M. incognita* occurred in one potato growth cycle with the J2 in the soil initially infesting the potato root, colonizing, growing and laying eggs in the roots. The second generation of J2 infested the roots and potatoes during the potato setting and maturity period, resulted in a rough potato surface and a decrease in the commercial rate. Therefore, the production of potatoes like cool, intolerant of high temperature habits, appropriate early planting or selection of early varieties to achieve time to avoid or reduce the second generation of J2 damage to potato tubers.

Key words: Potato; Root-knot nematodes; Species identification; Infection dynamics

马铃薯（*Solanum tuberosum* L.）是世界上仅次于小麦、水稻、玉米的第四大粮食作物[1]，中国马铃薯种植面积和总产量均居世界第一位[2]。植物寄生线虫是马铃薯生长发育过程中重要限制性病原物之一[3,4]，在全球每年可造成超过 1 730 亿美元的损失[5]，我国每年因植物寄生线虫病害造成的损失达 700 亿元。其中，根结线虫是威胁全球农业生产的十大病原线虫之一，哥伦比亚根结线虫（*M. chitwoodi*）、伪哥伦比亚根结线虫（*M. fallax*）、爪哇根结线虫（*M. javanica*）、北方根结线虫（*M. hapla*）、南方根结线虫（*M. incognita*）和花生根结线虫（*M. arenaria*）是全球马铃薯生产中危害最大的 6 种根结线虫[6,7]，被欧洲植物保护组织（EPPO）列入检疫名录[8]。根结线虫以 2 龄幼虫（J2）侵入马铃薯根部和发育中的块茎，在根系上产生瘤状根结，块茎于采收时或储藏期内呈现凹凸不平的粗糙表面[9-11]，其商品化率严重下降[12]，不适于销售的块茎比例可达 30%~40%[13]。

如今，根结线虫已在我国部分马铃薯产区报道发生[14]，南方根结线虫（*M. incognita*）是我国根结线虫的优势种之一[15]，山东省首次在滕州市的马铃薯田中发现南方根结线虫（*M. incognita*）[16]。随着我国 2015 年正式开启马铃薯主粮化发展战略[17]，马铃薯种植面积保持平稳走高[18]，根结线虫对马铃薯的危害也日趋严重。不同根结线虫种群间寄主和致病性方面存在差异，因此明确根结线虫的种类是针对性防治根结线虫病的必要前提。本研究系统鉴定了马铃薯根结线虫种类及在马铃薯上的动态发育过程，可对有效防治马铃薯根结线虫病提供有用参考。

1　材料与方法

1.1　线虫样品的采集与分离

1.1.1　样品采集

马铃薯整个生育期内调查取样，样品采集自山东省潍坊市坊子区胜邦基地、青岛市平度市仁兆镇马铃薯田。每 7d 采集一次马铃薯地下组织及根际土壤样本，每块地随机取样 5 棵。

1.1.2 线虫分离

将采集的马铃薯根系用自来水洗净，采用直接解剖法[19]分离根结线虫雌虫，将采集的土壤样品充分混匀后，称取 500g 进行淘洗过筛（40-80-400 目），使用改良贝曼漏斗法[19]分离土壤样品中的病原线虫。

1.2 马铃薯根内线虫分离与染色

采用次氯酸钠-酸性品红法[20]对马铃薯根系内的线虫进行染色观察。将洗净的马铃薯根系称重后放入装有 50mL 蒸馏水和适量 5% NaClO 的烧杯中，4min 后用流水充分冲洗根系 45s 后，于蒸馏水中浸泡 15min 后转移至另一个盛有 30～50mL 的蒸馏水的烧杯中，在上述烧杯中加入 1mL 酸性品红并置于沸水 10min，冷却后用水漂洗，然后将根组织放入 30mL 甘油中煮沸使根系褪色，解剖镜下观察、记录不同龄期的根结线虫数量。

1.3 线虫形态鉴定

1.3.1 线虫杀死固定

采用温和热杀死法[21]进行，将线虫悬浮液静置 2h，吸走上清液，保留底部 5mL 左右，将试管放置于 55～60℃的水浴锅中 10min 后取出，待冷却后装入青霉素瓶中，并加入等体积的 5%福尔马林固定液进行固定备用。

1.3.2 形态学鉴定

用挑针挑取土壤中分离得到的 J2，置于载玻片水滴中制成临时玻片。解剖根结以获得雌虫 20 条，置于 45%乳酸中，将制作的会阴花纹置于有一滴甘油的载玻片上，加盖玻片。利用 ZEISS 显微镜观察线虫形态特征，并运用 De Man 公式对形态数据进行计算。以刘维志等人对线虫分类特征的描述为参照，鉴定线虫种类[22]。

1.4 线虫分子生物学鉴定

1.4.1 线虫 DNA 提取

采用彭德良等[23]线虫 DNA 提取方法。在体式显微镜下挑取不同马铃薯病根组织内的雌虫并置于装有 8μL ddH_2O 的 200μL 离心管中，反复冻融 3 次（液氮处理 1min，65℃水浴 1min）后置于冰上，然后在离心管中分别加入 1μL 蛋白酶 K 和 1μL 10×PCR buffer（Mg^{2+} Plus）。涡旋混匀后放置于 PCR 仪中，65℃温育 30min，95℃下 10min，反应结束即可得到单条线虫 DNA。

1.4.2 ITS-rDNA 区段 PCR 扩增

以 1.4.1 中提取的线虫 DNA 为模板，采用线虫通用引物 TW81（5′-GTTTCCGTAGGTGAACCTGC-3′）、AB28（5′-ATATGCTTAAGTTCAGCGGGT-3′）对目标线虫 rDNA-ITS 区序列进行 PCR 扩增。本研究采用 25μL 反应体系：DNA 2μL，10×Ex Tap Buffer 2.5μL，Ex Tap 0.5μL，dNTP Mix 2μL，上下游引物（10μmol/L）各 2μL，ddH_2O 14μL。PCR 扩增程序参考薛清等[24]方法。

1.4.3 PCR 产物的 T-A 克隆

根据 pMD™19-T Vector 使用手册，配置 5μL 反应体系：pMD™19-T Vector 1μL，PCR 纯化片段 2μL，ddH_2O 2μL，加入等体积的（5μL）Solution Ⅰ，16℃反应 16h。载体 pMD™19-

T 转化至 Trelief™5α 感受态细胞（北京擎科生物科技有限公司）中，37℃培养过夜。挑取单菌落培养后进行 PCR 鉴定。取 5μL PCR 产物于 1%琼脂糖凝胶电泳检测，条带大小检测无误后，阳性菌液送至北京擎科青岛分公司测序。使用 DNA man 进行序列拼接，BLAST 比对后，从 GenBank 中下载同源性较高的序列。利用 MEGA 7.0 的 Muscle 算法进行多重序列比对后，基于邻接法构建系统发育树。

1.4.4 南方根结线虫特异性引物检测

根据 ITS-rDNA 序列分析结果，采用南方根结线虫特异性引物[25] IncK-14-F（5′-GGGATGTGTAAATGCTCCTG-3′）、IncK-14-R（5′-CCCGCTACACCCTCAACTTC-3′）对提取的线虫 DNA 和南方根结线虫群体 DNA 模板进行 PCR 扩增。扩增程序为：95℃预变性 3min，95℃变性 30s，58℃退火 30s，72℃延伸 1min，34 个循环后，72℃终延伸 5min，最后 4℃保存备用。取 5μL PCR 产物于 1%琼脂糖凝胶电泳检测，紫外灯下观察、拍照。

2 结果与分析

2.1 马铃薯根结线虫病症状观察

根结线虫主要危害马铃薯地下组织（根系和块茎，图 1）。马铃薯生长发育初期，以具有迁移性侵染能力的 J2 侵染根部，一般从须根开始发病，侵染初期，根部无明显症状，随着根结线虫的生长发育，在根部可见隆起的根结，直径 2~5cm，影响根系吸收和运输能力。危害后期，根结线虫发育成雌虫后产卵，于根结末端可见根结线虫的卵块。适宜条件下，卵再次孵化形成 J2，其侵染马铃薯块茎后，薯块表面粗糙，危害严重时可见明显的虫瘿。

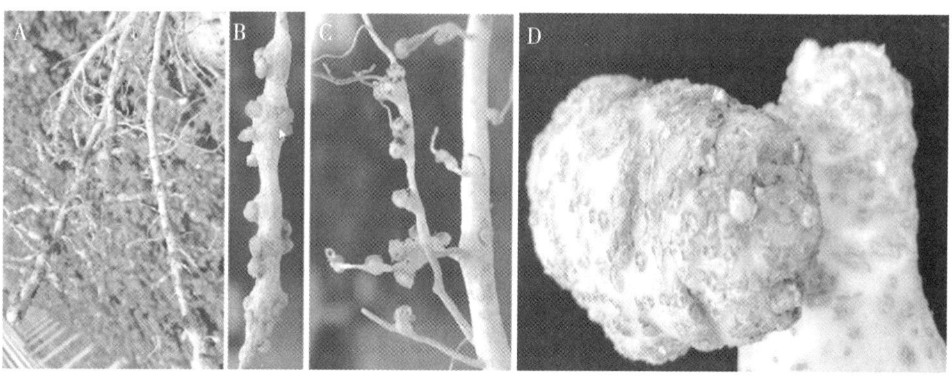

A-C. 根结线虫危害马铃薯根部症状；D. 根结线虫侵染马铃薯块茎症状

图 1 根结线虫危害马铃薯症状

2.2 形态学鉴定结果

马铃薯根结线虫 J2 呈线形，蠕虫状；头部无明显突起，可见不完整的环纹，口针纤细，约为 12.65μm，口针基部球小，尾部呈长锥形且末端具有清晰的透明区。J2 体长约 427.3μm，DGO 约为 2.13μm，尾长约 50.8μm。雌虫虫体呈梨形，乳白色，具一明显的短颈，口针纤细，约为 14.9μm。体长约 801.0μm，体宽约 559.6μm，DGO 约为 3.8μm。雌虫会阴花纹呈圆形或椭圆形，背弓高，方形或梯形，由平滑到波浪形的不连续线纹组成，稀疏，较粗；侧区不明显，线纹发生轻微弯曲；阴门窄，边缘光滑；阴门与肛门之间无线纹，肛门区无线纹（图2）。马铃薯根结线虫形态鉴定特征与已报道的多个南方根结线虫的形态特征一致。因此，初步将危害山东省潍坊市和山东省平度市马铃薯的根结线虫种类判定为南方根结线虫。

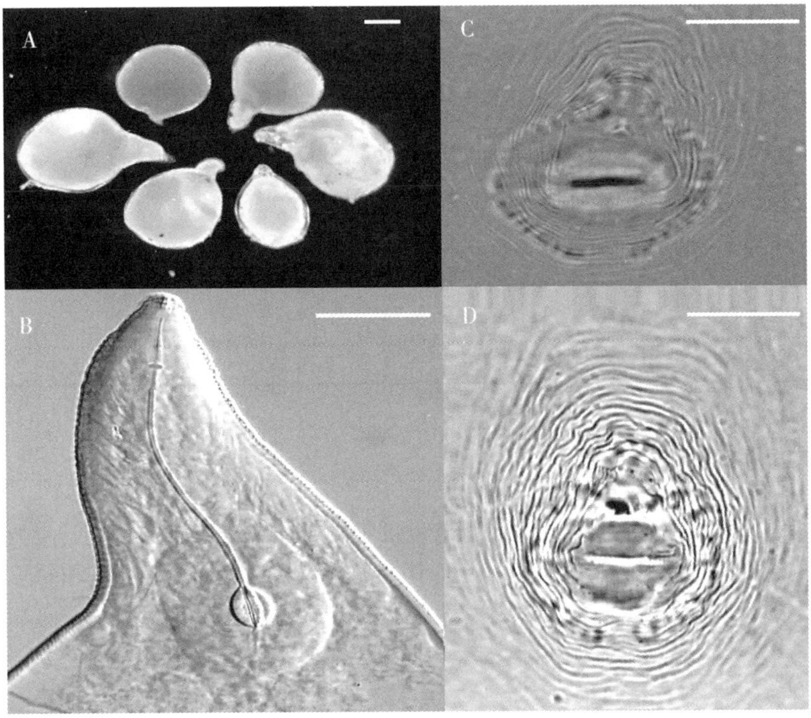

A. 整体虫态；B. 雌虫头部；C-D. 雌虫会阴花纹

（比例尺 A-D 均为 50μm）

图2　南方根结线虫雌虫虫态

测量数据详见表1和表2。

表1　南方根结线虫雌虫测量数据　　单位：μm

项目	数值	项目	数值
L	801. 0±100. 7（613. 4~951. 5）	MEVL	13. 3±1. 1（11. 2~14. 9）
W	559. 6±99. 8（351. 8~659. 9）	MEVW	11. 6±1. 3（10. 1~13. 9）
a	1. 47±0. 2（1. 1~1. 7）	MEVL/W	1. 2±0. 2（1. 0~1. 5）
ST	14. 9±1. 249（13. 3~17. 0）	MBL/W	0. 9±0. 1（0. 8~1. 0）
StK	3. 4±0. 4（2. 8~3. 8）	H-EP	42. 1±2. 2（38. 2~43. 9）
StH	1. 5±0. 3（1. 0~1. 8）	VSL	25. 0±3. 0（18. 9~31. 2）
StK/H	2. 3±0. 4（1. 8~2. 9）	V. a	17. 0±2. 3（11. 4~21. 0）
DGO	3. 8±0. 5（3. 0~4. 6）	A-TE	10. 4±1. 3（8. 5~13. 4）
MBL	47. 2±3. 3（42. 4~53. 9）	A-PH	9. 5±2. 1（5. 3~13. 2）
MBW	54. 2±6. 3（46. 4~64. 2）	PH-PH	27. 2±2. 5（23. 0~34. 3）
AM	92. 1±14. 3（76. 7~113. 2）		

表 2　南方根结线虫 2 龄幼虫测量数据　　单位：μm

项目	数值	项目	数值
L	427. 3±26. 2（382. 0~496. 7）	MBW	7. 5±0. 8（6. 3~9. 3）
W	14. 9±1. 7（12. 2~18. 5）	MBH	12. 7±1. 7（10. 0~16. 5）
a	29. 1±3. 8（20. 9~34. 4）	H-Ep	82. 5±5. 1（72. 9~90. 6）
ST	12. 7±0. 6（11. 7~14. 2）	b	3. 9±0. 6（2. 9~5. 8）
DGO	2. 1±0. 2（1. 7~2. 5）	b′	3. 3±0. 5（2. 4~4. 7）
StK	2. 5±0. 3（1. 9~3. 2）	Tail	50. 8±5. 8（40. 9~60. 8）
StH	1. 5±0. 4（1. 0~2. 6）	c	8. 5±1. 0（6. 8~10. 6）
M	39. 3±3. 7（32. 7~45. 9）	TTL	15. 9±2. 6（11. 6~20. 1）
LW	5. 6±0. 3（5. 2~6. 1）	ABW	9. 8±1. 6（7. 2~13. 1）
LH	2. 9±0. 4（2. 2~3. 9）	c′	5. 1±0. 6（4. 2~6. 3）
AM	60. 8±3. 8（53. 3~68. 6）	Tail/TTL	3. 3±0. 9（2. 3~6. 0）

2. 3　分子生物学特征

2. 3. 1　线虫 rDNA-ITS 序列分析

利用线虫通用引物 TW81/AB28 对根结线虫 DNA 样品进行 PCR 扩增，经纯化、克隆，双向测通后，获得根结线虫 ITS 区序列。经 BLAST 比对，发现从潍坊市和平度市仁兆镇分离得到的根结线虫与已知的南方根结线虫种群序列的同源性最高，为 99%。

2. 3. 2　线虫 rDNA-ITS 序列聚类分析

将得到的线虫种群 rDNA-ITS 区序列（OQ923660）在 NCBI 数据库中进行 Blast 比对，并从 GenBank 数据库中下载与目标序列相似性高的 10 个南方根结线虫序列及其 8 个近缘种的 rDNA-ITS 区序列，松材线虫作为种外群，采用 NJ 法同供试线虫构建系统发育树。结果表明：马铃薯根结线虫种群与 GenBank 中 10 个分别来自我国（福建、江苏、河南、台湾），以及印度等地的南方根结线虫种群（MT490905. 1、MT490194. 1、MT490922. 1、MT490926. 1、MT490915. 1、MT490894. 1、MT159698. 1、MT159701. 1、KT869139. 1、MT490908. 1、KP233823. 1、KP233824. 1）序列聚集在同一个分支上，其支持率为 99%（图 3）。另外，马铃薯根结线虫与象耳豆根结线虫、西班牙根结线虫、北方根结线虫、拟禾本科根结线虫、较小根结线虫、伪根结线虫、奇氏根结线虫、苹果根结线虫等近缘种聚类在不同分支上，南方根结线虫与西班牙根结线虫亲缘关系较近，但与其显著区分，进一步证实该根结线虫为南方根结线虫（*Meloidogyne incognita*）。

2. 3. 3　南方根结线虫特异性引物检测

采用 IncK-14F/IncK-14R 对所分离得到的根结线虫与已确定的南方根结线虫（*M. incognita*）同时进行 PCR 扩增检测。结果如图 4 所示，供试线虫和南方根结线虫（*M. incognita*）均扩增出 399bp 长的目标 DNA 片段（图 4），进一步说明了分离自马铃薯的根结线虫为南方根结线虫（*M. incognita*）。

2. 4　南方根结线虫侵染马铃薯动态

2. 4. 1　马铃薯根中南方根结线虫的群体动态

2021 年所调查马铃薯田南方根结线虫 J2 最早均出现于 4 月 18 日（马铃薯种植 15d），

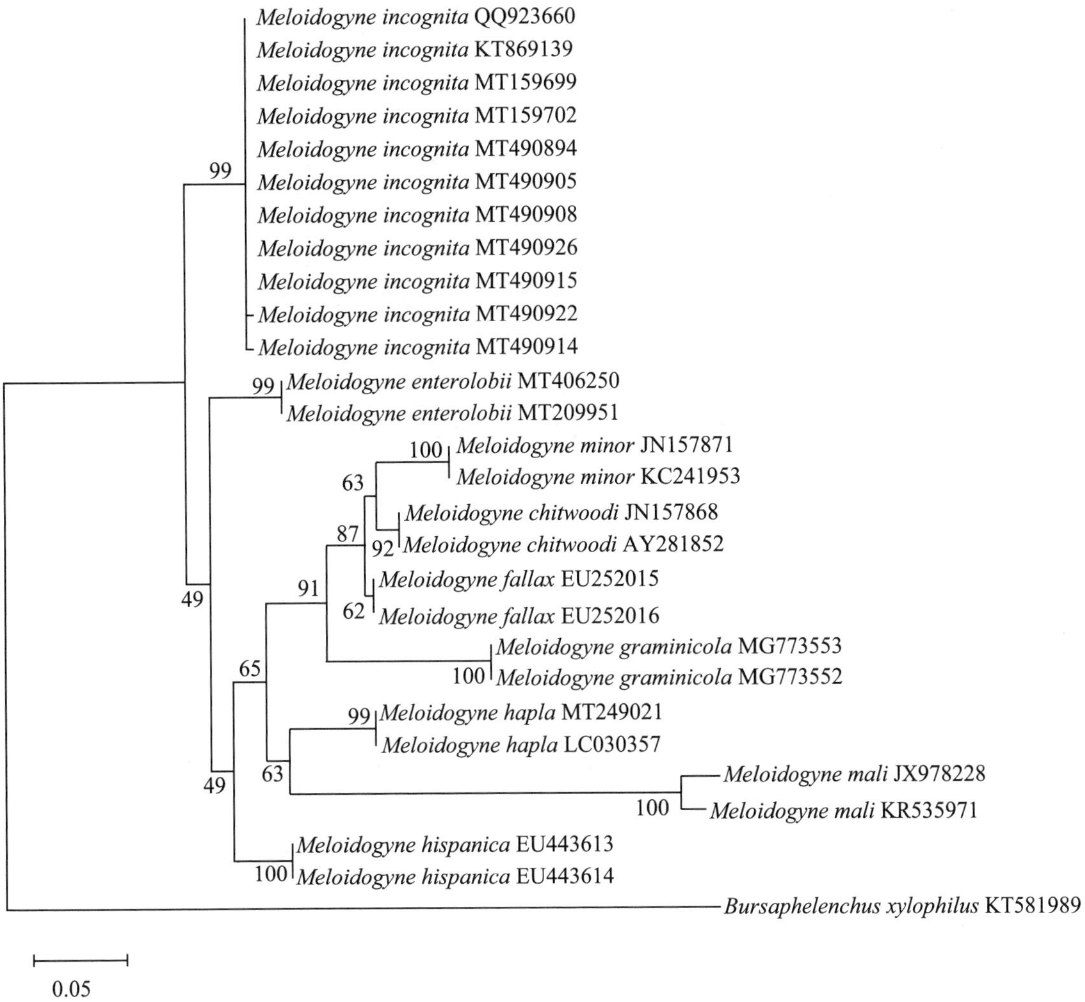

图 3　基于根结线虫 rDNA-ITS 区序列构建的系统发育树

之后侵入量明显呈上升趋势。侵入根后，J2 随即确立取食位点，根部可见隆起的根结。4 月 25 日马铃薯根内部分根结线虫发育成 3、4 龄幼虫（J3、J4），田块-1 和对照田块的马铃薯根内 J3、J4 分别占根内线虫总数的 33.3%和 60%，后经 14d 左右可见马铃薯根内线虫均发育成 J3、J4，其中田块-1 中马铃薯根内 4 龄幼虫最早于 5 月 16 日出现，由此判断，南方根结线虫完成幼虫阶段的发育（图 5）。

在调查的两块马铃薯田中，南方根结线虫雌虫最早于 5 月 23 日出现，因此判断南方根结线虫发育至成虫阶段。在随后的调查取样中发现逐渐有雌虫开始产卵，其中田块-1 最早于 5 月 30 日可见成熟雌虫产生，其密度为 48.4%。经 14d 后，马铃薯根部的根结线虫雌虫均已产卵。因此推断，南方根结线虫完成第一代生活史。

田块-1 和对照田中，南方根结线虫 J2 于 7 月 11 日再次出现，此时 J2 的密度分别为 6.78%和 23.57%，结果表明，第二代南方根结线虫 J2 开始危害。马铃薯收获时（7 月 11 日）调查取样发现，第二代南方根结线虫已发育成雌虫，分别占根部根结线虫总数的

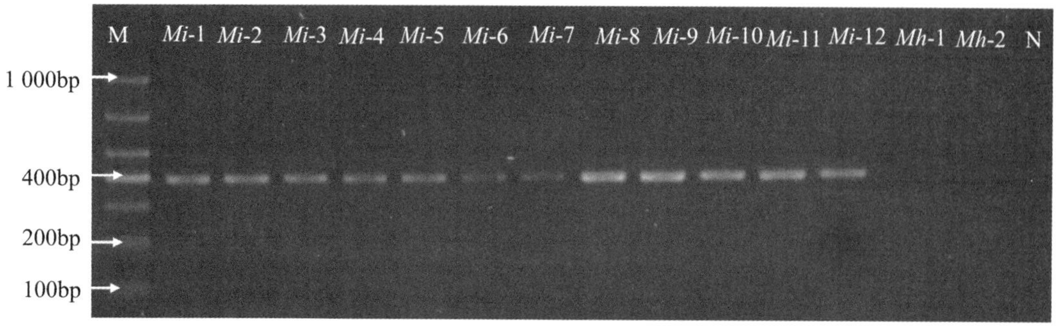

M 为 Marker；*Mi*-1 至 *Mi*-10 为雌虫；*Mi*-11 至 *Mi*-12 为阳性对照（*Mi*）；

Mh-1 至 *Mh*-2 为阴性对照；N 为 ddH_2O

图 4　南方根结线虫特异性引物扩增结果

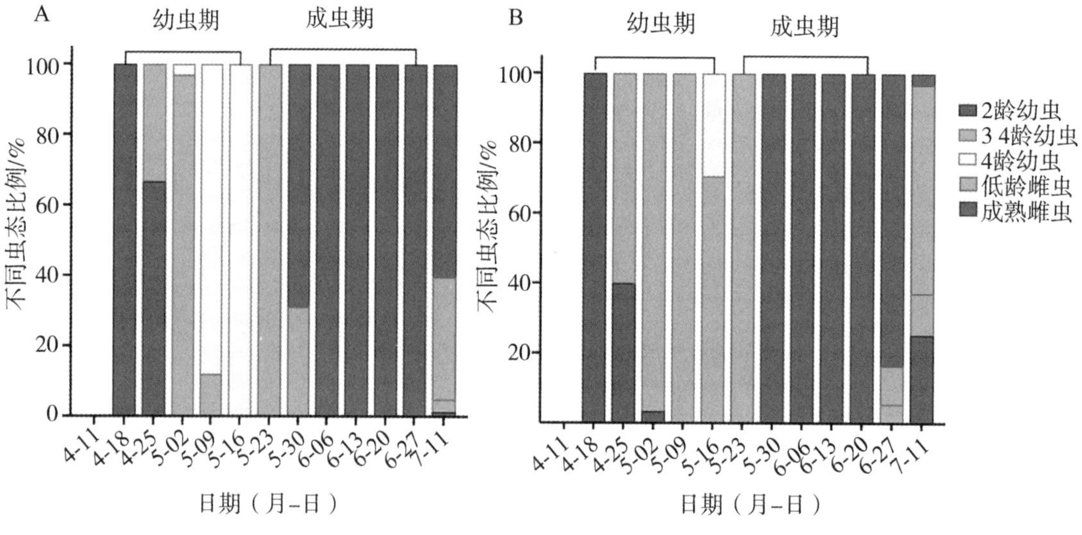

A. 田块-1；B. 对照田块

图 5　马铃薯根中南方根结线虫的群体动态

34. 86%和 59. 62%。

2. 4. 2　土壤中南方根结线虫 2 龄幼虫群体动态

2021 年 4 月 11 日（马铃薯种植后 15d）所调查的 3 块马铃薯种植田土壤中均已出现少量的南方根结线虫 J2，其密度分别为 18 条/500g、7 条/500g 和 18 条/500g，随着 J2 侵入马铃薯根系，其群体密度基本均呈下降趋势，并于 5 月 16 日和 23 日下降至最低点——500g 土壤中未分离到 J2，此时土壤中的南方根结线虫已侵入马铃薯的根中。随着第一代南方根结线虫卵的孵化，5 月 30 日后土壤中的根结线虫 J2 再次出现，且 3 块马铃薯种植田土壤中的 J2 数量逐渐上升，6 月 20 日对照田土壤中 J2 群体密度达到峰值，为 402 条/500g，田块-1 和田块-2 则于 7 月 11 日达到峰值。随着根结线虫 J2 侵入马铃薯块茎，对照田土壤中根结线虫 J2 的密度明显下降（图 6）。

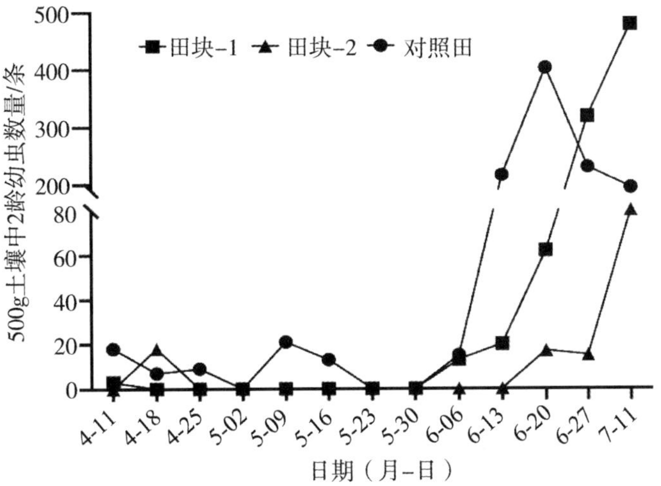

图 6　马铃薯根际土壤中 J2s 群体动态

3　结论与讨论

本研究通过形态学鉴定和分子生物学相结合的方法，发现山东省潍坊市和平度市的马铃薯根结线虫为南方根结线虫，与 Mao 等[26]、马海艳等[16]相继在云南省、山东省报道南方根结线虫危害马铃薯的研究结果基本一致。通过在马铃薯生育期内开展取样调查，发现 J2 于种植 15d 后侵入马铃薯根系，40~45d 后完成第一代生活史。于马铃薯种植后第 54 天根内雌虫数量最多且土壤中 J2 数量开始回升，收获时发现薯块内根结线虫已发育成 J3、J4，由此推测马铃薯薯块受第二代南方根结线虫的危害。

根结线虫生活史分为卵、幼虫和成虫 3 个阶段，J2 侵入植物根系，于根中柱维管束附近与植物建立宿主-寄生关系后诱导形成巨型细胞，最终在该位点定殖，蜕皮 3 次后发育成 J3、J4，最后发育成成虫、繁殖[27,28]。南方根结线虫生活史的长短与温度关系密切[28]，适宜温度和非致死温度均可生长、繁殖，但会引起生育期延长的情况[29]，冯志新[30]研究根结线虫的低温忍耐力发现 *M. chitwoodi* > *M. hapla* > *M. incognita* > *M. arenaria* > *M. javanica*。陈立杰等[31]研究认为，番茄上南方根结线虫 J2 正常存活的温度是 15~20℃，卵孵化的适宜温度为 15~30℃。Triantaphyllou 等[32]发现 29℃条件下，南方根结线虫侵入番茄根后 19~21d 出现第一个产卵雌虫。胡庆丽等[33]研究发现，南方根结线虫在马铃薯上繁殖的最适宜温度为 24~28℃，12℃可少量侵染，但是发育缓慢。本研究试验结果发现第一代经 35d 发育成雌虫，第二代发育历时有所缩短，可能与春、夏当地气温和降水有关。

本研究通过对马铃薯根结线虫的发育动态进行监测，表明南方根结线虫在马铃薯生育期内发生两代，第一代南方根结线虫危害马铃薯根系，卵孵化后第二代 J2 则对马铃薯块茎的品质造成影响。因此，在生产上结合马铃薯喜冷凉、不耐高温的习性[34,35]，可适当提前种植或选种早熟品种，实现时间上躲避或减轻第二代南方根结线虫 J2 对马铃薯块茎的危害。

参考文献

[1] HERMANSEN A, LU D, FORBES G. Potato production in China and Norway: similarities, differences and future challenges [J]. Potato Res, 2012, 55 (3-4): 197-203.

[2] 张烁. 中国马铃薯种植区划研究 [D]. 北京：中国农业科学院, 2021.

[3] MEDINA I L, GOMES C B, CORREA V R, *et al.* Genetic diversity of *Meloidogyne* spp. Parasitising potato in Brazil and aggressiveness of *M. javanica* populations on susceptible cultivars [J]. Nematology, 2017, 19: 69-80.

[4] ŽIBRAT U, GERIČ STARE B, KNAPIČ M, *et al.* Detection of Root-Knot Nematode *Meloidogyne luci* infestation of potato tubers using hyperspectral remote sensing and real-time pcr molecular methods [J]. Remote Sensing, 2021, 13 (10): 1996.

[5] 彭焕，赵薇，姚珂，等. 植物寄生线虫基因组学研究进展 [J]. 生物技术通报, 2021, 37 (7): 3-13.

[6] LIMA F S O, MATTOS V S, SILVA E S, *et al.* Nematodes affecting potato and sustainable practices for their management [M]. Potato-From Incas to All over the World. IntechOpen, 2018.

[7] ONKENDI E M, KARIUKI G M, MARAIS M, *et al.* The threat of root-knot nematodes (*Meloidogyne* spp.) in Africa: a review [J]. Plant Pathology, 2014, 63: 727-737.

[8] EPPO Global Database: *Meloidogyne chitwoodi.* EPPO datasheets on pests recommended for regulation. Available online: https: //gd. eppo. int (accessed on 9 April 2021).

[9] BALI S, ZHANG L, FRANCO J, *et al.* Biotechnological advances with applicability in potatoes for resistance against root-knot nematodes [J]. Current Opinion in Biotechnology, 2021, 70: 226-233.

[10] JATALA P. A review of the research program on the development of resistance to the root-knot nematodes *Meloidogynes* spp. [C] //Report of the 2nd Nematode Planning Conference, 1978. The International Potato Center, 1978.

[11] JATALA P, BOOTH R H, WIERSEMA S G. Development of *Meloidogyne incognita* in stored potato tubers [J]. Journal of Nematol. 1982, 14 (1): 142-143.

[12] INGHAM R E, HAMM P B, WILLIAMS R E, *et al.* Control of *Meloidogyne chitwoodi* in potato with fumigant and nonfumigant nematicides [J]. Journal of Nematol, 2000, 32: 556-565.

[13] REGMI H, DESAEGER J. Integrated management of root-knot nematode (*Meloidogyne* spp.) in Florida tomatoes combining host resistance and nematicides [J]. Crop Protection, 2020, 134: 105-170.

[14] 毛彦芝，牛若超，孙继英，等. 马铃薯田线虫病害发生及防治 [J]. 土壤与作物, 2022, 11 (1): 104-114.

[15] JINLING L, HAN J, LONGHUA S, *et al.* Identification of species and race of root-knot nematodes on crops in southern China [J]. Journal of Huazhong Agricultural University, 2003, 22: 544-548.

[16] 马海艳，安修海，邢佑博，等. 山东首次发现南方根结线虫危害马铃薯 [C] //马铃薯产业与美丽乡村, 2020: 604-607.

[17] 谷悦. 马铃薯主粮化为国家粮食安全战略重要一步：农业部公开解答关于马铃薯主粮化的问题 [J]. 中国食品, 2015 (3): 36-39.

[18] 徐宁，张洪亮，张荣华，等. 中国马铃薯种植业现状与展望 [J]. 中国马铃薯, 2021, 35 (1): 81-96.

[19] 赵洪海，梁晨，刘维志，等. 采自烟台市的南方根结线虫的描述 [J]. 莱阳农学院学报, 2000 (2): 81-85.

[20] 刘维志. 中国检疫性植物线虫 [M]. 北京：中国农业科学技术出版社，2004.
[21] 倪春辉，李惠霞，刘永刚，等. 甘肃党参中花生茎线虫的鉴定与记述 [J]. 华南农业大学学报，2022，43 (4)：99-105.
[22] 谢辉. 植物线虫分类学 [M]. 2 版. 北京：高等教育出版社，2005.
[23] 彭德良，SUBBOTIN S，MOENS M. 小麦禾谷孢囊线虫 (*Heterodera avenae*) 的核糖体基因 (rDNA) 限制性片段长度多态性研究 [J]. 植物病理学报，2003，33e：7.
[24] 薛清，杜虹锐，薛会英，等. 苜蓿滑刃线虫线粒体基因组及其系统发育研究 [J]. 生物技术通报，2021，37 (7)：98-106.
[25] RANDIG O，BONGIOVANNI M，CARNEIRO R M，*et al*. Genetic diversity of root-knot nematodes from Brazil and development of SCAR markers specific for the coffee-damaging species [J]. Genome，2002，45：862-870.
[26] MAO Y Z，HU Y F，LI C J，*et al*. Molecular identification of Meloidogyne species isolated from potato in China and evaluation of the response of potato genotypes to these isolates [J]. Nematology，2019，21：847-856.
[27] MOENS M，PERRY R N，STARR J L. Meloidogyne species-a diverse group of novel and important plant parasites [J]. Root Knot Nematodes，2009. DOI：10. 1079/9781845934927. 0001.
[28] EISENBACK J D，TRIANTAPHYLLOU H H. Root-knot nematodes：Meloidogyne species and races [M] //Manual of agricultural nematology. CRC Press，2020：191-274.
[29] 刘维志. 植物病原线虫学 [M]. 北京：中国农业出版社，2000.
[30] 冯志新. 植物线虫学 [M]. 北京：中国农业出版社，2001.
[31] 陈立杰，魏峰，段玉玺，等. 温湿度对南方根结线虫卵孵化和二龄幼虫的影响 [J]. 植物保护，2009，35 (2)：48-52.
[32] TRIANTAPHYLLOU A C，HIRSCHMANN H. Post-infection development of *Meloidogyne incognita* Chitwood 1949 (Nematoda：Heteroderidae) [J]. Ann. Inst. Phytopathol.，Benaki，1960，3：3-11.
[33] 吴庆丽，王鲜，廖金铃，等. 不同光温条件对马铃薯繁殖根结线虫效果的影响 [J]. 植物保护，2006 (6)：27-29.
[34] 薄秀娟. 马铃薯高产高效种植技术的应用 [J]. 农家参谋，2022 (13)：45-47.
[35] 苗起萃，王靖东. 地膜覆盖马铃薯高产高效种植技术 [J]. 农家致富顾问，2021 (4)：47.

中国线虫学研究（第九卷）Nematology Research in China（Vol. 9）：32-33

食用菌菌渣防治番茄根结线虫病研究初探*

覃丽萍[1,2,3]**，黄金玲[1,2,3]，李工芳[1,2,3]，刘峥嵘[4]，陆秀红[1,2,3]***
（[1]广西壮族自治区农业科学院植物保护研究所，南宁　530007；
[2]广西作物病虫害生物学重点实验室，南宁　530007；
[3]农业农村部华南果蔬绿色防控重点实验室，南宁　530007；[4]广西大学，南宁　530005）

Preliminary Study on Use of Edible Fungi Residues to Control Root-knot Nematode Disease of Tomato*

Qin Liping[1,2,3]**, Huang Jinling[1,2,3], Li Gongfang[1,2,3], Liu Zhengrong[4], Lu Xiuhong[1,2,3]***
([1] *Plant Protection Research Institute, Guangxi Academy of Agricultural Science, Nanning* 530007, *China*; [2] *Guangxi Key Laboratory of Biology for Crop Diseases and Insect Pests, Nanning* 530007, *China*; [3] *Key Laboratory of Green Prevention and Control on Fruits and Vegetables in South China Ministry of Agriculture and Rural Affairs, Nanning* 530007, *China*; [4] *Guangxi University, Nanning* 530005, *China*)

摘　要：根结线虫（*Meloidogyne* spp.）是番茄（*Lycopersicon esculentun*）的重要病原线虫，在全世界广泛分布，可引起番茄产量和品质下降，造成严重经济损失。目前番茄线虫病以化学药剂防治为主，具有简便、高效的优点，但化学杀线虫剂的长期使用易造成环境污染、农药残留超标等问题，在农药、化肥双减的大趋势下，生产上迫切需要研究、应用无公害技术来防治根结线虫病。

有研究发现食用菌菌渣对线虫有良好的防治效果，且对作物有增产作用；还有研究提示，菌渣对线虫的防治效果可能与所栽培的食用菌品种有关。为找准利用对象，提高利用效率，笔者通过室内盆栽开展了秀珍菇（*Pleurotus pulmonarius*）菌渣覆土对番茄根线虫病的防治效果研究。结果显示：秀珍菇菌渣覆土 2 个月后番茄的根结指数为 29. 63，对照的根结指数为 48. 62，二者差异达 5%显著水平，秀珍菇菌渣覆土处理对番茄根结线虫病的防治效果为 39. 29%；秀珍菇菌渣覆土处理番茄植株的平均株高、单株总鲜重和茎径分别为 79. 51cm、93. 04g、7. 27mm，对照的则分别为 72. 53cm、82. 69g、6. 36mm，其中秀珍菇菌渣覆土处理的株高和茎径分别比对照提高了 9. 63%和 14. 36%，差异达 5%显著水平，秀珍菇菌渣覆土处理的单株总鲜重虽与对照相比差异未达显著水平，但仍比对照提高了 12. 52%。本研究表

* 基金项目：广西自然科学基金（2020GXNSFAA297076）；广西农业科学院科技发展基金（桂农科 2021YT062）

** 第一作者：覃丽萍，从事植物病害生物防治研究。E-mail：qlp961003@163.com
*** 通信作者：陆秀红，副研究员，从事植物线虫病防治研究。E-mail：447597587@qq.com

明秀珍菇菌渣可显著降低番茄根结线虫病的危害，同时还可促进番茄生长，提高生物量积累。利用食用菌菌渣防治根结线虫病相当有潜力、前景，值得深入研究并加以应用，以促进线虫病绿色防控目标的实现。

关键词：食用菌菌渣；番茄根结线虫；秀珍菇；绿色防控

中国线虫学研究（第九卷）Nematology Research in China（Vol. 9）：34-39

福建省水稻潜根线虫的种类鉴定及种群结构*

黄泓晶**，柯叶鑫，李　硕，肖　顺，程　曦，刘国坤***
（福建农林大学/生物农药与化学生物学教育部重点实验室，福州　350002）

摘　要：对福建省 15 个主稻区的 168 个水稻样本的潜根线虫种类进行了分离，所有样本均可分离出潜根线虫，线虫经形态学特征与分子生物学方法鉴定为细尖潜根线虫（*Hirschmanniella mucronata*）和水稻潜根线虫（*H. oryzae*），前者分离频率为 77.4%，后者分离频率为 45.8%，混合种群的分频率为 23.2%。该结果与 20 世纪 80 年代前报道的福建省稻田潜根线虫种群结构相比，发生了很大的改变。细尖潜根线虫成为优势种，可能与水稻栽培模式、水稻品种、水肥管理等改变有关。
关键词：水稻；细尖潜根线虫；水稻潜根线虫；种群结构

Species Identification and Structure of *Hirschmanniella* spp. on Rice in Fujian Province, China*

Huang Hongjing**, Ke Yexin, Li Shuo, Xiao Shun, Cheng Xi, Liu Guokun***
(*Key Laboratory of Biopesticide and Chemical Biology*, *Ministry of Education*, *Fujian Agriculture and Forestry University*, *Fuzhou* 350002, *China*)

Abstract: A survey of *Hirschmanniella* spp. in 15 main rice producing regions in Fujian Province, China. *Hirschmanniella* spp. were found in 100% of the root samples collected. *H. mucronata* and *H. oryzae have been* isolated and identified based on morphological and molecular characteristics, with separation frequency of 77.4% and 45.8% respectively, and the isolation frequency of the two mixed populations was 23.1%. The result is different from the species structures of *Hirschmanniella* reported in 1980s in Fujian Province. *H. mucronata has been becoming dominant species*, *which may* related to changes of rice varieties, cultivation models of rice, irrigation and fertilization management.
Key words: *Oryza sativa*; *Hirschmanniella mucronata*; *H. oryzae*; Population construction

潜根线虫（*Hirschmanniella* spp.）是水稻上的重要寄生线虫，通过侵入水稻根系后在皮层薄壁组织里迁移取食危害，但地上部多表现生长不良现象，无特异性症状，隐蔽性强。全世界估计 58%的水稻田被潜根线虫危害，可引起约 25%的产量损失[1]。潜根线虫在我国水稻产区广泛分布，共发现有 16 种潜根线虫[2]，主要以混合种群在田间发生，在调查田块中，

---

* 基金项目：福建省现代农业水稻产业技术体系建设项目（2019—2021）
** 第一作者：黄泓晶，硕士研究生，研究方向为植物线虫病害及线虫病害防治。E-mail：494973370@qq.com
*** 通信作者：刘国坤，教授，研究方向为植物线虫病害及线虫病害防治。E-mail：liuguok@126.com

水稻潜根线虫（*H. oryzae*）均为优势种群，其他常见混合发生种类主要有纤细潜根线虫（*H. gracilis*）、细尖潜根线虫（*H. mucronata*）、小结潜根线虫（*H. microtyla*）、贝尔潜根线虫（*H. belli*）等[3]。福建省此前报道主要以水稻潜根线虫（*H. oryzae*）为优势种，混合种群中占比 60%~80%，与小结潜根线虫、细尖潜根线虫、纤细潜根线虫等混合发生[4]。近 20 多年来，福建省水稻品种与栽培模式等发生了很大变化，作者对水稻潜根线虫种类调查研究中，发现潜根线虫种群结构发生了较大的改变，现将研究结果报道如下。

1 材料与方法

1.1 样本采集与线虫分离

2020—2021 年，在福建省福州市、三明市、南平市、泉州市、宁德市、漳州市等 7 个地区的 15 个主要的水稻种植区，随机采集分蘖末期或成熟期水稻根部样本，共 168 个。采集的样本根系经自来水冲洗，将根剪成 2mm 小段，采用浅盘法[5]分离潜根线虫。

1.2 潜根线虫形态学鉴定

挑取分离的线虫制作临时玻片[5]，在尼康显微镜（Eclipse Ni-U 931609）下观察，用配套 NikonDS-Ri1 相机拍照，用配套软件（NIS-Elements BR）测量，根据 de Man 公式[5]进行形态测计。

1.3 潜根线虫的分子生物学鉴定

1.3.1 DNA 提取与扩增

采用冻融裂解法[6]对单条线虫进行 DNA 提取。采用 25μL 扩增反应体系：7.5μL ddH_2O，上下游引物各 1μL、3μL 线虫 DNA 模板，12.5μL EX Premix TaqTM（TAKARA）。

ITS 区通用引物对为 TW81/AB28（5′-GTTTCCGTAGGTGAACCTGC-3′；5′-ATATGCTTAAGTTCAGCGGGT-3′）[7]，扩增反应条件为 94℃，4min；94℃，30s；55℃，30s；72℃，60s（35 个循环）；72℃，10min。

28S D2-D3 区通用引物对为 D2A/D3B（5′-ACAAGTACCGTGAGGGAAAGTTG-3′；5′-TCGGAAGGAACCAGCTACTA-3′）[8]，扩增反应条件为 94℃，3min；94℃，30s；54℃，40s；72℃，2min（35 个循环）；72℃，10min。

1.3.2 序列处理、提交与系统发育树构建

PCR 产物用 1%的琼脂糖凝胶电泳检测确定为目标条带后，送至上海生工生物公司进行测序，获得的序列上传至 Genbank，获取序列号。采用模型 G-INS-I[8]进行多重序列比对，除去无效点位，转换为 nexus 与 PHYLIP 格式，使用 MrBayes 3.2.7 on XSEDE 程序，构建贝叶斯树[9]，使用 RAxML-HPC2 on XSEDE 程序构建极大似然树[10]，最后在贝叶斯 50%多数原则一致的树上用 Adobe Illustrator CC 软件对自展值（BS）和后验概率（PP）进行标注。

2 结果分析

2.1 2 种潜根线虫的形态特征

在所分离的水稻根系样本中，发现 2 种潜根线虫，根据形态特征，分别鉴定为水稻潜根线虫与细尖潜根线虫。

水稻潜根线虫：虫体细长，头部低平，具有 3~5 个环纹，口针小于 19μm，口针基部球

圆。雌虫阴门位于体中部偏后，卵巢对生。雄虫交合刺发达，交合伞覆盖尾部超过 2/3，但未延伸至尾尖；雌雄虫尾部长锥形，末端有一针状的腹面尾尖突（图 1）。

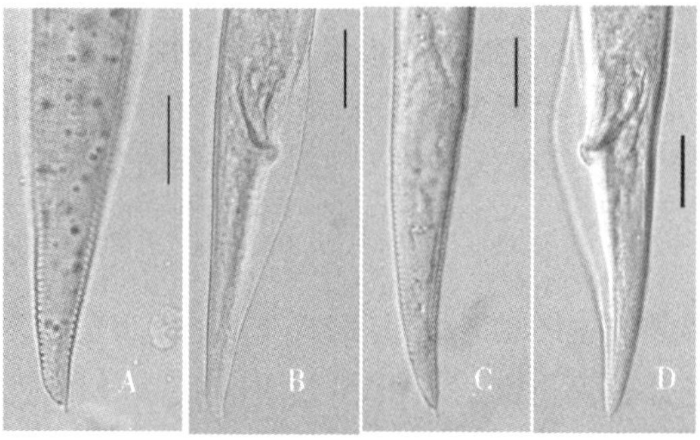

A-B. 水稻潜根线虫尾部，A 为雌虫，B 为雄虫；
C-D. 细尖潜根线虫尾部，C 为雌虫，D 为雄虫（比例尺：20μm）

图 1　水稻潜根线虫（*H. oryzae*）与细尖潜根线虫（*H. mucronata*）尾部形态

细尖潜根线虫：虫体细长，头部高圆，呈半球形，无缢缩，唇区体环小于等于 5 个。口针通常大于 25μm，口针基部球大且圆。雌虫阴门位于虫体中部，双生殖腺对伸。雄虫交合伞覆盖尾部超过 2/3，未延伸至尾尖。雌虫虫尾部尖锥形，尾尖突尖细，位于尾的中央（图 2）。

2.2　分子生物学鉴定

对采集分离的福建省 7 个市 15 个稻区的多个潜根线虫种群进行 ITS 区、28S D2D3 区测序，所得序列结果上传 Genbank，获得序列号（图 2，图 3）。

将水稻潜根线虫种群的 ITS 区序列进行数据库比对，其与广东（MT704933）、台湾（DQ309588）等地的 *H. oryzea* 种群一致性达 98%以上。细尖潜根线虫种群 ITS 区序列与广东（MT704948）、台湾（DQ309589）等地的 *H. mucronata* 种群一致性达 97%以上。系统发育树（图 2）表明，所有的 *H. mucronata* 聚集在同一分支（PP = 0. 99，BS = 60），所有的 *H. oryzea* 聚在同一大分支（PP = 1，BS = 100）。

将水稻潜根线虫种群 28S D2D3 区序列进行数据库比对，其与云南（MN445998）、伊朗（JX291141）等地的 *H. oryzea* 种群一致性均达 99%以上。细尖潜根线虫种群与广东（KY424320）、泰国（MT597913）等地的 *H. mucronata* 种群一致性均达 99%以上。系统发育树（图 3）表明，所有 *H. mucronata* 聚集在同一大分支（PP = 1，BS = 92），所有的 *H. oryzea* 聚在同一大分支（PP = 1，BS = 28）。

2.3　潜根线虫种类的分离频率

所采集的福建省 15 个稻区的 168 个水稻根系样本中均分离到潜根线虫，经形态学观测与分子生物学鉴定，77 个样本分离到水稻潜根线虫，分离频率为 45. 8%，130 个样本分离到细尖潜根线虫，占比 77. 4%；另外 39 个样本为混合种群，分离到 2 个种，占比 23. 2%。

3　讨论

本次对福建省水稻主产区潜根线虫的调查鉴定中，只发现 2 种潜根线虫，分别为水稻潜

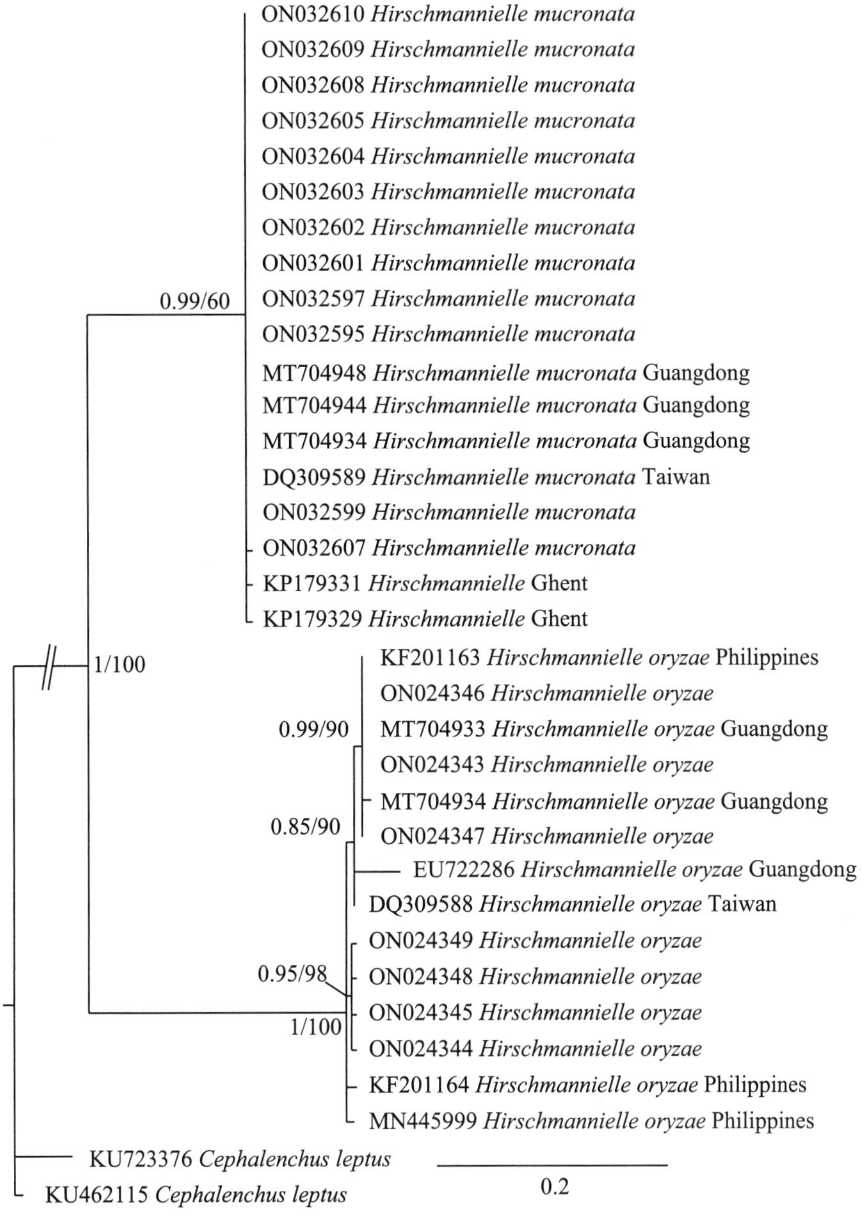

图 2　基于 ITS 区以贝叶斯法和极大似然法整合构建的系统发育树（分支数值 PP/BS）

根线虫和细尖潜根线虫，水稻潜根线虫种群单独分离频率为 22.6%，细尖潜根线虫种群单独分离频率为 54.2%，混合种群分频率为 23.2%。1981 年尹淦鏐首次报道中国南方稻区普遍发生水稻潜根线虫病，随后在全国各地水稻产区调研过程中，发现均有潜根线虫危害，大多为混合种群，其中水稻潜根线虫为优势种群[11]。福建省稻区在 80 年代调查过程中，水稻潜根线虫和小结潜根线虫为常见种，分离频率分别为 81% 和 91%，其余种群以细尖潜根线虫等 6 种为次要种，分离频率仅达 9%～27% 不等[12]。2004—2007 年，谢志成调查福州金山地区水稻，发现水稻潜根线虫为优势种，在群体中占比 79.6%，细尖潜根线虫、小结潜根

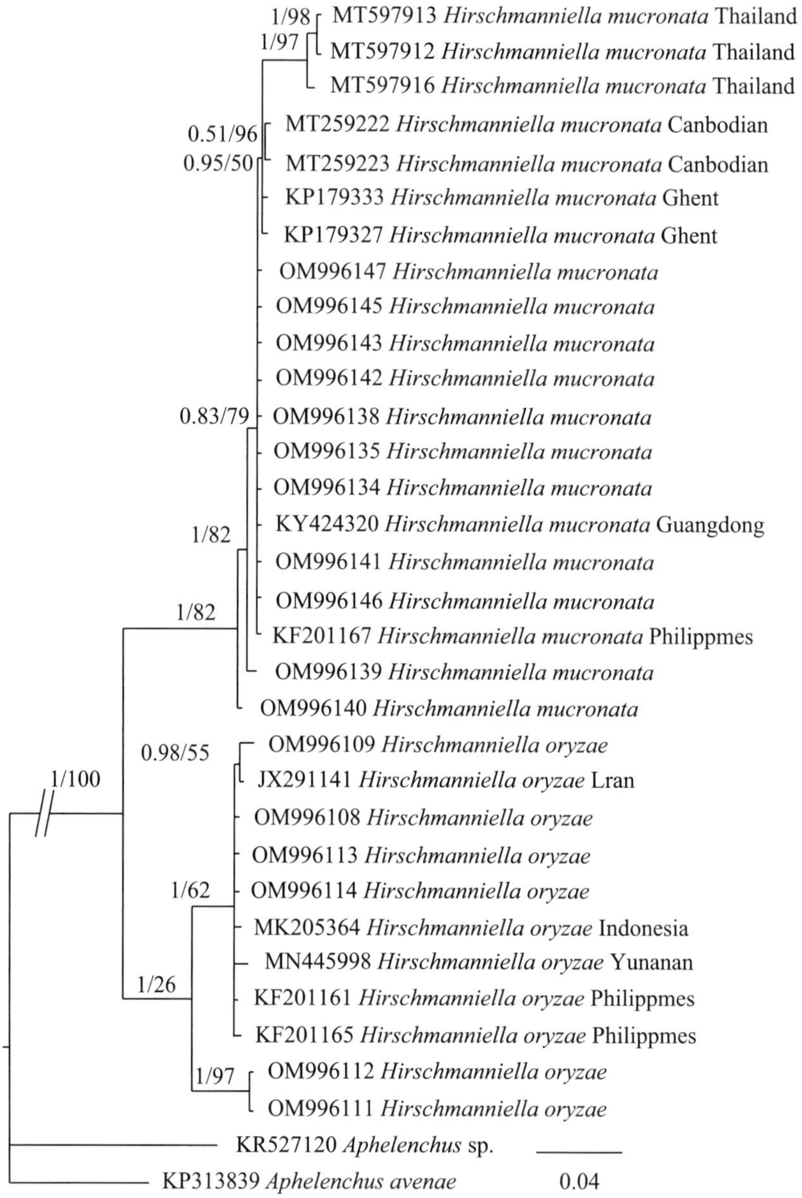

图 3　基于 28S D2D3 区序列构建的系统发育树（分支数值 PP/BS）

线虫等为次要种[4]。在本次调查涉及 7 个地市的水稻田块中，均可分离到潜根线虫，但种群结构已发生改变，在许多田块中，细尖潜根线虫与水稻潜根线虫可为单一种群，只有 23.2%样本中为混合种群；细尖潜根线虫由次要种变为优势种，水稻潜根线虫由优势种变为次优势种。福建省潜根线虫种群结构发生改变可能同水稻品种、栽培模式、水肥管理等有关，蔬菜轮作、烟草轮作可使水稻潜根线虫数量分别下降 88.1%和 82.9%[13]。谢志成[4]的研究表明，所有水稻品种均会受潜根线虫侵染，但不同水稻品种对潜根线虫有一定的选择性，所以潜根线虫的种群结构有差异性。前人研究更多集中于水稻潜根线虫种的研究，但近年来一些国家和地区报道发现细尖潜根线虫为稻田优势种，危害性大，如在泰国巴吞他尼府（Pathum Thani）水稻田间分离的优势种经形态与分子鉴定均为细尖潜根线虫[14]；我国在江苏省农业科学院试验地分离到的潜根线虫优势种为细尖潜根线虫[15]。因此研究潜根线虫种间竞争力与种群结构演替，对于选育出抗潜根线虫病的水稻品种具有重要意义。

参考文献

[1] 王义成，金晨钟，岳再阳．水稻潜根线虫病的危害损失和防治技术［J］．湖南农业科学，1988（2）：39-41.

[2] HU X Q，YU M，LIN L F，*et al.* Species and distribution of rice root nematode in Yunnan Province，China［J］．Journal of Integrative Agriculture，2004，3（8）：39-44.

[3] 殷友琴，李学文．水稻潜根线虫病发生和防治研究［J］．湖南农学院学报，1984，3：61-70.

[4] 谢志成．水稻根部线虫鉴定及潜根线虫根结线虫对水稻的致病性［D］．福州：福建农林大学，2007.

[5] 张绍升，刘国坤，肖顺．中国作物线虫病害研究与诊控技术［M］．福州：福建科学技术出版社，2021.

[6] LIU G K，Chen J，XIAO S，*et al.* Development of species-specific PCR primers and sensitive detection of the *Tylenchulus semipenetrans* in China［J］．Agricultural Sciences in China，2011，10（2）：252-258.

[7] SUBBOTIN S A，STURHAN D，CHIZHOV V N，*et al.* Phylogenetic analysis of *Tylenchida thorne*，1949 as inferred from D2 and D3 expansion fragments of the 28S rRNA gene sequences［J］．Nematology，2006，8（3）：455-474.

[8] KATOH K，STANDLEY D M. MAFFT multiple sequence alignment software version 7：improvements in performance and usability［J］．Molcula Biology and Evolution，2013，30（4）：772-780.

[9] RONQUIST F，HUELSENBECK J P. MrBayes 3：Bayesian phylogenetic inference under mixed models［J］．Bioinformatics，2003，19（12）：1572-1574.

[10] STAMATAKIS A，HOOVER P，ROUGEMONT J. A rapid bootstrap algorithm for the RAxML Web servers［J］．Systematic Biology，2008，57（5）：758-771.

[11] 尹淦镠，冯志新．农作物寄生线虫的初步调查鉴定［J］．植物保护学报，1981，3（2）：111-126.

[12] 张绍升．福建稻田潜根线虫（*Hirschmanniella* spp.）七个种鉴定初报［J］．福建农学院学报，1987（2）：155-159.

[13] 高学彪，周慧娟．几种农业措施对水稻潜根线虫病的防治作用及机理的研究［J］．华中农业大学学报，1998，17（4）：4.

[14] 山草梅，叶蕾，张连虎，等．水稻抗潜根线虫基因 OsRAI1 的克隆及功能分析［J］．生物技术通报，2021，37（7）：146-155.

[15] 冯辉，魏利辉，陈怀谷，等．细尖潜根线虫（*Hirschmanniella mucronata*）江苏分离群体形态学和分子特征描述［J］．植物病理学报，2016，46（4）：474-484.

A New Record of *Laimaphelenchus spiflatus* Gu *et al.*, 2020（Rhabditida：Aphelenchoididae）from Shanxi Province, with Proposal of *L. liaoningensis* Song *et al.*, 2020 as a Junior Synonym*

Wang Liyi[1**], Zhao Zengqi[2], Wang Jianming[1], Xu Yumei[1***]

（[1] *Laboratory of Nematology, Department of Plant Pathology, College of Plant Protection, Shanxi Agricultural University, Taigu* 030801, *China*；

[2] *Manaaki Whenua-Landcare Research, Private Bag* 92170, *Auckland Mail Centre, Auckland* 1142, *New Zealand*）

Abstract：This contribution reports a new finding of *Laimaphelenchus* species in the Shanxi Province of northern China. Furthermore, a comparison was made between *L. spiflatus* and *L. liaoningensis* using original morphological, morphometrical and molecular data. The shape and structure of the tail tip were considered as critical morphological character for *Laimaphelenchus* identification, and no differences were observed between *L. spiflatus* and *L. liaoningensis*. Specifically, the stalk lacked tubercles but had flat fused stacked structures with 8-12 finger-like appendages. The partial small subunit（SSU）and D2-D3 segments of large subunit（LSU）rDNA were also highly similar, differing in only 1-3 positions in alignment. Therefore, the taxonomic status of *L. liaoningensis* was determined to be a junior synonym of *L. spiflatus* in this study.

Key words：LSU；Morphometrics；SSU；Compare；Junior synonym

1 Introduction

Laimaphelenchus spiflatus was originally described in 2020 from the twig of a declining Chinese pine, *Pinus tabuliformis* in Beijing, China[1]. In the same year, *L. liaoningensis* was discovered in wood samples collected from dead pine trees（*P. armandii*）in Shenyang, Liaoning Province, China[2]. Jahanshahi Afshar *et al.*（2021）suggested that *L. liaoningensis* and *L. spiflatus* are likely the same species based on the identical D2-D3 sequences of both species[3].

Laimaphelenchus spiflatus has recently been identified in a survey of aphelenchid nematodes

* Funding：National Natural Science Foundation of China（31801958）；Central Project Guide Local Science and Technology for Development of Shanxi Province（YDZJSX2021A033）

** First author：Wang Liyi, Master student, mainly engaged in the study of nematology. E-mail：18235864767@163.com

*** Corresponding author：Xu Yumei, Professor, mainly engaged in the study of nematology. E-mail：ymxu@sxau.edu.cn associated with conifers in Shanxi Province. This study documents the presence of *L. spiflatus* in Shanxi Province and provides a discussion of the status of *L. liaoningensis* and *L. spiflatus* based on a comparison of their morphology, morphometric, and molecular data.

2 Material and Method

Bark samples were collected from Meyer spruce (*Picea meyeri* Rehder & E. H. Wilson) (GPS coordinates: 37° 25′ 19″ N, 112° 34′ 56″ E, 800m a. s. l.) and oleaster trees (*Elaeagnus angustifolia* L.) (GPS coordinates: 37°25′17″N, 112°35′9″E, 800m a. s. l.) on the campus of Shanxi Agricultural University, Taigu district, from Shanxi Province, China.

The methods for nematode extraction and specimens processing were described in detail by Wang *et al.* (2022)[4]. For DNA extraction, PCR, and sequencing, the protocol described by Xu *et al.* (2013)[7] were followed. The SSU and LSU D2-D3 rDNA were amplified using the primers 1096F/1912R, 1813F/2646R (Holterman *et al.* 2006)[5] and D2A/D3B (Nunn, 1992)[6], respectively. The resulting sequences have been deposited in GenBank under the accession numbers OQ925400 & OQ925401 for SSU and OQ925398 & OQ925399 for SSU, respectively. The protocols for sequence alignment and phylogenetic inference were outlined by Xu *et al.* (2018)[8].

3 Resultsand Discussion

Morphological and morphometric data of the two recovered isolates of *L. spiflatus* are presented in Table 1, Fig. 1, Fig. 2. The morphological and morphometric characteristics of the present populations of *L. spiflatus* were in agreement with the original descriptions by Gu *et al.* (2020)[1]. The majority of the range values of morphometric characteristics overlapped or were within the same interval for both *L. spiflatus* and *L. liaoningensis*, except for a few differences. The body length of females [tg13-b: 1 078 (1 008−1 130μm); tg91-b: 966 (873−1 034μm) *vs* 1 462 (1 252−1 722μm)], b' of females [tg91-b: 4.6 (4.1−5.0) *vs* 6.4 (5.2−7.1)], median bulb diameter of females [tg91-b: 12.7 (12−13) *vs* 17.0 (14.0−22.0)] and b' of males [tg13-b: 4.8 (4.6−4.9); tg91-b: 4.6 (4.2−5.1) *vs* 6.1 (5.6−6.6)] were slightly lower for the recovered isolates of *L. spiflatus* when compared with *L. liaoningensis* (Table 1).

Table 1 Morphometric data for *Laimaphelenchus spiflatus* Gu *et al.*, 2020 in Shanxi Province. All measurements are in μm and in the form: mean±s. d. (range)

Characters	*Picea meyeri* (tg13-b)		*Elaeagnus angustifolia* (tg91-b)		Gu *et al.* (2020)		Song *et al.* (2020)	
	Females	Males	Females	Males	Females	Males	Females	Males
n	6	5	14	11	15	15	15	15
L	1 078±51.1 (1 008~1 130)	1 053±27.9 (1 007~1 081)	966±51.5 (873~1 034)	933±63.9 (810~1 041)	1 150±108 (976~1 437)	1 092±78.6 (905~1 235)	1 462±139 (1 252~1 722)	1 206±117 (972~1 383)
a	48.1±2.7 (44.2~51.8)	52.3±2.7 (50.2~55.4)	47.4±2.5 (44.2~52.5)	52.5±2.2 (49.6~56.0)	53.9±4.1 (45.0~62.0)	55.5±3 (49.9~60.8)	39.0±3.3 (32.0~46.0)	44.0±4.5 (37.6~53.6)

(continued)

Characters	*Picea meyeri* (tg13-b)		*Elaeagnus angustifolia* (tg91-b)		Gu *et al.* (2020)		Song *et al.* (2020)	
	Females	Males	Females	Males	Females	Males	Females	Males
b	13.1±0.4 (12.7~13.9)	12.5±0.5 (11.9~13.0)	12.0±0.7 (10.8~12.9)	11.5±0.7 (10.3~12.7)	12.7±0.9 (11.1~14.7)	12.2±0.6 (10.5~13.5)	16.0±2.0 (12.0~20.0)	14.0±1.1 (12.0~16.0)
b′	5.0±0.5 (4.1~5.5)	4.8±0.1 (4.6~4.9)	4.6±0.3 (4.1~5.0)	4.6±0.2 (4.2~5.1)	5.4±0.3 (4.9~6.1)	5.3±0.3 (4.7~5.8)	6.4±0.6 (5.2~7.1)	6.1±0.3 (5.6~6.6)
c	19.7±1.5 (18.0~21.8)	18.9±0.6 (18.3~19.9)	19.8±1.2 (17.8~21.8)	18.4±1.0 (16.3~20.2)	21.0±1.1 (19.4~23.4)	20.2±1.2 (17.9~22.0)	24.0±1.4 (21.0~26.0)	20.0±1.0 (19.0~22.0)
c′	4.1±0.4 (3.7~4.7)	3.1±0.2 (2.9~3.4)	4.2±0.3 (3.7~4.6)	3.1±0.2 (2.9~3.6)	4.2±0.3 (3.8~4.9)	3.1±0.2 (2.8~3.5)	3.6±0.3 (3.1~4.1)	3.0±0.3 (2.6~3.4)
V/T	69.5 ±0.9 (68.1~70.3)	61.4±8.2 (53.0~73.9)	69.4±0.8 (68.3~71.0)	56.8±4.9 (48.3~63.0)	69.0±0.8 (67.5~70.5)	55.1±4.1 (46.5~62.4)	69.0±2.4 (61.0~72.0)	69.0±5.2 (57.0~75.0)
m	45.6±3.3 (39.7~49.8)	45.0±2.8 (41.1~48.1)	45.2±3.8 (39.1~49.9)	45.3±3.8 (39.0~53.3)	N/A	N/A	45.0±4.1 (37.0~54.0)	50.0±4.8 (41.0~56.0)
Stylet length	12.6±0.5 (12~13)	12.4±0.5 (12~13)	12.5±0.4 (12~13)	12.2±0.4 (12~13)	12.5±0.5 (11.9~13.3)	12.3±0.3 (11.7~13.0)	14.0±1.0 (12.0~16.0)	14.0±0.8 (12.0~16.0)
Lipregion diam.	6.5±0.4 (6~7)	6.6±0.3 (6~7)	6.3±0.3 (6~7)	6.2±0.3 (6~7)	7.0±0.1 (6.8~7.3)	6.8±0.3 (6.1~7.2)	7.6±0.5 (6.1~8.1)	7.3±0.4 (6.8~8.0)
Lipregion height	2.9±0.6 (2~4)	2.8±0.1 (2~4)	3.0±0.2 (2~4)	2.9±0.3 (2~3)	2.9±0.1 (2.7~3.2)	2.8±0.1 (2.6~3.2)	3.6±0.4 (3.0~4.5)	3.4±0.3 (2.9~3.9)
Excretory pore-anterior end	97.7±5.1 (89~104)	102.7±4.9 (98~108)	96.0±5.1 (88~107)	96.7±6.4 (87~105)	105±7.3 (90~114)	89±4.2 (82~98)	116.0±14.2 (101.0~151.0)	108.0±10.0 (92.0~126.0)
Nerve ring-anterior end	98.5±3.7 (93~104)	101.3±5.5 (94~109)	95.6±5.3 (85~105)	97.0±5.2 (88~104)	N/A	N/A	N/A	N/A
Median bulb length	18.0±0.8 (17~19)	17.8±0.7 (17~19)	17.7±0.9 (16~19)	16.9±0.8 (16~18)	18.1±0.6 (17.2~19.1)	17.4±0.5 (16.0~18.0)	21.0±2.5 (15.0~26.0)	18.0±1.0 (16.0~20.0)
Median bulb diam.	13.4±0.9 (12~14)	12.4±0.6 (12~13)	12.7±0.5 (12~13)	12.1±0.6 (11~13)	12.5±0.7 (11.6~13.8)	11.6±0.5 (10.0~12.2)	17.0±2.4 (14.0~22.0)	14.0±1.1 (12.0~16.0)
Pharynx length	82.1±4.1 (77~88)	84.3±4.3 (78~88)	80.3±3.8 (74~89)	81.1±3.5 (76~86)	N/A	N/A	N/A	N/A
PUS	104±8.3 (96~118)	—	96±12.3 (81~116)	—	116±25.9 (92~172)	—	141.0±19.6 (102.0~184.0)	—
Tail length	54.4±1.8 (52~56)	55.9±2.0 (53~58)	48.8±3.6 (44~55)	50.6±3.0 (44~55)	55±4 (48~66)	54±3.8 (47~59)	62.0±5.8 (53.0~70.0)	61.0±6.4 (48.0~70.0)
Anal body diam.	13.2±0.9 (12~14)	17.9±1.1 (17~19)	11.8±0.6 (11~13)	16.2±1.0 (14~17)	13.0±0.5 (12.0~13.6)	17.3±0.9 (14.9~18.7)	17.0±1.9 (13.0~20.0)	20.0±2.1 (17.0~25.0)
Spicules length	—	25.7±1.1 (24~27)	—	24.4±1.9 (21~26)	—	27.3±1.3 (23.4~28.8)	—	28.0±1.7 (24.0~30.0)

N/A: not available in the original paper.

The recovered isolates of *L. spiflatus* (SSU: OQ925400 & OQ925401; LSU: OQ925398 & OQ925399) showed 99% identity with the type population (SSU: MN401305; LSU: MN401306 & MN401307) and with *L. liaoningensis* (SSU: MN401305; LSU: MN401306 & MN401307),

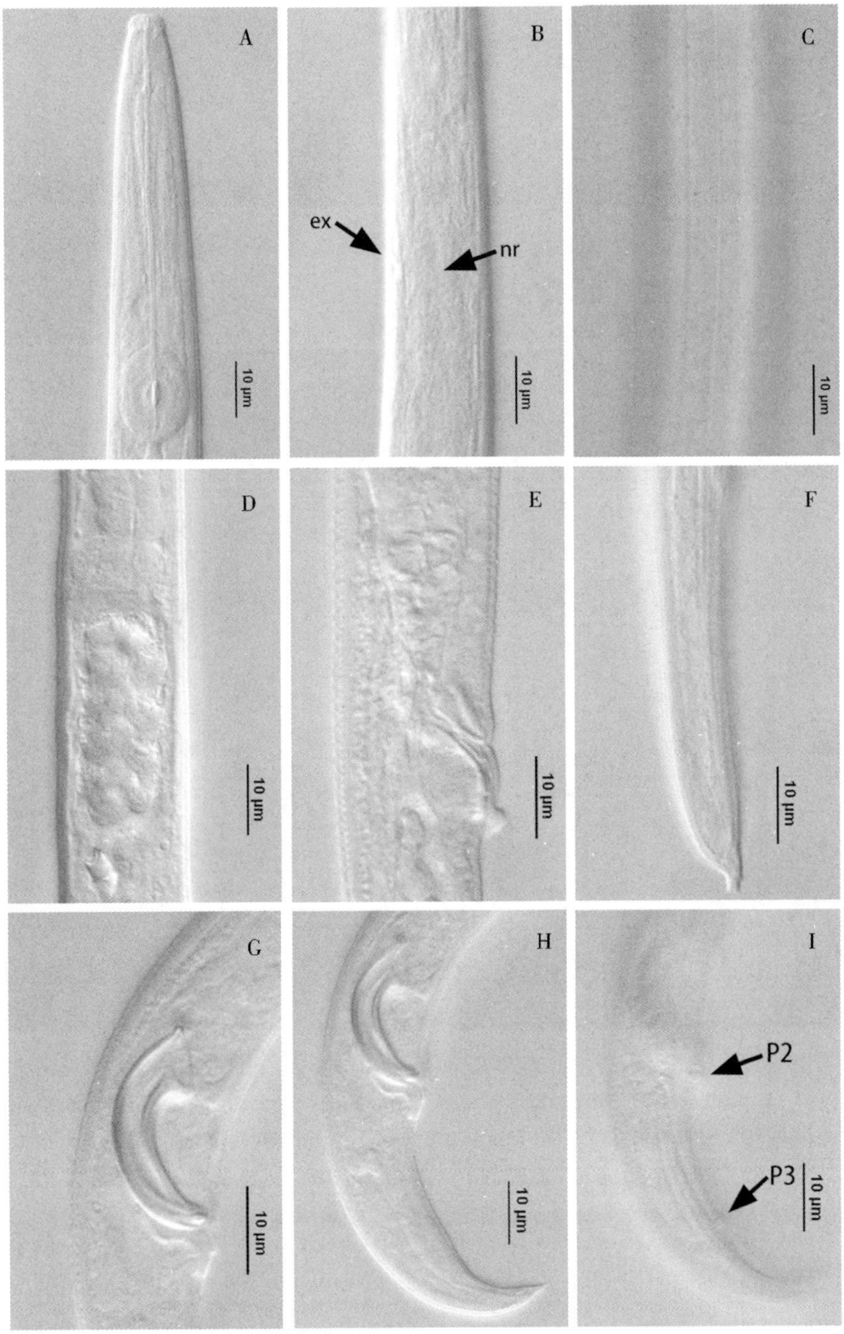

A-F. Female; G-I. Male

A. Anterior region; B. Excretory pore (arrow labeled *ex*) and nerve ring (arrow labeled *nr*); C. Lateral lines; D. Spermatheca; E. Vulval region; F-H. Tail; G. Spicules; I. Caudal papillae (arrows)

Fig. 1 Photomicrographs of *Laimaphelenchus spiflatus* Gu *et al.*, 2020 taken from glycerine-mounted specimens on permanent slides

A-E. Female; F. Male

A. Cephalic region, amphidial opening (arrow); B. Lateral lines; C. Vulval region (arrow); D. Anus (arrow); E. Tail tip; F. Caudal papillae (arrows)

Fig. 2 Scanning electron micrographs of *Laimaphelenchus spiflatus* Gu *et al.*, 2020

but had 1–3 nucleotide variations. Phylogenetic analysis (Fig. 3, Fig. 4) demonstrated that the recovered isolates from Shanxi Province formed a clade with the type population of *L. spiflatus* and *L. liaoningensis* with high Bayesian phylogenetic probability values (1.00 for SSU and 0.94 for LSU D2-D3).

Regarding the morphological characteristics, *L. spiflatus* and *L. liaoningensis* shared all features, such as relatively long female body, four lines in the lateral field, and vulva with a well-developed anterior flap. Both species also had male two pairs of caudal papillae, except for the shape and structure of the tail tip. Specifically, *L. spiflatus* had a tail with a stalk ending in 8–12 finger-like projections, while *L. liaoningensis* had a stalk-like terminus and two tubercles with 4–6

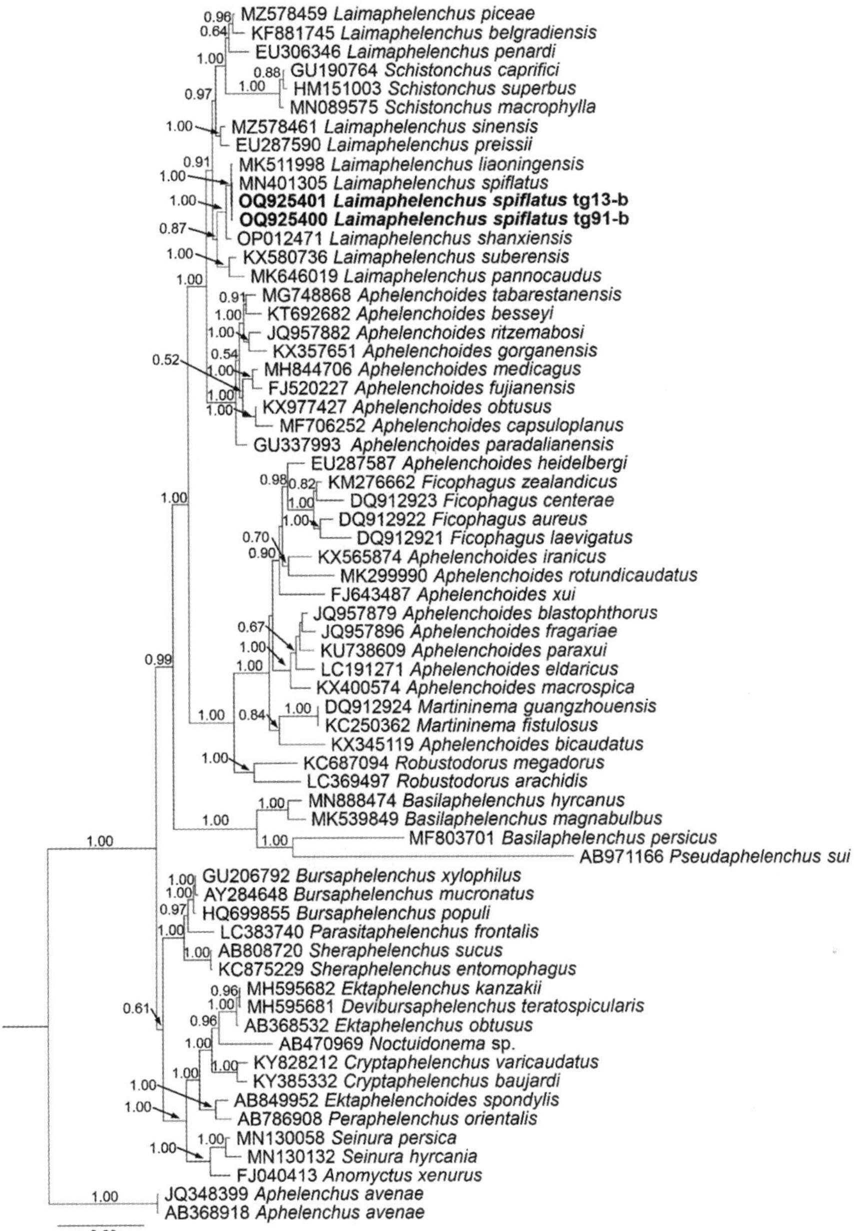

Fig. 3 Bayesian consensus tree inferred from SSU sequences of the recovered populations of *Laimaphelenchus* Fuchs, 1937 under TrN+I+G model. Posterior probability values exceeding 0.50 are given on appropriate clades. Newly obtained sequences in this study are in bold letters

finger-like protrusions. However, according to Jahanshahi Afshar *et al.* (2021)[3] and their detailed classification and description of the tail tip shape of the genus and the SEM photos, the tail tip shapes of the two species were found to be the same, with both having a tail that lacked tubercles and had flat fused stacked structures with 8–12 finger-like appendages.

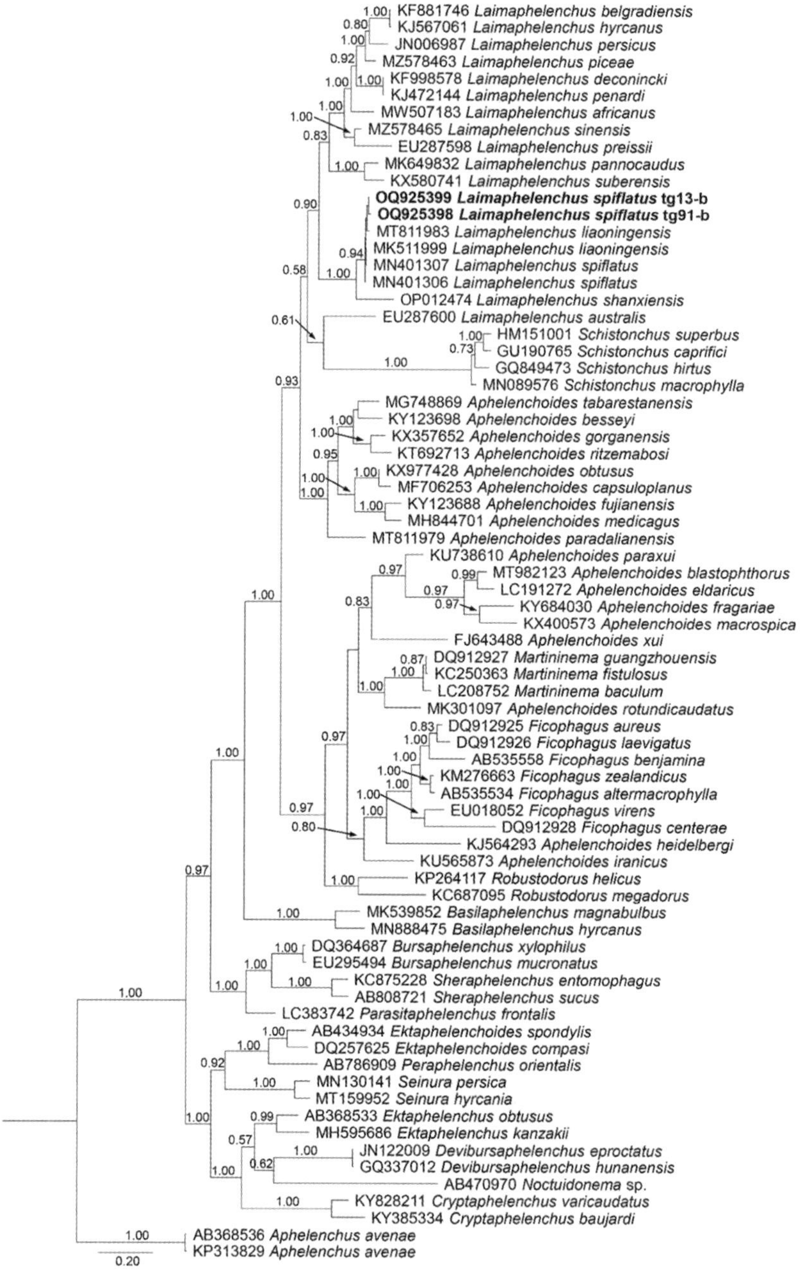

Fig. 4 Bayesian consensus tree inferred from LSU D2−D3 sequences of the recovered populations of *Laimaphelenchus* Fuchs, 1937 under GTR+I+G model. Posterior probability values exceeding 0.50 are given on appropriate clades. Newly obtained sequences in this study are in bold letters

The morphometric data of type population of *L. spiflatus* and *L. liaoningensis* showed that the majority of the range values overlapped or were within the same interval, with the only notable difference being a slightly lower median bulb diameter in *L. spiflatus*. However, this difference can be explained by intraspecific variability and preparation for microscopy. Molecular analysis showed that the SSU and LSU D2-D3 sequences of *L. spiflatus* and *L. liaoningensis* were highly similar, differing only by one or a few nucleotides. Phylogenetic analysis supported that they are a monophyletic group and are conspecific. Therefore, according to the principle of priority in taxonomy, *L. spiflatus* has priority over *L. liaoningensis*, and the latter should be considered a junior synonym of *L. spiflatus*.

References

[1] GU J F, MARIA M, FANG Y W, *et al.* Molecular and morphological characterisation of *Laimaphelenchus spiflatus* n. sp. (Nematoda: Aphelenchoididae) from China [J]. Nematology, 2020a, 22 (8): 843-853.

[2] SONG Y T, LIU R J, JIANG Y, *et al.* Molecular and morphological characterization of *Laimaphelenchus liaoningensis* n. sp. (Nematoda: Aphelenchoididae) in China [J]. Journal of Nanjing Forestry University (Natural Sciences Edition), 2020, 44 (4): 93-101.

[3] JAHANSHAHI AFSHAR F, RASHIDIFARD M, ABOLAFIA J, *et al. Laimaphelenchus africanus* n. sp. (Tylenchomorpha: Aphelenchoididae) from South Africa, a morphological and molecular phylogenetic study, with an update to the diagnostics of the genus [J]. Journal of Nematology, 2021, 53: 1-14.

[4] WANG L, YANG Y, ZHAO Z, *et al.* Description of *Laimaphelenchus piceae* sp. n., a new evidence of cryptic speciation, and *L. sinensis* Gu *et al.*, 2020 (Rhabditida: Aphelenchoididae), a record from Shanxi province, north China [J]. Nematology, 2022, 24 (6): 601-616.

[5] HOLTERMAN M, VAN DER WURFF A, VAN DEN ELSEN S, *et al.* Phylum-wide analysis of SSU rDNA reveals deep phylogenetic relationships among nematodes and accelerated evolution toward crown clades [J]. Molecular Biology and Evolution, 2006, 23 (9): 1792-1800.

[6] NUNN G B. Nematode molecular evolution [D]. University of Nottingham, Nottingham, UK, 1992.

[7] XU Y M, ZHAO Z Q, WANG J M, *et al.* A new species of the genus *Tripylina* Brzeski, 1963 (Nematode: Enoplida: Trischistomatidae) from Shanxi Province, China [J]. Zootaxa, 2013, 3630 (3): 561-570.

[8] XU Y M, DANIEL L, YE W, *et al.* Description of *Tripylella jianjuni* sp. n. (Nematoda: Tripylidae) from New Zealand [J]. Nematology, 2018, 20 (8): 795-810.

中国线虫学研究（第九卷）Nematology Research in China（Vol. 9）: 48-53

Advances in Morphological and Molecular Identification of PWN*

Gu Jianfeng[1**], Fang Yiwu[1], Ma Xinxin[2], Lü Xiaoling[1], Duan Weijun[1]

([1]*Ningbo Key Laboratory of Port Biological and Food Safety Testing*; *Technical center of Ningbo Customs*; *Ningbo Inspection and Quarantine Science Technology Academy*, *Ningbo* 315100, *China*; [2]*Ningbo Zhongsheng product testing Co.*, *Ltd*, *Ningbo* 315100, *China*)

Abstract: Pine Wood Nematode is a quarantine nematode of great concern in the world, and it does great harm to pine forests worldwide. This paper introduced the latest progress of morphological and molecular identification, and suggested a PWN detection and identification procedure at the port. In the future, the rapid and convenient AI identification and isothermal amplification methods is supposed to be more convenient, efficient and widely accepted.

Key words: *Bursaphelenchus xylophilus*; Identification; Morphological; Molecular; Quarantine

The Pine Wood Nematode (PWN), *Bursaphelenchus xylophilus* (Steiner & Buhrer) Nickle, 1970 was first found in 1929 from longleaf pine (*Pinus palustris*) in Texas, USA[1]. In 1972, it was confirmed to cause Pine wilt disease[2]. PWN is originated in North America and commonly found in coniferous trees in the United States and Canada, but local pine wilt only occurs sporadically and causes little damage[3-6]. When PWN was introduced into Asian and European countries through pine logs and wood packaging exportation, it caused great damage due to changes in host, vector insects, environment and natural enemies, and even the species of commensal bacteria and fungi. PWN is a plant quarantine pest in China and many other countries.

This article mainly introduces the latest research progress in the morphological and molecular identification technology of PWN.

1 Morphological identification

1.1 Groups with in Genus *Bursaphelenchus*

Based on the number of lateral lines as the basic characteristics, Braasch proposed the grouping by combining male spicules, the number and arrangement of caudal papillae, and the presence or non－presence of female vulval flap and female tail shape[7]. Later Braasch *et al.* combined molecular phylogenetic analysis and morphological studies of more than 40 species and established a grouping system, Which divides the 76 species into 14 groups[8].

* Funding: National key research and development plan NQI project (2022YFF0608804); Science and Technology Plan Project of Ningbo Zhongsheng Product Testing Co., Ltd. (2022ZS004)

** First author: Gu Jianfeng. E-mail: jeffgu00@qq.com

The main morphological characteristics of *xilophilus* group are as follows: 4 lateral lines; the typical shape of male spicules which was long and arched, with sharp rostrum and disc-like process at the distal end; seven male caudal papillae present, P3 and P4 were adjacent and located at the beginning of the bursa, and a 7-11 μm long vulval flap. Among the above features, the lateral lines and male caudal papillae are not easy to observe under the light microscope, so the most important features are the male spicules shape and long vulval flap in females.

Fifteen nematodes have been reported in the *xylophilus* group, including *B. xylophilus*, *B. fraudulentus* Rühm, 1956 (J. B. Goodey, 1960), *B. mucronatus* Mamiya & Enda, 1979, *B. conicaudatus* Kanzaki, Tsuda & Futai, 2000, *B. baujardi Walia* Negi, Bajaj & Kalia, 2003, *B. luxuriosae* Kanzaki & Futai, 2003, *B. doui* Braasch, Gu, Burgermeister & Zhang, 2004, *B. singaporensis* Gu, Zhang, Braasch & Burgermeister, 2005, *B. macromucronatus* Gu, Zheng, Braasch & Burgermeister, 2008, *B. populi* Tomalak & Filipiak, 2010, *B. paraluxuriosae* Gu, Wang & Braasch, 2012, *B. firmae* Kanzaki, Maehara, Aikawa & Matsumato, 2012, *B. koreanus* Gu, Wang & Chen, 2013, *B. gillanii* Schönfeld, Braasch, Riedel & Gu, 2013, and *B. acaloleptae* Kanzaki, Ekino, Maehara, Aikawa & Giblin-Davis, 2019.

1.2 Morphological identification of PWN

Morphological methods have many advantages over molecular methods such as rapid, low cost, simple operation. Morphological methods will complement molecular methods for a long time. In addition, the morphological identification of PWN may be the easiest among all the nematode identification methods, only that experience and training is required.

The female tail shape is the key to the identification of PWN. Generally, the female tail of PWN is cylindrical or subcylindrical, with no mucron. Occasionally, some females in some population had very short mucron, but no more than 2μm. However, there are exceptions. Studies by many scholars have confirmed that the tail shape of PWN changes depending on host, environment and other conditions. For example, most of the PWN isolated from *Pinus* tree of Liaoning had long or short tail mucron with blunt rounded or sharp ends, and the length was 1.8μm (0.3-3.2μm). However, after cultured on *Botrytis cinerea*, all females had bluntly rounded tail without mucron[9].

Anyhow, the above situation (almost all female PWN have mucron, or the length is more than 2μm) is relatively rare. In general, PWN have typical morphological characteristics: the female tail is nearly cylindrical, the end is wide and round, and there is no mucron. Some females of the population had short tail mucron, but the length did not exceed 2μm. In addition, individuals without mucron were usually found in the population. So in suspicious cases, contact a specialist or use molecular methods to assist detection is suggested.

In addition, it should be noted that dispersive larvae (also known as persistent larvae or durable larvae) of PWN were obtained from long horn beetles or wood, with thin needle-like bodies ($a \geq 45$), and the lip area was nearly hemisphical (dome-like), with degenerated stylet, which is easy to miss identification.

1.3 Identification by artificial intelligence (AI)

With the development of AI technology, the intelligent nematode image recognition system is expected to solve the nematode morphology identification problem of, which can greatly improve the customs control ability and work efficiency.

Females of PWN have special identification characteristics, which can be used as the main basis for identification. However, the tails of female nematodes alone was difficult to distinguish accurately based on the female tails. So the female tail can combine with the head and the tail of the male (spicules), which can achieve better identification results. Our laboratory has rich experience in the identification of nematodes of the genus *Bursaphelenchus*. We have the world's leading specimen library, and have accumulated a large number of characteristic pictures of PWN and closely related species, which lays a foundation for the recognition of AI. At present, our team determined that the micrographs taken with a 40x objective lens were used as the base gallery, and the nematode head, female tail, and male tail were the main recognition targets. The intelligent recognition system of nematode microscopic images was initially constructed by combining the deep learning classification network and decision tree method. According to the female tail alone, the recognition accuracy of PWN was up to 96%. For example, the recognition accuracy of head and female tail was up to 98%. In addition, the head could be used to identify the genera *Heterodera*, *Meloidogyne*, *Xiphinema* and *Longidorus*, and the comprehensive recognition accuracy reached more than 94%. A mobile online nematode recognition APP was also developed, which, together with WIFI microscope, could realize real-time display of nematode micrographs and intelligent recognition after photography on a mobile phone or tablet, which is suitable for application at ports.

2 Molecular identification of PWN

2.1 DNA extraction

Any molecular assay, whether for individual nematodes, nematode fluids, wood chips or sawdust, sufficient DNA is required. The current methods for DNA extraction mainly include proteinase K method[10], kit extraction method, alkali lysis method[11], and various nematode lysate methods[12]. However, the proteinase K method needs more than 1 h and requires liquid nitrogen freezing or ultra-low temperature freezing process, which limits its use. The nematode DNA extraction kit of Ningbo Baichuan Biology can be completed in less than 10 minutes. Only one lysate and one heating step is needed, and the operation is simple, rapid and stable. It is a good alternative to proteinase K method.

2.2 Quantitative real-time PCR (qPCR) method

In 1989, Choo *et al.* developed real-time PCR, also known as qPCR[13]. Add fluorescent dyes in the PCR reaction system or with the fluorescent probe of the groups, using double fluorescent signal after each expansion cycle of accumulation, real-time monitoring of the PCR process, to avoid the cross contamination caused by PCR post-processing, match with corresponding fluorescence PCR instrument, more sensitive than conventional polymerase chain reaction (PCR), and more rapidly. Chen Fengmao *et al.* constructed a real-time PCR method for the detection of pine nematode nematode by targeting ITS region[14]. Specific primer combination F11/R11 and probe Taqman-11 could distinguish PWN from similar species in dead pines. It has the advantages of sensitivity and accuracy, but the disadvantages are that the experimental process is relatively time-consuming, and the related reagents and equipment are expensive.

2.3 Recombinase Polymerase Amplificatio (RPA)

RPA is a technology developed by TwistDx Inc in 2006[15]. This technique mainly relies on recombinase T4 UvsX, SSB, and strand displacement DNA polymerase, which can bind single-stranded nucleic acid (oligonucleotide primers), to achieve exponential amplification of the target region on the template. RPA does not require thermal denaturation, nor does it require temperature control equipment. The amplification process can be completed within 20 min under the conditions of 23-45℃ to achieve the rapid detection target. Cha *et al.* used portable optical isothermal device (POID) combined with DAP buffer to further simplify the detection procedure, and could complete the diagnosis of PWN disease within 25min in the field[16]. However, this method requires mixing fluorescent dyes after isothermal amplification and then judging based on the photometric value, which has the risk of false positives and is inconvenient to operate. Fang *et al.* (2021) also developed a dual RPA detection method for PWN and *B. mucronatus*.

2.4 Enzyme-mediated Duplex Exponential Amplification (EmDEA)

EmDEA is developed by Suzhou Jingrui Biotechnology Co., LTD. By double amplification of the target nucleic acid and fluorescent probe through multi-enzyme cooperation, EmDEA achieves extremely high detection sensitivity and specificity in a short time. Based on the analysis of specific gene segments in ITS region of PWN, upstream and downstream primers and fluorescent probes were designed for isothermal nucleic acid amplification and rapid fluorescence detection. At the same time of exponential rapid amplification of the target sequence, the binding of the probe to the target sequence initiates enzyme-mediated probe hydrolysis, the fluorescence group on the probe is separated from the quenched group, and the fluorescence signal is generated and accumulated. The rapid identification is performed according to the characteristics of the fluorescence curve and the time when the fluorescence value exceeds the set threshold. This method is suitable for the rapid detection of various samples such as sawdust, single nematode, nematode fluid, etc. It requires less than 30min and is easy to operate (nematode lysate has been preloaded; the activation solution has been preloaded into the tube cap). Only one pipetting process is required, and the sensitivity can reach 3-5 copies/reaction.

2.5 Sequencing or DNA barcoding methods

DNA barcoding is a standardized molecular diagnostic technique for species identification using a relatively short, accepted standard DNA fragment in the genome. At its core is DNA sequencing and sequence analysis. In 2016, He *et al.* suggested that 28S and ITS regions could be used as candidate barcoding genes in PWN group due to their certain genetic distance intervals and relatively high species recognition rate[17].

3 Looking forward

During the quarantine of imported logs or wood packages at customs ports, or the investigation or monitoring of PWN by relevant domestic forestry departments, the operation is required to be convenient, portable, fast and accurate. Artificial intelligence identification is an extension of morphological identification, and molecular biological methods such as isothermal amplification are the trend. In the future, the intelligent recognition software will be more accurate and convenient, and the operation method or equipment of molecular biology will be more rapid and accurate.

In our suggestion, PWN can be identified using either conventional detection methods (which usually take more than 24 hours) or rapid detection methods (which usually take about 30 minutes) (Figure 1). The conventional detection method uses the modified funnel method to isolate nematodes, then morphological identification is performed first. When female and male adults were found, they could be identified based on morphological characteristics or intelligent recognition software. In case of suspicious cases, molecular biological methods such as EmDEA or DNA barcoding should be further used for identification, or contact experts.

In the rapid detection method, nematode DNA was extracted from sawdust by drilling, and then detected by isothermal fluorescence method like EmDEA. If the result is negative, it is qualified to be released. If the result is positive, it needs to be further checked by the above routine detection method (because the dead nematodes in samples may cause positive results), which can greatly accelerate the clearance speed.

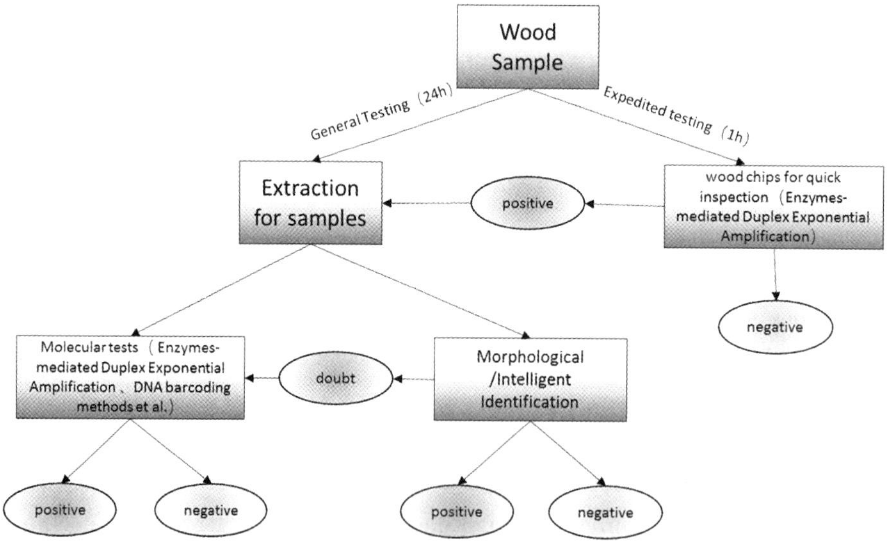

Figure 1 Flow chart of PWN quarantine

References

[1] NICKLE W R, GOLDEN A M, MAMIYA Y, *et al.* On the taxonomy and morphology of the PWN,

Bursaphelenchus xylophilus (Steiner &Buhrer 1934) Nickle 1970. Journal of nematology, 1981, 13 (3): 385-392.

[2] MAMIYA Y. PWN, *Bursaphelenchus lignicolus* Mamiya and Kiyohara, as a causal agent of pine wilting disease. Review of Plant Protection Research, 1972, 5: 46-60.

[3] ROBBINS K. Distribution of the pinewood nematode in the United States//Proceedings 1982 National Pine Wilt Disease Workshop. Illinois Natural History Survey, 1982.

[4] RUTHERFORD T A, MAMIYA Y, WEBSTER J M. Nematode-induced pine wilt disease: factors influencing its occurrence and distribution. Forest Science, 1990, 36 (1): 145-155.

[5] Bowers W W, Hudak J, Raske A G, *et al.* Host and vector surveys for the pinewood nematode, *Bursaphelenchus xylophilus* (Steiner and Buhrer) Nickle (Nematoda: Aphelenchoididae) in Canada. Newfoundland and Labrador Region, Forestry Canada, 1992.

[6] SUTHERLAND J R, PETERSON M J. The pinewood nematode in Canada: history, distribution, hosts, potential vectors and research//Sustainability of pine forests in relation to pine wilt and decline. Proceedings of International Symposium, Tokyo, Japan, 27-28 October, 1998. Shokado Shoten, 1999: 247-253.

[7] BRAASCH H. *Bursaphelenchus* species in conifers in Europe: distribution and morphological relationships. EPPO Bulletin, 2001, 31: 127-142.

[8] BRAASCH H., BURGERMEISTER W., Gu J F. Revised intra-generic grouping of *Bursaphelenchus* Fuchs, 1937 (Nematoda: Aphelenchoididae). Journal of Nematode Morphology and Systematics, 2009, 12: 65-88.

[9] GU JIANFENG, FANG YIWU, LIU LELE, *et al.* Identification of a female PWN with tail apiculata. Chinese Journal of Plant Protection, 2021, 48 (2): 434-441.

[10] Wang Jiangling, Zhang Jiancheng, Gu Jianfeng. DNA extraction from single strand nematodes. Plant Quarantine, 2011, 25 (2): 32-35.

[11] STANTON J M, MCNICOL C D, STEELE V. Non-manual lysis of second-stage Meloidogyne juveniles for identification of pure and mixed samples based on the polymerase chain reaction. Australasian Plant Pathology, 1998, 27 (2): 112-115.

[12] WAEYENBERGE L, RYSS A, MOENS M, *et al.* Molecular characterisation of 18 Pratylenchus species using rDNA restriction fragment length polymorphism. Nematology, 2000, 2 (2): 135-142.

[13] CHOO Q, KUO G, WEINER A, *et al.* Isolation of a cDNA clone derived from a blood-borne non-A, non-B viral hepatitis genome. Science, 244 (4902): 359-362.

[14] CHEN FENGMAO, YE JIANREN, TANG JIAN, *et al.* RAPD detection of PWN and Nematodes Pseudopine. Journal of Nanjing Forestry University: Natural Science Edition, 2005, 29 (4): 25-28.

[15] PIEPENBURG O, WILLIAMS C H, STEMPLE D L, *et al.* DNA detection using recombination proteins. PLoS Biology, 2006, 4 (7): e204.

[16] CHA D, KIM D, CHOI W, *et al.* Point-of-care diagnostic (POCD) method for detecting *Bursaphelenchus xylophilus* in pinewood using recombinase polymerase amplification (RPA) with the portable optical isothermal device (POID). PLoS ONE, 2020, 15 (1): e0227476.

[17] HE JIE, GU JIANFENG. DNA barcoding screening in PWN group. Plant Quarantine, 2016, 30 (1): 14-20.

Cloning and Characterization of microRNA396 and Its Targets in *Cucumis metuliferus* Resistant to *Meloidogyne incognita*[*]

Ye Deyou[1**], Bao-Lam Huynh[2], Qi Yonghong[3]

([1] *Institute of Vegetables, Gansu Academy of Agricultural Sciences, Lanzhou* 730070, *China*; [2] *Department of Nematology, University of California, Riverside CA*, 92521, *USA*; [3] *Institute of Plant Protection, Gansu Academy of Agricultural Sciences, Lanzhou* 730070, *China*)

Abstract: MicroRNAs (miRNAs), a class of non-coding small RNAs, play an important role in biotic and abiotic stress responses. Previously, great progress has been made in miRNA-guided gene regulation in plant-parasitic cyst nematodes interactions. However, little is known about miRNAs regulation of nematode stress response in *Cucumis metuliferus*, which is a relative of cultivated cucumber or melon and highly resistant to the root-knot nematode *Meloidogyne incognita*. On the basis of our investigation recently, five cmu-miR396 precursors were firstly identified and characterized from *C. metuliferus* in this study. Subsequently, a total of 12 genes were predicted to be potential targets of cmu-miR396 and their function involved in transcription factors, metabolism, signal transduction and stress response. Moreover, three cmu-GRFs genes with significant differential expression in roots of *C. metuliferus* were successfully amplified and confirmed by RT-PCR experimentally. Additionally, the analysis was performed for the phylogeny and subcellular localization of the cmu-GRFs protein, and three transcription factors of cmu-GRFs were confirmed as the candidate targets of cmu-miR396 through 5′RLM-RACE assays. These data laid a foundation for further functional analyses to reveal the roles played by miR396 during the interaction between *C. metuliferus* and *M. incognita*, providing a better understanding of the mechanisms and underlying regulatory networks of miR396-mediated resistance to *M. incognita* in *C. metuliferus*.

Key words: Cloning; Characterization; microRNA396; Targets; *Cucumis metuliferus*; *Meloidogyne incognita*

* Funding: National Natural Science Foundation of China (31760508); Key R & D plan of Gansu Academy of Agricultural Sciences (2022GAAS34)

** First author: Ye Deyou. E-mail: ydy287@163.com

Development and Application of Recombinase Polymerase Amplification assay for rapid and visual Detection of *Pratylenchus coffeae*

Wu Caiyun*, Fan Linjuan, Xu Xueliang, Liu Zirong, Yao Jian, Yao Yingjuan**

(*Institute of Agricultural Applied Microbiology, Jiangxi Academy of Agricultural Sciences, Nanchang* 330200, *China*)

Abstract: *Pratylenchus coffeae* is a migratory plant-parasitic nematode, which can cause root-rot disease in many crops and serious economic losses every year. Rapid and visual detection of *P. coffeae* is essential for more effective prevention and control. Recombinase polymerase amplification (RPA) is a novel isothermal DNA amplification approach that has been used to detect a variety of animal and plant pathogens. In our study, we developed RPA assays for the specific detection of *P. coffeae*. Specific primers and probes were designed based on the conserved sequences of the rDNA-ITS sequence. The RPA amplification step could be completed at 38℃ in 25min and recombinase Polymerase Amplification Assay combined with lateral flow dipstick (LFD-RPA) incubation time of approximately 8-12min without a thermal cycling instrument. The RPA method was capable for specific detection of *P. coffeae* from populations of different geographical origins, single individual, the tuber of Chinese yam and infected soil. The detection limit of *P. coffeae* in RPA assay was 0.05ng/μL genomic DNA. The results indicated that the developed RPA assay was a sensitive, rapid, and practicable approach for the detection and molecular diagnosis of *P. coffeae*.

Key words: *Pratylenchus coffeae*; Recombinase polymerase amplification; Lateral flow dipstick; Detection

* First author: Wu Caiyun, master, probationer, integrated pest management. E-mail: wucaiyunwy@163.com

** Corresponding author: Yao Yingjuan. E-mail: yaoyingjuan2008@163.com

Effects of long-term Consecutive Monoculture of Yam on the Fungal Community and Function in the Rhizospheric Soil

Yao Jian[1]*, Yuan Mengyu[2], Wu Caiyun[1], Fan Lingjuan[1], Liu Zirong[1], Xu Xueliang[1], Yao Yingjuan[1]**

(*[1]Institute of Agricultural Applied Microbiology, Jiangxi Academy of Agricultural Sciences, Nanchang* 330200, *China*; *[2]College of Plant Science and Technology, Huazhong Agricultural University, Wuhan* 430070, *China*)

Abstract: Long-term consecutive monoculture of yam is easy to cause replant diseases, which seriously affects the growth of yam. However, little is known about the consecutive monoculture of yam affects the rhizospheric soil fungal community structure. In this study, the effects of consecutive monoculture on rhizospheric soil characteristics, fungal diversity and community structure were investigated in the Yongfeng yam fields under monoculture of 1, 10 and 20 years. Long-term monoculture caused soil acidification, increased the concentration of available potassium and available phosphorus, increased the relative abundance of Basidiomycota, but decreased the relative abundance of Ascomycota. Furthermore, monoculture of Yongfeng yam for 10 years showed the lowest fungal richness and diversity while monoculture of Yongfeng yam for 20 years showed the highest fungal richness and diversity. The fungal network of 1 year cultivation showed the most complex while the network of consecutive monoculture 10 years showed the lowest complex, which may be related to lower complex rhizospheric soil bacterial network structure and the most complex rhizospheric soil bacterial network structure under consecutive monoculture for 1 year and 10 years, respectively.

Key words: Consecutive monoculture; Rhizospheric soil; Fungal community; Network

---

* First author: Yao Jian. E-mail: yaojian417@ 163. com

** Corresponding author: Yao Yingjuan. E-mail: yaoyingjuan2008@ 163. com

GmLecRKs-GmCDL1-GmMPK3/6 通路在大豆孢囊线虫抗性中的功能研究

张　磊*，谭远华，祝　群，汪　瑞，邓苗苗，郭晓黎**
（华中农业大学植物科学技术学院，农业微生物学资源发掘与利用全国重点实验室，武汉　430070）

Functional Analysis of GmLecRKs-GmCDL1-GmMPK3/6 Pathway in Regulating Soybean Cyst Nematode Resistance

Zhang Lei*, Tan Yuanhua, Zhu Qun, Wang Rui, Deng Miaomiao, Guo Xiaoli**
(*State Key Laboratory of Agricultural Microbiology*, *College of Plant Science and Technology*, *Huazhong Agricultural University*, *Wuhan* 430070, *China*)

摘　要：大豆孢囊线虫（*Heterodera glycines* Ichinohe，SCN）为大豆根部重要病害，严重危害我国大豆种植区；虽然前期研究表明模式识别受体与共受体参与植物与孢囊线虫互作，但是目前对大豆如何感知和响应孢囊线虫侵染仍了解较少，相关信号转导组分有待进一步挖掘。

本研究基于孢囊线虫侵染前后感病品种 Williams 82 和抗病品种 PI88788 根部定量磷酸化蛋白质组与生化试验，发现大豆孢囊线虫侵染和损伤处理激活大豆根部 MAPK 信号通路；Co-IP、GST pull-down 和萤光素酶互补试验结果显示 GmMPK3/6 与细胞质类受体激酶 GmCDL1 在植物体内和体外均互作；体外磷酸化试验表明 GmMPK3/6 磷酸化 GmCDL1 Thr372，且其磷酸化依赖 GmMKK4 活化。Thr372 磷酸化增强 GmCDL1 蛋白稳定性，但不影响其激酶活性、亚细胞定位及与 GmMPK3/6 互作。功能分析发现 GmMPK3/6 介导 GmCDL1 Thr372 磷酸化进一步增强 MAPK 磷酸化与大豆孢囊线虫抗性，形成正反馈调节。SCN 侵染、损伤和 flg22 处理诱导 GmCDL1 Thr372 磷酸化水平升高，增强 GmCDL1 蛋白稳定性。通过 IP-MS、Co-IP、GST pull-down 和萤光素酶互补试验发现 GmCDL1 与 L 型凝集素受体激酶 GmLecRK02g/08g 在植物体内互作，且损伤处理后其互作减弱；功能分析发现 GmCDL1 激活下游免疫反应依赖 GmLecRK02g/08g。综上，本研究发现 GmLecRKs-GmCDL1-GmMPK3/6 通路激活大豆孢囊线虫基础抗性，且细胞质类受体激酶与 MAPK 之间存在正反馈调节。

关键词：大豆；孢囊线虫；细胞质类受体激酶；MPK3/6；磷酸化

---

* 第一作者：张磊，博士研究生，从事作物抗线虫基础研究。E-mail：zlei@mail.hzau.edu.cn
** 通信作者：郭晓黎，教授，从事作物与植物线虫互作研究。E-mail：guoxi@mail.hzau.edu.cn

中国线虫学研究（第九卷）Nematology Research in China（Vol. 9）: 58-59

Host-induced Silencing of a Nematode Chitin Synthase Gene Decreases Abundance of Rhizosphere Fungal Community While Enhancing *Heterodera glycines* Resistance of Soybean

Tian Shuan[1*], Shi Xue[1*], Qu Baoyuan[2], Kang Houxiang[1], Huang Wenkun[1], Peng Huan[1], Peng Deliang[1], Wang Jiajun[3], Liu Shiming[1**], Kong Ling'an[1**]

(*[1] Institute of Plant Protection, Chinese Academy of Agricultural Sciences, Beijing* 100193, *China*; *[2] State Key Laboratory of Plant Genomics, Institute of Genetics and Developmental Biology, Chinese Academy of Sciences, Beijing* 100101, *China*; *[3] Institute of Soybean Research, Heilongjiang Academy of Agricultural Sciences, Harbin* 150086, *China*)

Abstract: A transgenic variety of soybean [*Glycine max* (L.) Merr.], H57, has been developed from wild-type variety Jack, with host-induced gene silencing of a chitin synthase gene (*CHS*) in soybean cyst nematode (SCN, *Heterodera glycines* Ichinohe), a devastating pathogen in soybean. Infection with SCN was assessed at 60 days after planting of H57 and Jack into SCN-infected soil by examining recovered cysts from rhizosphere soil and comparing with an infected bulk soil control. 16S and ITS amplicons were identified by high-throughput sequencing to analyze rhizosphere microbial communities (bacterial and fungal), and bioinformatic analysis was used to define operational taxonomic units. Alpha diversity, using five indexes, and relative abundance were determined. Soybean H57 showed significantly enhanced and heritable resistance to SCN compared with Jack. The diversity and richness (abundance) of the bacterial community of H57 and Jack were significantly and similarly increased relative to the bulk soil. The fungal community of H57 had considerably lower abundance than both other treatments, and lower diversity than the bulk soil. The relative abundance of only two bacterial phyla (Acidobacteria and Actinobacteria) and one fungal phylum (Glomeromycota), and three bacterial genera (Candidatus _ Solibacter, Candidatus _ Udaeobacter and Bryobacter) and one fungal genus (Aspergillus), differed significantly between rhizosphere soils of H57 and Jack. This study established a basis for interaction

---

* First authors: Tian Shuan; Shi Xue

** Corresponding authors: Liu Shiming, Professor, mainly engaged in research on plant nematode molecular biology and nematode disease management. E-mail: liushiming01@caas.cn

Kong Ling'an, Associate Professor, mainly engaged in the research on plant nematode molecular biology and nematode disease management. E-mail: konglingan@caas.cn research between soybean with *SCN-CHS* host-induced gene silencing and the rhizosphere microbial community, and for potentially planting soybean H57 to manage SCN.

Key words: Chitin synthase; Host-induced gene silencing; Rhizosphere microbial community; SCN resistance; Soybean; Transgenic soybean H57

Identification of key MicroRNAs in *Cucumis metuliferus* under *Meloidogyne incognita* Stress*

Ye Deyou[1]**, Qi Yonghong[2]

([1]*Institute of Vegetables*, *Gansu Academy of Agricultural Sciences*, *Lanzhou*, *Gansu*, 730070, *China*; [2]*Institute of Plant Protection*, *Gansu Academy of Agricultural Sciences*, *Lanzhou* 730070, *China*)

Abstract: MicroRNAs (miRNAs) are important transcriptional and post-transcriptional modulators of gene expression that play crucial roles in the responses to diverse stresses. Although significant progress has been made recently on miRNA-mediated gene regulation in plant-nematode interactions, none has been reported on root-knot nematode (*Meloidogyne* spp.) infection in *Cucumis metuliferus*, which is a relative of cucumber with resistance to *M. incognita*. To gain insights into the regulatory roles of miRNAs for resistance to *M. incognita*, expression profiles were created for miRNAs and their targets in *C. metuliferus*. Ten miRNAs were identified from our miRNAs sequencing data of *C. metuliferus* for expression analysis through quantitative reverse transcription-PCR (qRT-PCR) in this study. The results indicated that *M. incognita* infection had a significant effect on both miRNAs expression and their corresponding targets in either resistant or susceptible plants but with differential expression. Moreover, four out of ten selected miRNA-target pairs, miR156-SBP, miR390-ARF3, ath-miR159a-MYB104 and aly-miR827-3p-PTI, exhibited inverse expression patterns between miRNAs and their targets. This study demonstrates that miRNA-mediated gene regulation was involved in *C. metuliferus*-*M. incognita* interactions. Four miRNAs were identified as having negative correlation in expression with their corresponding targets. These modules represent components of a candidate regulatory system and lay a foundation for investigating resistance mechanisms induced in *C. metuliferus* by *M. incognita* infection. Further functional analyses will be required to investigate the biological significance of these regulatory modules in *C. metuliferus*-*M. incognita* interactions.

Key words: *Cucumis metuliferus*; *Meloidogyne incognita*; miRNAs; Target genes

* Funding: National Natural Science Foundation of China (31760508); Key R & D plan of Gansu Academy of Agricultural Sciences (2022GAAS34)

** First author: Ye Deyou. E-mail: ydy287@163.com

Isolation and Characterization of Streptomycetes Strains JXGZ01 with Nematicidal Activity Against Root-knot Nematode, *Meloidogyne incognita*

Xu Xueliang*, Fan Linjuan, Wu Caiyun, Liu Zirong, Yao Jian**, Yao Yingjuan**

(*Institute of Agricultural Applied Microbiology, Jiangxi Academy of Agricultural Sciences, Nanchang* 330200, *China*)

Abstract: Root-knot nematodes are the most economically destructive class of exclusively endoparasitic pathogenic nematodes that can cause yield loss in many cash and food crops. In this study, we evaluated the control effect of *Meloidogyne incognita* by a novel *Streptomyces* strain JXGZ01 screened from yam field soil. The results showed that the mortality of second-stage juveniles (J2s) treated with undiluted culture filtrate of JXGZ01 for 12h and 24h was 70. 86% and 85. 58%, respectively, and 66. 41% and 81. 55% for 12h and 24h for 2-fold dilution, respectively. Both strong acids and alkali significantly reduced the nematicidal activity of the culture filtrate, but neither temperature nor ultraviolet irradiation had any significant effect. In addition, 2-fold dilution, 4-fold dilution and 8-fold dilutions inhibited the hatching of *M. incognita* eggs by 91. 25%, 83. 67% and 75. 69%, respectively. Finally, the results of the pot experiments showed that the groups treated with different dilutions of culture filtrate significantly reduced the number of galls by 65. 87% to 97. 30%, the number of egg masses and hatching J2s by 38. 17% to 92. 84% and 37. 57% to 93. 82%, respectively, and the hatching rate of eggs by 27. 92% to 45. 27% relative to the control group. This indicates that the strain JXGZ01 has great potential and application value for development as a biocontrol agent for *M. incognita*.

Key words: *Meloidogyne incognita*; Streptomycetes; Culture filtrate; Nematicidal activity; Hatching inhibition rate

* First author: Xu Xueliang, master. E-mail: xuxueliang@ 126. com

** Corresponding authors: Yao Jian. E-mail: yaojian417@ 163. com

Yao Yingjuan. E-mail: yaoyingjuan2008@ 163. com

中国线虫学研究（第九卷）Nematology Research in China（Vol. 9）：62-63

Population Dynamics of *Meloidogyne graminicola* in Soil in Different Types of Rice Agroecosystems in Hunan Province, China

Yang Zhuhong[1,2*], Zhang Lu[1*], Li Xinwen[3], Lin Yufeng[3], Ye Shan[1,2], Ding Zhong[1,2**]

（[1] *College of Plant Protection, Hunan Agricultural University, Changsha* 410128, *China*; [2] *Hunan Provincial Engineering and Technology Research Center for Biopesticide and Formulation Processing, Changsha* 410128, *China*; [3] *Agriculture and Rural Department of Hunan Province, Plant Protection and Inspection Station, Changsha* 410005, *China*）

Abstract: Rice is an important staple food for a large part of the world's population and a model plant for studying the interaction between plants and pathogens. As a major rice production country, China contains approximately 19% of the world's rice planting area and accounts for 32% of the world's rice output (FAO, http://www.fao.org/faostat/zh/#data). The rice root-knot nematode *Meloidogyne graminicola* is increasingly widely distributed in China and cause a severe incidence in the Hunan province. A full understanding of the population dynamics of *M. graminicola* within and between crop cycles facilitates management decisions and the development of new or more accurate management practices. Therefore, it is necessary to investigate its population dynamics in paddy fields. This study was conducted to ascertain the effect of rice agroecosystems on the population dynamics of *M. graminicola* and root gall development in rice. The results indicated that the population density of *M. graminicola* in soil was markedly influenced by agroecosystem, rainfall and temperature. The population density of *M. graminicola* J2s and eggs in the soil and root galls were significantly higher in the dry aerobic rice agroecosystem and in the rainfed upland agroecosystem than in the lowland double-rice cropping sequence agroecosystem. Rainfall, as it can affect soil moisture, was the key factor affecting the density of nematodes in both the rainfed upland agroecosystem and the dry aerobic rice agroecosystem. Field flooding was still an effective way to reduce the population density of *M. graminicola*. In addition, we observed that *M. graminicola* can lay eggs outside rice roots under laboratory conditions. Therefore, we propose a hypothesis that *M. graminicola* lays egg masses within roots when the soil moisture is high but lays eggs outside when

* First authors: Yang Zhuhong; Zhang Lu

** Correspondence author: Ding Zhong. E-mail: dingzh@hunau.net

soil moisture is suitable. This study clarified the population dynamics of *M. graminicola* in different types of rice agroecosystems, which is conducive for controlling rice root-knot nematodes.
Key words: *Meloidogyne graminicola*; Population density; Agroecosystem; Soil moisture; Root gall

Effect of Different Initial Population Densities of *Meloidogyne graminicola* on Growth and Yield of Upland Rice cv. Hanyou73

Yang Zhuhong[1,2]*，Zhang Lu[1]*，Li Xinwen[3]，Lin Yufeng[3]，Ye Shan[1,2]，Ding Zhong[1,2]**
（[1] *College of Plant Protection*，*Hunan Agricultural University*，*Changsha* 410128，*China*；[2] *Hunan Provincial Engineering and Technology Research Center for Biopesticide and Formulation Processing*，*Changsha* 410128，*China*；[3] *Agriculture and Rural Department of Hunan Province*，*Plant Protection and Inspection Station*，*Changsha* 410005，*China*）

Abstract：*Meloidogyne graminicola*，commonly known as rice root-knot nematode，is one of the most economically damaging pathogens in rice cultivation areas，particularly in Asia. The rice yield loss caused by *M. graminicola* may vary depending on the severity of disease，and it is closely related to the environmental conditions and agro-ecosystem. The present investigation was conducted to ascertain the effect of initial *M. graminicola* population of soil on nematode multiplication，root gall development，rice growth and yield under screenhouse condition. After seed germination，rice cv. Hanyou73 were grown in soil containing eggs and second stage juveniles（J2）of *M. graminicola* at different levels i. e. 0，0. 02，0. 04，0. 06，0. 1，0. 5，1，and 2 eggs + J2/cm^3 soil，the ratio of eggs to J2 in soil was 1 to 1. Maximum number of galls was recorded in 0. 5 eggs + J2/cm^3 soil. Rice fertility was unaffected，but with the increase in density levels of *M. graminicola*，root length，root weight，plant height，tiller number，dry shoot weight，panicle number，panicle length，grain number，grain weight，and 1000-seed weight progressively decreased. The relationship between initial *M. graminicola* densities and relative grain yield fitted to the Seinhorst model was $Y = 0.24 + (0.76)(0.3252)^{P_i}$. Nematode reproduction in roots was higher at low initial densities and inhibited at high initial densities，whereas reproduction in soil was uncorrelated with initial densities. The reproduction factor of nematode decreased with increasing initial population densities.

Key words：*Meloidogyne graminicola*；initial population density；rice；plant growth and yield parameter

* First authors：Yang Zhuhong；Zhang Lu
** Correspondence author：Ding Zhong. E-mail：dingzh@ hunau. net

Recombinase Polymerase Amplification Coupled with CRISPR-Cas 12a Technology for Rapid and Highly Sensitive Detection of *Heterodera avenae* and *Heterodera filipjevi**

Shao Hudie[1,2**], Huang Wenkun[1], Kong Ling'an[1], Li Chuanren[2], Peng Deliang[1***], Peng Huan[1***]
([1] *Institute of Plant Protection, Chinese Academy of Agricultural Sciences, Beijing* 100193, *China*; [2] *College of Agriculture, Yangtze University, Jingzhou* 434025, *China*)

Abstract: The cereal cyst nematodes, *Heterodera avenae* and *Heterodera filipjevi* are recognized as cyst nematodes that infect cereal crops and cause severe economic losses worldwide. Rapid, visual detection of cyst nematodes is essential for more effective control of this pest. In this study, recombinase polymerase amplification (RPA) combined with clustered regularly interspaced short palindromic repeats (CRISPR)/Cas12a (formerly known as cpf1) was developed for the rapid detection of *H. avenae* and *H. filipjevi* from infested field samples. The RPA reaction was performed at a wide range of temperatures from 35 to 42℃ within 15min. There was no cross-reactivity between *H. avenae*, *H. filipjevi* and the common closely related plant-parasitic nematodes, indicating the high specificity of this assay. The detection limit of RPA-Cas12a was as low as 10^{-4} single second-stage juvenile (J2), 10^{-5} single cyst, and 0.001 ng of genomic DNA, which is 10 times greater than that of RPA-LFD detection. The RPA-Cas12a assay was able to detect 10^{-1} single J2 of *H. avenae* and *H. filipjevi* in 10g of soil. In addition, the RPA-LFD assay and RPA-Cas12a assays both could quickly detect *H. avenae* and *H. filipjevi* from naturally infested soil, and the entire detection process could be completed within 1h. These results indicated that the RPA-Cas12a assay developed herein is a simple, rapid, specific, sensitive, and visual method that can be easily adapted for the quick detection of *H. avenae* and *H. filipjevi* in infested fields.

Key words: *Heterodera avenae*; *Heterodera filipjevi*; Recombinase polymerase amplification (RPA); CRISPR/Cas12a

---

* Funding: National Natural Science Foundation of China (31972247, 31672012); Science and Technology Communication Project of Chinese Academy of Agricultural Sciences (2060302-51)

** First author: Shao Hudie, Ph. D., mainly engaged in genetic diversity research of plant nematodes

*** Corresponding authors: Peng Huan, Ph. D., Associate Researcher, mainly engaged in the research of plant nematode molecular biology and nematode disease management. E-mail: hpeng83@ 126. com

Peng Deliang, Ph. D., Researcher, mainly engaged in plant nematode classification and Research on the treatment of nematode diseases. E-mail: pengdeliang@ caas. cn

Resistance to *Heterodera filipjevi* in Wheat: An Emphasis on Classical and Modern Management Approaches*

Neveen Atta Elhamouly[1,2]**, Peng Deliang[1]***

([1]*State Key Laboratory for Biology of Plant Diseases and Insect Pests, Institute of Plant Protection, Chinese Academy of Agricultural Sciences, Beijing* 100193, *China*;
[2]*Department of Botany, Faculty of Agriculture, Menoufia University, Shibin El-Kom, Egypt*)

Abstract: Cereal cyst nematodes (CCNs) have the potential to produce significant economic yield losses when paired with other biotic and abiotic factors. When these nematodes are found in a disease complex, the devastation they wreak can be severe, particularly in areas prone to water stress. *Heterodera avenae*, *Heterodera filipjevi*, and *Heterodera latipons* are the three most economically significant CCN species in diverse parts of the world. Unfortunately, when numerous species and pathotypes coexist in nature, breeding for resistance becomes extremely difficult. Ineffective breeding methods and lengthy screening processes, as well as a lack of expertise and recognition of CCNs as a factor limiting wheat production potential, hinder genetic advancement for CCN resistance. Several initiatives have been launched in order to boost wheat resistance to CCNs. The majority of these efforts are focused on screening and selecting appropriate parents for breeding operations, followed by the application of genetic markers linked with this resistance. Using biotechnological techniques such as gene silencing, exploitation of nematode effector genes, proteinase inhibitors, chemodisruptive peptides, and a combination of one or more of these approaches. Furthermore, genome editing techniques such as CRISPR-Cas9 may be beneficial for boosting wheat resistance to CCN. In this review study, we shed light on the most recent research breakthroughs and the state of diagnostic methodologies and procedures for *Heterodera filipjevi*, as well as current strategies to control this nematode employing genetic resistance, biological agents, and chemical procedures.

Key words: *Heterodera filipjevi*; Distribution; Life cycle; Host range; Diagnosis; Pathotypes; Resistance genes; Chemical and biological management strategies

* 基金项目：国家自然科学基金（32072398）；中国农业科学院科技创新工程（ASTIP-02-IPP-15）

** First author: Neveen Atta Elhamouly. E-mail: neven. atta001@ agr. menofia. edu. eg

*** Corresponding author: Peng Deliang. E-mail: pengdeliang@ caas. cn

Thirty Years of Plant Parasitic Nematode Research in China

Lizzete Dayana Romero Moya*, Peng Deliang**
(*State Key Laboratory for Biology of Plant Diseases and Insect Pests. Institute of Plant Protection*, *Chinese Academy of Agricultural Sciences*, *Beijing* 100193, *China*)

Abstract: Phytoparasitic nematodes are a major problem for global food security, as they are present in nearly every important agricultural crop and cause significant yield losses amounting to an estimated US $ 157 billion annually (Jones *et al.*, 2013). To address this challenge, knowledge of plant diseases is crucial. Plant pathologists worldwide work to identify and understand diseases caused by plant-parasitic nematodes (Mesa *et al.*, 2020). This analysis aims to examine the state of research on nematology in China, including its main lines of work, priorities, and evolution from 1993 to 2023. A synthetic bibliometric analysis of studies on plant-parasitic nematodes published in China during this period was conducted using documents from Scopus, Web of Science, and Science Direct. The analysis characterized the progression of scientific outputs by topics, years, and authors, revealing a significant increase in the number of publications per year, particularly up to 2022. Among all provinces and municipalities in China, Beijing has one of the highest numbers of publications on nematodes. China, following the United States, has made the greatest contribution to agricultural nematology research worldwide. Within the keywords analyzed, the lines of research and genders that received the most attention were biological control, resistance, *Meloidogyne* sp. and *Heterodera* sp., respectively. The study of phytoparasitic nematodes in China have generated important advances in plant pathology. Although nematology has increased the number of its scientists and research institutions in recent years, the number of research documents in the field is expected to continue growing.

Key words: Nematology; Phytoparasitic; Bibliometry; Bibliographic data bases; Scientific research

* First author: Lizzete Dayana Romero Moya, Doctoral candidate. E-mail: 2020y90100067@caas.cn
** Corresponding author: Peng Deliang. E-mail: dlpeng@ippcaas.cn

中国线虫学研究（第九卷）Nematology Research in China（Vol. 9）：68-71

miRNA 在植物与线虫互作及细菌诱导植物免疫中的研究进展*

杨　帆[1**]，范海燕[1]，赵　迪[2]，王媛媛[3]，刘晓宇[4]，朱晓峰[1]，段玉玺[1]，陈立杰[1***]
（[1]沈阳农业大学植物保护学院，沈阳　110866；[2]沈阳农业大学分析测试中心，沈阳　110866；
[3]沈阳农业大学生物技术学院，沈阳　110866；[4]沈阳农业大学理学院，沈阳　110866）

摘　要：植物线虫病是世界上严重危害农作物的土传病害，可对作物生产造成巨大的经济损失，生物防治作为有效的防治手段之一，已经越来越受到人们的关注。miRNA（微小 RNA，microRNA）是一类广泛的小型非编码内源性 RNA。miRNA 介导的基因沉默是基因表达的基本调控机制[1]。超过 30%的蛋白编码基因可能受到 miRNAs 的调控[2]。因此，miRNA 在植物中几乎所有关键的生物过程中都发挥着重要的调控作用，并以多种方式成为基因的调控因子[3]。此外，一些 miRNA 在植物与线虫互作中的功能已被证实，如 miRNA 合成途径的阻断可增强植物对根结线虫的免疫力[4]，因此，miRNA 在植物线虫的免疫调节中发挥着重要作用。

关键词：植物线虫；miRNA；植物免疫调节

Research Advances on miRNA in Plant-nematode Interactions and Bacteria-induced Plant Immunity*

Yang Fan[1**]，Fan Haiyan[1]，Zhao Di[2]，Wang Yuanyuan[3]，Liu Xiaoyu[4]，Zhu Xiaofeng[1]，Duan Yuxi[1]，Chen Lijie[1***]
（[1] *College of Plant Protection*，*Shenyang Agriculture University*，*Shenyang*　110866，*China*；
[2] *Analysis and Testing Center*，*Shenyang Agriculture University*，*Shenyang*　110866，*China*；
[3] *College of Biotechnology*，*Shenyang Agriculture University*，*Shenyang*　110866，*China*；
[4] *College of Science*，*Shenyang Agriculture University*，*Shenyang*　110866，*China*）

Abstract：Plant nematode disease is one of the most serious soil-borne diseases in the world，which can cause huge economic losses to crop production. As one of the effective control methods，biological control has been paid more and more attention. miRNA（microRNA）is a broad class of small non-coding endogenous RNA. miRNA-mediated gene silencing is the basic regulatory mechanism of gene expression[1]. More than 30% of protein-coding genes may be regulated by miRNAs[2]. Therefore，miRNAs play an important regulatory role in almost all key biological processes in plants and act as regulatory factors of genes in a variety of ways[3]. In addition，the functions of some miRNAs in

* 基金项目：国家科技基础资源调查专项（2018FY100300）；辽宁省教育厅青年科技人才“育苗”项目（LSNQN201902）；辽宁省博士科研启动基金计划项目（2019-BS-210）；国家寄生虫资源库（NPRC-2019-194-30）
** 作者简介：杨帆，博士研究生，从事植物线虫研究。E-mail yangjingdong2333@ 163. com
*** 通信作者：陈立杰，教授，从事植物线虫研究。E-mail chenlj-0210@ syau. edu. cn

plant-nematode interactions have been confirmed, for example, blocking of miRNA synthesis pathway can enhance plant immunity against root-knot nematodes[4]. Therefore, miRNAs play an important role in immune regulation of plant nematodes.

Key words: Plant nematodes; miRNA; Plant immune regulation

1 miRNA 响应植物线虫胁迫的研究进展

miRNA 是一类由内源基因编码，长度约为 22 个核苷酸的非编码单链 RNA 分子，它们在动植物中参与转录后基因表达调控。miRNA 目前在植物线虫领域研究较多，也是最早被鉴定到参与免疫反应的非编码 RNA。miRNA 的主要功能是通过引起靶基因 mRNA 的降解或者阻止靶基因 mRNA 的翻译，进而抑制靶基因功能[5]。植物寄生线虫如孢囊线虫和根结线虫会在寄主根系形成特殊的取食细胞如合胞体和巨细胞，是寄主细胞的基因被额外诱导表达形成的新型取食细胞，其中的 miRNA 发挥了作用[6]。miR319/TCP4 在根结线虫的影响下调节番茄中茉莉酸的生物合成[7]。miR159 不但自身可以受到生长素诱导，它还可以调控 *MYB*33 和 *MYB*65 的表达[8]，并参与调控寄主根结中靶基因 *MYB*33 的表达从而限制南方根结线虫的早期侵染[9]。Cabrera 等证明 miR390a 和 Tas3 在巨细胞中可共表达，根结形成过程需要 miR390/TAS3/ARF3 分子模型调控[10]。Díaz-Manzano 等研究发现 miRNA172/TOE1/FT 基因调控模型在爪哇根结线虫侵染拟南芥的过程中，通过生长素而起到调节巨细胞形成和发育的作用[11]。miR396 在拟南芥和大豆中靶向生长调控因子（GRF，growth-regulating factors），在植物线虫侵染寄主形成巨细胞时通过调节 GRF1、GRF3、GRF6 和 GRF9[12,13] 等基因表达而影响巨细胞的发育。拟南芥接种线虫后，miRNA 及其靶基因呈负调节关系，其中 miR156、miR159、miR172 和 miR396 表达量下调[14]。Pan 等分析 miR159-MYB、miR319-TCP4 和 miR167-ARF8 三对基因调控模型的逆表达方式可能响应了南方根结线虫侵染棉花根系的胁迫反应[15]。

尽管 miRNA 在植物与线虫互作的机理已有研究，但 miRNA 在生防细菌诱导植物对线虫产生抗性的作用机制的研究仍然相对较少。

2 miRNA 响应细菌诱导植物免疫的研究进展

促进植物生长和保护植物免受根结线虫侵害的根际细菌称为线虫生防细菌；目前 miRNA 在生防细菌诱导植物抵抗线虫侵染的机制研究较少，但在其他病害研究报道较多。已有报道表明，接种假单胞菌 MLR6 对细胞膜的稳定性有积极的调控作用，它可以减少细胞膜电解质渗漏，促进活性氧的积累[16]；铜绿假单胞菌能够从浮游的生活方式向固定的生活方式转变，是因为 GacS/GacA 双组分系统激活了两种小型 RNA 分子 RsmY 和 RsmZ 的产生[17]。拟南芥 miR393 是第一个在植物免疫应答过程中起关键作用的 miRNA，被丁香假单胞菌鞭毛蛋白 flg22 诱导激发，miR393、miR160 和 miR167 通过抑制生长素信号途径而影响细菌的侵染[18]。Flg22 被用于 PTI（PAMP-triggered immunity，PAMP 触发免疫）体系中重要的元素可诱导胼胝质沉积，但是人们对 PTI 在根系上的作用，尤其是在植物与线虫互作中作用知之甚少。Li 等证明了来源于丁香假单胞菌的 RNA 激发了拟南芥的先天免疫反应，细菌侵染拟南芥后 miR398a 和 miR773 表达量降低，调控靶基因 *SOD* 上调表达来抵抗细菌侵染[19]。miR858-MYB83 调控系统被证明具有平衡甜菜孢囊线虫对拟南芥致病性的优异分子功能[20]。生防芽孢杆菌 AR156 诱导了拟南芥抗 *Pst* DC3000 的侵染，是 miR825 和 miR472 受到抑制从而增强 *NBS-LRR* 基因家族 CNLs 介导基础免疫而实现的[21]。恶臭假单胞菌 Sneb821 可以通过调控番茄 miR482d 的合成，来抵抗南方根结线虫的侵染[22]。现阶段的研究表明，miRNA 确实可以参与到生防菌诱导植物抵抗病原物侵染的过程中，但是 miRNA 在生防菌诱抗途径中所扮演的分子调控角色有待于进一步的探究。

3 问题与展望

随着植物免疫学的发展，植物病害生物防治的研究正在不断深入，而生防菌在诱导植物对病原物产生抗性或免疫的过程中，有多种多样的激发途径发挥着重要作用。尽管 miRNA 在生防菌诱抗过程中的分子机制取得了一定的认知，但这些探究仍然是不够系统全面的，仍有很多工作需要进一步展开。首先，miRNA 是否与 lncRNA（长链非编码 RNA，long non-coding RNA）或者 circRNA（环状 RNA，circular RNA）有着一定的互作关系，且 RNA 分子间的具体结合位点需加以验证。其次，是否有些 miRNA 可以直接作用在靶基因或者关键蛋白上尚不可知，这对了解 miRNA 的作用机制将有更全面的帮助。另外，生防菌是如何调控这些非编码 RNA 的表达以及哪些生防菌的代谢产物可以调控这些 RNA 的合成？生防菌可能通过一些鞭毛蛋白或者一些植物激素来诱导植物相关非编码 RNA 的生成，此类工作有待于进一步展开。最后，miRNA 的靶基因是如何调控植物相关抗病途径，来抵御病原物的侵染，以及这些基因会不会诱导抗病基因下游的抗病蛋白的表达等等，诸多未知细节仍有待进一步考证。

参考文献

[1] LEWIS B，BURGE C，BARTEL D. Conserved seed pairing，often flanked by adenosines，indicates that thousands of human genes are microRNA targets [J]. Cell，2005，120：15-20.

[2] XIE X，LU J，KULBOKAS E，*et al.* Systematic discovery of regulatory motifs in human promoters and 3′UTRs by comparison of several mammals [J]. Nature，2005，434：338-345.

[3] ZHANG B，WANG Q，PAN X. MicroRNAs and their regulatory roles in animals and plants [J]. Journal of Cellular Physiology，2007，210：279-289.

[4] RUIZFERRER V，CABRERA J，MARTINEZARGUDO I，*et al.* Silenced retrotransposons are major rasiRNAs targets in*Arabidopsis* galls induced by *Meloidogyne javanica* [J]. Molecular Plant Pathology，2018，19：2431-2445.

[5] YANG F，ZHAO D，FAN H Y，*et al.* Functional analysis of long non-coding RNAs reveal their novel roles in biocontrol of bacteria-induced tomato resistance to*Meloidogyne incognita* [J]. International Journal of Molecular Sciences，2020，21：911-929.

[6] JAUBERT-POSSAMAI S，NOUREDDINE Y，FAVERY B. MicroRNAs，new players in the plant-nematode interaction [J]. Frontiers in Plant Science，2019，10：1-8.

[7] ZHAO W C，LI Z L，FAN J W，*et al.* Identification of jasmonic acid-associated microRNAs and characterization of the regulatory roles of the miR319/TCP4 module under root-knot nematode stress in tomato [J]. Journal of Experimental Botany，2015，66（15）：4653-4667.

[8] ZHANG W, GAO S, ZHOU X, *et al.* Bacteria-responsive microRNAs regulate plant innate immunity by modulating plant hormone networks [J]. Plant Molecular Biology, 2011, 75: 93-105.

[9] MEDINA C, DA ROCHA M, MAGLIANO M, *et al.* Characterization of microRNAs from *Arabidopsis* galls highlights a role for miR159 in the plant response to the root-knot nematode *Meloidogyne incognita* [J]. New Phytologist, 2017, 216: 882-896.

[10] CABRERA J, BARCALA M, CARCIA A, *et al.* Differentially expressed small RNAs in *Arabidopsis* galls formed by *Meloidogyne javanica*: a functional role for miR390 and its TAS3-derived tasiRNAs [J]. New Phytologist, 2016, 209: 1625-1640.

[11] DíAZ-MANZANO F E, CABRERA J, RIPOLL J, *et al.* A role for the gene regulatory module microRNA172/Target of Early Activation Tagged1/Flowering Locust (miRNA172/TOE1/FT) in the feeding sites induced by *Meloidogyne javanica* in *Arabidopsis thaliana* [J]. New Phytologist, 2018, 217: 813-827.

[12] HEWEZI T, BAUM T J. Complex feedback regulations govern the expression of miRNA396 and its GRF target genes [J]. Plant Signal Behavior, 2012, 7: 749-751.

[13] NOON J B, HEWEZI T, BAUM T J. Homeostasis in the soybean miRNA396-GRF network is essential for productive soybean cyst nematode infections [J]. Journal of Experimental Botony, 2019, 70: 1653-1668.

[14] JAUBERT-POSSAMAI S, NOUREDDINE Y, FAVERY B. MicroRNAs, new players in the plant-nematode interaction [J]. Frontiers in Plant Science, 2019, 10: 1-8.

[15] PAN X, NICHOLS RL, LI C, *et al.* MicroRNA-target gene responses to root knot nematode (*Meloidogyne incognita*) infection in cotton (*Gossypium hirsutum* L.) [J]. Genomics, 2019, 111: 383-390.

[16] RABHI N, SILINI A, CHERIFSILINI H, *et al.* *Pseudomonas knackmussii* MLR6, a rhizospheric strain isolated from halophyte, enhances salt tolerance in *Arabidopsis thaliana* [J]. Journal of Applied Microbiology, 2018, 125: 1836-1851.

[17] CHAMBONNIER G, ROUX L, REDELBERGER D, *et al.* The hybrid histidine kinase lads forms a multicomponent signal transduction system with the GacS/GacA two-component system in*Pseudomonas aeruginosa* [J]. PLoS Genetics, 2016, 12 (5): e1006032.

[18] NAVARRO L, DUNOYER P, JAY F, *et al.* A plant miRNA contributes to antibacterial resistance by repressing auxin signaling [J]. Science, 2006, 312: 436-439.

[19] LI Y, ZHANG Q Q, ZHANG J G, *et al.* Identification of microRNAs involved in pathogen-associated molecular pattern-triggered plant innate immunity [J]. Plant Physiology, 2010, 152: 2222-2231.

[20] PIYA S, KIHM C, RICE J H, *et al.* Cooperative regulatory functions of miR858 and MYB83 during cyst nematode parasitism [J]. Plant Physiology, 2017, 174: 1897-1912.

[21] JIANG C H, FAN Z H, LI Z J, *et al.* *Bacillus cereus* AR156 triggers induced systemic resistance against *Pseudomonas syringae* pv. tomato DC3000 by suppressing miR472 and activating CNLs-mediated basal immunity in Arabidopsis [J]. Molecular Plant Pathology, 2020, 10: 1-17.

[22] YANG F, DING L, ZHAO D, *et al.* Identification and functional analysis of tomato microRNAs in the biocontrol bacterium *Pseudomonas putida* induced plant resistance to *Meloidogyne incognita* [J]. Phytopathology, 2022, 112 (11): 2372-2382.

Phased 的 T2T 基因组揭示异源多倍体线虫起源模式为未减数配子与单倍体配子的杂交

代大东*，解传帅，周雅艺，张书荣，彭东海，郑金水，孙　明**
（华中农业大学/农业微生物资源发掘与利用全国重点实验室，武汉　430007）

摘　要：据估计，植物寄生线虫（PPN）每年给全球作物生产造成 800 亿~1 730 亿美元的经济损失，并对全球粮食安全构成威胁。根结线虫（RKNs）是最具破坏性的 PPN，因为它们可以感染几乎所有的维管植物。RKN 展示了从双性生殖到专性孤雌生殖和从二倍体到四倍体的多倍性水平及多种繁殖策略。例如，*M. incognita*（Mi）、*M. arenaria*（Ma）和 *M. javanica*（Mj），通常以专性有丝分裂孤雌生殖和异源多倍体为特征。然而，在具有无性繁殖的动物中多倍体化的形成和后果在很大程度上仍然未知。

大多数动物、植物和真菌通过有性繁殖产生后代，而无性繁殖是真核生物中一种罕见但广泛分布的特征。虽然为什么大多数生物体保持有性繁殖仍然是一个悬而未决的问题，但一些假设，例如比无性物种更快的适应性、更快的进化速度和消除有害突变等是有经验支持的假设。然而，少数真核生物保持专性孤雌生殖，杂交和多倍体化在孤雌物种中产生独特的遗传变异组合。由于缺乏重组，与它们的有性亲属相比，孤雌生殖物种在环境适应过程中通常表现出较弱的竞争力，并且被认为通过一种称为 Muller 棘轮的机制积累了有害突变。因此，无性繁殖被认为是进化的死胡同。但是，也有一些例外。例如多倍体 RKN 显示出比其二倍体性亲属更强的生存优势。

在本研究中，笔者组装了 4 个多倍体和 1 个二倍体 RKN 基因组，但没有发现典型的端粒结构。通过进一步分析确定了两个重复序列，它们取代了典型的端粒序列，并且可能充当多倍体和二倍体 RKN 的功能性端粒，同时，它定义了染色体的边界。经过两轮组装后，笔者获得了 4 个 phased 的多倍体 RKN 端粒到端粒（T2T）基因组和 1 个单倍体 *M. graminicola*（Mg）T2T 基因组，并将多倍体的基因组结构解析为 AAB 三倍体和 AABB 四倍体。亚基因组的系统发育揭示了多倍体 RKN 起源模式是单倍体配子和未减数配子之间的杂交。笔者的结果表明现存的花生根结线虫同时存在三倍体和四倍体两个谱系。笔者还观察到多倍化后广泛的染色体融合和同源基因表达减少，这可能抵消克隆繁殖的缺点并增加多倍体 RKN 的适应性。笔者的研究结果阐明了专性孤雌动物的进化机制，并为 PPN 的研究和控制提供了准确和完整的基因组资源。

关键词：根结线虫；异源多倍体；进化机制；T2T 基因组

---

* 第一作者：代大东，博士后，从事植物寄生线虫多组学研究。E-mail：daidadong@mail.hzau.edu.cn
** 通信作者：孙明，教授，从事微生物病虫害防控及新型微生物农药创制相关研究。E-mail：m98sun@mail.hzau.edu.cn

Volutella ciliate Q7 对马铃薯腐烂茎线虫的作用*

马　娟**，李秀花，高　波，王容燕，陈书龙***
（河北省农林科学院植物保护研究所/河北省农业有害生物综合防治工程技术研究中心/农业农村部华北北部作物有害生物综合治理重点实验室，石家庄　050031）

Control Effect of *Volutella ciliate* Q7 on *Ditylenchus destructor**

Ma Juan**，Li Xiuhua，Gao Bo，Wang Rongyan，Chen Shulong***
（*Institute of Plant Protection*，*Hebei Academy of Agricultural andForestryg Sciences/IPM centre of Hebei Province/Key Laboratory of Integrated Pest Management on Crops in Northern Region of North China*，*Ministry of Agriculture*，*Shijiazhuang*　050031，*China*）

摘　要：由马铃薯腐烂茎线虫（*Ditylenchus destructor*）引起的甘薯茎线虫病是我国北方薯区甘薯生产上最严重的病害之一。该线虫在甘薯整个生育期均能寄生危害，主要危害薯块，也危害茎蔓及秧苗。当前我国实际生产中化学农药仍然是防治马铃薯腐烂茎线虫的重要手段，但部分地区杀线剂用量大、频次高，不仅危害人类健康，污染环境，还造成线虫抗药性水平逐渐增加。研发高效、低毒、低残留的防控方法是目前甘薯生产中急需解决的紧迫问题。*Volutella ciliate* Q7 由甘薯薯块中分离得到，对马铃薯腐烂茎线虫具有很强的诱集作用，通过菌的诱集而降低线虫对植株的危害，是一条甘薯茎线虫病防控新途径。通过水琼脂平板法测试不同真菌对茎线虫的诱集作用，发现处理 24h 后 Q7 对线虫诱集率达到 16%，显著高于 *Fusarium tricinctum*、*F. oxysporum*、*F. moniliforme* 、*F. avenaceum* 及 *F. graminearum* 等。通过盆栽试验和田间试验测试 *V. ciliate* Q7 对甘薯茎线虫病的诱控效果。盆栽试验表明，未用药对照中每株薯苗内侵入的线虫数量高达 289 条，使用 *V. ciliate* Q7 处理后侵入甘薯内的线虫数量下降到 36 条，显著低于对照甘薯苗内的线虫数量。田间调查结果发现未用药对照区线虫危害较重，茎线虫病情指数为 35. 03。使用失活的 Q7 处理后甘薯薯块被害率和病情指数显著降低，对茎线虫防控效果为 51. 73%。10%噻唑膦颗粒剂施用量为 2 kg/亩时对茎线虫防效达到 81. 7%，其使用量降低防效显著下降，当用量为 1. 2 kg/亩时对茎线虫防效降至 56. 76%。将 *V. ciliata* Q7 与 10%噻唑膦颗粒剂 1. 2 kg/亩组合使用后，对茎线虫的防控效果可达 87. 22%，优于 10%噻唑膦 2 kg/亩防治效果。*Volutella ciliate* Q7 在田间施用后持效期长，稳定性好，还可与低剂量杀线剂组合使用，极大延缓线虫抗药性的产生。

关键词：甘薯茎线虫；*Volutella ciliate* Q7；诱集率；田间试验

---

* 基金项目：国家现代农业产业技术体系（CARS-10-B16）；河北省农林科学院科技创新专项（2022KJCXZX-ZBS5）

** 第一作者：马娟，研究员，从事线虫学研究。E-mail：majuan_206@126. com
*** 通信作者：陈书龙，研究员，从事线虫学研究。E-mail：chenshulong65@163. com

百岁兰曲霉对水稻干尖线虫的作用机制研究

贾建平[1*]，邓龙飞[2]，李忠彩[2]，于敬文[1]，赵津田[1]，余曦玥[1]，彭德良[1]，刘世名[1]，黄文坤[1**]
（[1]中国农业科学院植物保护研究所植物病虫害生物学国家重点实验室，北京 100193；
[2]湖南省汉寿县农业农村局，常德 415900）

Interaction Mechanism of Nematicidal Fungus *Aspergillus welwitschiae* Against the White Tip Nematode *Aphelenchoides besseyi* in Rice

Jia Jianping[1*], Deng Longfei[2], Li Zhongcai[2], Yu Jingwen[1], Zhao Jintian[1], Yu Xiyue[1], Peng Deliang[1], Liu shiming[1], Huang Wenkun[1**]
（[1]*State Key Laboratory for Biology of Plant Diseases and Insect Pests*, *Institute of Plant Protection*, *Chinese Academy of Agricultural Sciences*, *Beijing* 100193, *China*;
[2]*Agriculture and Rural Affairs Bureau of Hanshou*, *Changde* 415900, *China*）

摘　要：水稻干尖线虫（*Aphelenchoides besseyi*）是危害水稻较为严重的植物病原寄生线虫，影响亚洲地区及中国水稻安全生产。由于化学杀线剂大多毒性高、挥发性强，不利于环境保护，也易使线虫产生抗药性，因此，亟待筛选出绿色环保且易于推广的防控方法，以期为新型、高效、安全的微生物源杀线剂开发奠定基础。

利用室内生物测定，对百岁兰曲霉孢子悬浮液防治水稻干尖线虫的效果进行评价。结果显示，2 倍、4 倍、8 倍、16 倍和 32 倍稀释的百岁兰曲霉孢子悬浮液均对水稻干尖线虫具有一定的毒杀作用。稀释 2 倍的曲霉孢子悬浮液处理 48h 后，水稻干尖线虫的校正死亡率达 96.07%，4 倍稀释液施用以后，干尖线虫卵孵化抑制率可达 35.19%。通过盆栽试验，发现百岁兰曲霉处理后，实穗率增加了 23.25%，千粒重增加了 13.99%，防治效果可达 42.32%，说明百岁兰曲霉的施用可以有效防治水稻干尖线虫，并提高水稻产量。

百岁兰曲霉生防真菌能有效控制水稻干尖线虫的危害，提高水稻叶绿素含量，增加水稻产量，并诱导水稻病程相关蛋白的表达，对水稻干尖线虫具有较好的生防潜力。

关键词：水稻干尖线虫；百岁兰曲霉；生物防治；互作机制；蛋白质组学

* 第一作者：贾建平，博士研究生，从事植物线虫病害研究。E-mail：jjianping1997@ 163. com
** 通信作者：黄文坤，研究员，从事植物线虫病害致病机理及综合防控技术研究。E-mail：wkhuang2002@ 163. com

3株不同来源淡紫紫孢菌对植物线虫的活性

吴　艳[1*]，张　会[1]，江兆春[2]，刘明睿[1]，杨再福[1**]
（[1]贵州大学农学院植物病理教研室，贵阳　550025；[2]贵州省植保植检站，贵阳　550001）

Nematicidal Activities of Three Strains of *Purpureocillium lilacinum* from Different Sources Against Plant Nematodes

Wu Yan[1*], Zhang Hui[1], Jiang Zhaochun[2], Liu Mingrui[1], Yang Zaifu[1**]
（[1] *Department of Plant Pathology, College of Agriculture, Guizhou University, Guiyang　550025 China*; [2] *Guizhou Station of Plant Protection and Quarantine, Guiyang　55001, China*）

摘　要：淡紫紫孢菌（*Purpureocillium lilacinum*）分布于土壤、植物组织、线虫体内等多种环境中，是植物线虫重要的生防真菌。

本研究从马铃薯金线虫（*Globodera rostochiensis*）孢囊、南方根结线虫（*Meloidogyne incognita*）雌虫、刺梨（*Rosa roxburghii*）根组织内分离真菌，通过形态学和多基因系统比对鉴定到3株淡紫紫孢菌，编号分别为P3、M51、E162。通过发酵液原液和稀释液处理水稻干尖线虫（*Aphelenchoides besseyi*）、腐烂茎线虫（*Ditylenchus destructor*）和孢子悬浮液处理马铃薯金线虫卵，测试其对植物线虫的活性。

研究结果表明，菌株P3发酵原液、2倍稀释液48h对水稻干尖线虫的校正死亡率分别为95.48%、24.77%，对腐烂茎线虫为100%和42.25%；菌株M51发酵原液、2倍稀释液48h对水稻干尖线虫校正死亡率为40.78%和21.10%，对腐烂茎线虫为54.28%和36.70%；菌株E162发酵原液、2倍稀释液48h对水稻干尖线虫校正死亡率分别为55.58%和17.44%，对腐烂茎线虫为87.94%和25.40%。3株菌孢子悬浮液处理马铃薯金线虫卵，3d后寄生率菌株P3为91.08%、M51为84.58%、E162为0%。结果显示，不同来源淡紫紫孢菌发酵液对2种植物线虫的致死率存在一定的差异，其中菌株P3活性最高；孢子悬浮液对马铃薯金线虫卵的寄生率差异明显，从马铃薯金线虫孢囊和南方根结线虫雌虫上分离的菌株（P3、M51）表现出较高的寄生率，而从刺梨根组织内分离的菌株（E162），则未表现出寄生性。

关键词：淡紫紫孢菌；植物线虫；杀线活性

* 第一作者：吴艳，硕士研究生，从事植物线虫病害研究。E-mail：18164816527m@ sina. cn
** 通信作者：杨再福，讲师，从事植物线虫病害研究。E-mail：zfyang@ gzu. edu. cn

7 种药剂对粗茎秦艽根结线虫病的防治效果

李云霞*，杨艳梅，刘福祥，李乾坤，胡先奇**
（云南农业大学植物保护学院/云南生物资源保护与利用国家重点实验室，昆明 650201）

Control Effect of Seven Nematicides on Root-knot Nematodes of *Gentiana crassicaulis*

Li Yunxia*，Yang Yanmei，Liu Fuxiang，Li Qiankun，Hu Xianqi**
（*State Key Laboratory for Conservation and Utilization of Bio-Resources in Yunnan/ College of Plant Protection*，*Yunnan Agricultural University*，*Kunming* 650201，*China*）

摘　要： 粗茎秦艽（*Gentiana crassicaulis*）是龙胆科多年生草本植物，以干燥根入药。云南省丽江市为粗茎秦艽道地药材产区，具有悠久的规模化栽培生产粗茎秦艽的历史。近年来，丽江市鲁甸乡大面积种植的粗茎秦艽普遍受根结线虫危害，平均发病率 84. 30%，病情指数为 34. 36，致使药材产量大幅度下降、品质降低，严重地块甚至绝收。经调查及鉴定确定，其病原线虫是北方根结线虫（*Meloidogyne hapla*）。

为找到高效、低毒、低残留的绿色防控药剂，2021 年 3—11 月笔者在丽江市鲁甸乡进行了田间药剂防效试验，选用 4 种高效低毒农药（9%寡糖・噻唑膦颗粒剂、5%阿维・噻唑膦颗粒剂、1%阿维菌素颗粒剂、10%噻唑膦颗粒剂）、1 种生物活性有机肥（红土运生物有机肥）、2 种微生物菌剂（2 亿孢子/g 哈茨木霉可湿性粉剂、线虫克星）进行田间小区试验。结果显示，7 种药剂对粗茎秦艽根结线虫病均有防治效果，采收末期（药剂处理后 255d），各药剂相对防效依次为：5%阿维・噻唑膦颗粒剂 73. 12%、10%噻唑膦颗粒剂 69. 08%、9%寡糖・噻唑膦颗粒剂 62. 81%、1%阿维菌素颗粒剂 40. 26%、2 亿孢子/g 哈茨木霉可湿性粉剂 40. 12%、线虫克星 32. 01%、红土运生物有机肥 19. 29%。其中，5%阿维・噻唑膦颗粒剂、10%噻唑膦颗粒剂、9%寡糖・噻唑膦颗粒剂的相对防效显著（$P \leq 0.05$）高于其余 4 种药剂，微生物菌剂 2 亿孢子/g 哈茨木霉可湿性粉剂、线虫克星的防治效果与当地常年施用的 1%阿维菌素颗粒剂相当。由于中药材使用性质的特殊性，人们对其绿色安全无残留有着更高标准，微生物菌剂在安全性、环保性及减缓根结线虫抗药性等方面都优于化学药剂，故推荐在粗茎秦艽根结线虫病的防治过程中可选用线虫克星、2 亿孢子/g 哈茨木霉可湿性粉剂，在生产实践中可以适当增加单位面积的施用量，从而保障粗茎秦艽的产量和品质。

关键词： 粗茎秦艽；根结线虫病；药剂；防效

---

* 第一作者：李云霞，博士研究生，从事植物线虫病害研究。E-mail：yx1160725@ 163. com
** 通信作者：胡先奇，教授，从事植物线虫病害研究。E-mail：xqhoo@ 126. com

孢囊线虫效应蛋白 Hg11576 靶向大豆 GmHIR1 抑制寄主免疫的分子机制研究*

姚　珂**，彭德良，彭　焕***
（中国农业科学院植物保护研究所/植物病虫害综合治理全国重点实验室，北京　100193）

Molecular Mechanism of Cyst Nematode Effector Hg11576 Target Soybean GmHIR1 to Suppress Host Immunity*

Yao Ke**，Peng Deliang，Peng Huan***
(*State Key Laboratory for Biology of Plant Diseases and Insect Pests*, *Institute of Plant Protection*, *Chinese Academy of Agricultural Sciences*, *Beijing*　100193, *China*)

摘　要：大豆孢囊线虫（*Heterodera glycines*）侵染引起的大豆孢囊线虫病（Soybean Cyst Nematode，SCN）在全世界的大豆（*Glycine max*）主产区均有分布和危害，防治难度大、防治费用高，全球每年由大豆孢囊线虫造成的经济损失高达数十亿美元。植物寄生线虫通过口针向寄主体内分泌大量的效应蛋白，这些效应蛋白在线虫维持取食位点以及抵抗寄主防卫反应中都发挥着关键作用。笔者通过对大豆孢囊线虫基因组效应蛋白的非编码区进行分析，发现了200多个启动子含有由“ATGCCA”组成的DOG-box（Dorsal Gland box）的新效应蛋白，其中大部分的功能暂无相关报道。通过对这些潜在效应蛋白进行筛选，得到了10个能抑制BAX或GPA2/RBP1介导的过敏性坏死的候选蛋白，其中包括*Hg*11576。*Hg*11576的表达部位主要位于SCN食道腺；*Hg*11576在烟草叶片中的瞬时表达能够抑制BAX激发的过敏性坏死反应；体外介导*Hg*11576基因沉默后，大豆孢囊线虫二龄幼虫的侵染率下降69%，单株白雌虫量减少47%；采用酵母双杂交发现Hg11576与大豆过敏性诱导蛋白GmHIR1互作。HIR蛋白是一类与植物过敏性反应相关的蛋白，能参与植物的免疫反应。在烟草叶片中瞬时表达*GmHIR*1能够引起过敏性坏死反应。以上结果表明，大豆孢囊线虫效应蛋白Hg11576通过靶向大豆过敏诱导蛋白GmHIR1，抑制寄主免疫反应，促进SCN的寄生，但Hg11576的生物学功能及与GmHIR1互作抑制寄主免疫反应的作用机制还在研究中。

关键词：大豆孢囊线虫；效应蛋白；植物与线虫互作；致病性；免疫

---

* 基金项目：公益性行业（农业）科研专项（201503114）；国家自然科学基金（31672012，31972247）
** 第一作者：姚珂，博士研究生，从事植物线虫致病机理研究。E-mail：11816092@zju.edu.cn
*** 通信作者：彭焕，研究员，从事植物与线虫互作机制研究。E-mail：hpeng83@126.com

马铃薯主栽品种及育种资源抗金线虫 *H*1 基因分子鉴定*

江　如**，彭　焕，彭德良***
（中国农业科学院植物保护研究所/植物病虫害综合治理全国重点实验室，北京　100193）

Molecular Identification of the *H*1 Gene for Resistance to *Globodera rostochiensis* in Major Potato Cultivar and Breeding Resources*

Jiang Ru**，Peng Huan，Peng Deliang***
（*State Key Laboratory for Biology of Plant Diseases and Insect Pests*，*Institute of Plant Protection*，*Chinese Academy of Agricultural Sciences*，*Beijing*　100193，*China*）

摘　要：马铃薯金线虫（*Globodera rostochiensis*）是国际公认的重要检疫性有害生物，严重危害马铃薯生产。马铃薯金线虫的防治措施中，种植抗病品种被认为是最具经济效益和生态效益的防治方法。随着分子标记技术的发展和广泛应用，包括抗马铃薯金线虫基因在内的许多植物抗病基因被分子定位和作图。利用与抗病基因紧密连锁的分子标记对抗病资源进行检测已经成为抗病基因鉴定快速而有效的方法。本研究利用 2 个与已知马铃薯抗马铃薯金线虫 *H*1 基因连锁的分子标记对我国 43 份马铃薯主栽品种及 13 份育种材料进行基因型鉴定。结果表明，8 个主栽马铃薯品种（会薯 15 号、宣薯 5 号、宣薯 6 号、云薯 505、云 304、大西洋、陇 14、冀张薯 12）和 5 份育种材料（cau-1、cau-2、cau-3、1047、1049）在两个分子标记中均为阳性。主栽马铃薯品种 P14-187-3 及育种材料 1048、1050 在一个分子标记上显示阳性条带。综上，本研究筛选鉴定 9 个马铃薯主栽品种和 7 份育种材料含抗马铃薯金线虫 Ro1 和 Ro4 型的 *H*1 基因，为马铃薯金线虫抗病品种的培育提供重要参考。

关键词：马铃薯金线虫；*H*1 基因；主栽品种；育种资源；筛选鉴定

* 基金项目：国家自然科学基金（32072398）；政府购买服务项目（15190025）；中国农业科学院科技创新工程（ASTIP-02-IPP-15）

** 第一作者：江如，博士研究生，从事植物线虫分子生物学研究。E-mail：jiangruby@126. com
*** 通信作者：彭德良，研究员，从事植物线虫研究。E-mail：pengdeliang@caas. cn

中国线虫学研究（第九卷）Nematology Research in China（Vol. 9）：79

8 种植物挥发物对马铃薯金线虫卵孵化的抑制效果

姚汉央*，杜　霞，杨艳梅，李云霞，李　艳，邓春菊，段锦凤，尹艳蝶，胡先奇**
（云南农业大学植物保护学院/云南生物资源保护与利用国家重点实验室，昆明　650201）

Hatching Inhibition of Eight Plant Volatile Organic Compounds for *Globodera rostochiensis*

Yao Hanyang*，Du Xia，Yang Yanmei，Li Yunxia，Li Yan，Deng Chunju，
Duan Jinfeng，Yin Yandie，Hu Xianqi**
（*State Key Laboratory for Conservation and Utilization of Bio-Resources in Yunnan/ College of Plant Protection*，*Yunnan Agricultural University*，*Kunming*　650201，*China*）

摘　要：植物挥发物（Volatile organic compounds，VOCs）是植物次生代谢产物，在植物的防御以及与其他植物、动物和微生物的相互作用中起着重要作用并具有广泛的生物学功能。在植物病虫害防控方面，植物挥发物被证实有多种功能，包括调节植物生长、吸引天敌昆虫、毒杀病原物或抑制其生长、增进植物防御反应、诱导植物抗性和缓解非生物胁迫压力等。在植物线虫防控中，植物挥发物主要影响线虫的趋化性。对根结线虫（*Meloidogyne* spp.）有作用的挥发物有 α-蒎烯、柠檬烯、玉米素等，对孢囊线虫（*Globodera* spp.、*Heterodera* spp. 等）有吸引作用的挥发物有乙烯利、水杨酸和甘露醇等，对线虫有趋避作用的挥发物有软脂酸亚油酸、2,6-二叔丁基对甲酚、邻苯二甲酸二丁酯等。另外，植物挥发物也可以毒杀线虫的幼虫或者抑制卵孵化，增强寄主植物的防御反应，诱导植物产生抗性从而降低线虫危害。本研究选择了 8 种常见的植物挥发物 α-蒎烯、β-蒎烯、β-石竹烯、丁香酚、茴香脑、柠檬烯、1-辛烯-3 醇、3-己醇，这些挥发物对植物线虫的影响在根结线虫、松材线虫（*Bursaphelenchus xylophilus*）等的研究中已有报道。本研究以马铃薯金线虫（*Globodera rostochiensis*）为对象，评估了 8 种挥发物对马铃薯金线虫卵孵化抑制效果。结果显示，柠檬烯、1-辛烯-3-醇、3-己醇 3 种植物挥发物能显著（$P \leqslant 0.05$）抑制马铃薯金线虫卵的孵化，抑制率分别为 67.90%、95.60%、95.27%，表明柠檬烯、1-辛烯-3-醇、3-己醇可以作为马铃薯金线虫绿色防控药剂新的选择，为新型熏蒸剂的研发提供选材。

关键词：马铃薯金线虫；植物挥发物；抑制孵化；新型熏蒸剂

* 第一作者：姚汉央，博士研究生，从事植物线虫病害研究。E-mail：406590843@ qq. com
** 通信作者：胡先奇，教授，从事植物线虫病害研究。E-mail：xqhoo@ 126. com

马铃薯金线虫群体遗传分化及毒性特征分析*

江　如**，彭　焕，刘世名，彭德良***
（中国农业科学院植物保护研究所/植物病虫害综合治理全国重点实验室，北京　100193）

Genetic Differentiation and Virulence Characterization of *Globodera rostochiensis* in China*

Jiang Ru**，Peng Huan，Liu Shiming，Peng Deliang***
(*State Key Laboratory for Biology of Plant Diseases and Insect Pests*，*Institute of Plant Protection*，*Chinese Academy of Agricultural Sciences*，*Beijing*　100193，*China*)

摘　要：马铃薯孢囊线虫（Potato cyst nematode，PCN），包括马铃薯金线虫（*Globodera rostochiensis*）以及马铃薯白线虫（*G. pallida*），是我国禁止进境的一类植物检疫危险性线虫，现纳入全国农业植物检疫性有害生物名单。其主要寄生茄科近90种植物，危害的农作物主要有马铃薯、茄子和番茄。自2018年我国发现马铃薯金线虫在贵州省发生以来，进一步调查发现其在云南省、贵州省和四川省等多地发生危害，全国扩散风险极高。然而，对于马铃薯金线虫在我国传播与危害，目前却无有效的早期检测和监测技术、无有效的根除办法且其传播来源以及我国群体的毒性特征等科学问题尚不明确。本研究基于基因组重测序，收集我国云贵川地区29个代表群体以及和国外及不同致病型群体包括欧洲6个群体、北美洲加拿大2个群体、美国55个群体以及东南亚14个群体进行共同分析，以期明确我国群体遗传构成，推测地理区域间群体亲缘关系和迁移事件，解析马铃薯金线虫在我国的传入及传播过程，追溯其可能的传播来源。利用鉴别寄主对我国马铃薯金线虫群体致病型进行鉴定，针对不同毒性群体筛选谱系特异性区域，开发致病型相关分子标记。进一步筛选候选效应子并明确其生物学功能，以期明确我国群体的毒性特征，为防止马铃薯金线虫进一步传播扩散和精准防控提供技术支撑。

关键词：马铃薯金线虫；群体遗传分化；毒性特征分析

---

* 基金项目：国家自然科学基金（32072398）；政府购买服务项目（15190025）；中国农业科学院科技创新工程（ASTIP-02-IPP-15）

** 第一作者：江如，博士研究生，从事植物线虫分子生物学研究。E-mail：jiangruby@126.com
*** 通信作者：彭德良，研究员，从事植物线虫研究。E-mail：pengdeliang@caas.cn

中国线虫学研究（第九卷）Nematology Research in China（Vol. 9）：81

马铃薯金线虫（*Globodera rostochiensis*）生物学特征研究

邓春菊*，杨艳梅，杜　霞，姚汉央，李云霞，李　艳，尹艳蝶，段锦凤，胡先奇**
（云南农业大学植物保护学院/省部共建云南生物资源保护与利用国家重点实验室，昆明　650201）

Study on Biological Characteristics of *Globodera rostochiensis*

Deng Chunju*，Yang Yanmei，Du Xia，Yao Hanyang，Li Yunxia，
Li Yan，Yin Yandie，Duan Jinfeng，Hu Xianqi**
(*State Key Laboratory for Conservation and Utilization of Bio-Resources in Yunnan/ College of Plant Protection*，*Yunnan Agricultural University*，*Kunming*　650201，*China*)

摘　要：马铃薯金线虫（*Globodera rostochiensis*）是马铃薯生产田间极为重要的检疫性有害生物，包括我国在内的很多国家已将其列为对外检疫对象。近年来，在我国西南的个别地区已在田间发现了马铃薯金线虫。为了认识传入我国的马铃薯金线虫在发生区的生物学特性，对其胚胎发育过程、卵孵化条件、生活史、致病性及寄主范围进行了初步研究。结果表明，在23～31℃下，马铃薯金线虫2龄幼虫接种后，在马铃薯根系中发育为成熟孢囊大约需要52.5d，亦证明其侵染马铃薯能完成生活史，是危害马铃薯的病原；在25℃下，马铃薯金线虫的单胞卵发育到2龄幼虫需要12.5～14.0d；在25℃下，马铃薯金线虫在马铃薯、番茄根系分泌物中14d累积孵化率分别为45.8%、40.0%；在20℃、25℃下，14d累积孵化率分别为26.4%、32.3%；25℃下，在pH=7溶液中的14d累积孵化率为1.4%。对9科22种作物及田间杂草的人工接种和田间调查发现：马铃薯金线虫能在4种茄科植物（马铃薯、番茄、茄子、刺天茄）上完成生活史；可以侵染茄科（龙葵、辣椒、烟草）、蓼科（酸模叶蓼、野荞麦、荞麦），但是未发现形成雌虫（孢囊），推测马铃薯金线虫不能在这6种供试植物上完成生活史；在其他的供试植物上均未发现有该线虫侵染。

关键词：马铃薯金线虫；发育；致病性；田间寄主

---

* 第一作者：邓春菊，硕士研究生，从事植物线虫病害研究。E-mail：1092784410@qq.com
** 通信作者：胡先奇，教授，从事植物线虫病害研究。E-mail：xqhoo@126.com

马铃薯品种资源对马铃薯金线虫耐病性的筛选*

易　军[1**]，黄　润[1]，李　兵[2]，符慧娟[1]，胡建军[2]，许秉智[3]，李星月[1***]
（[1]四川省农业科学院植物保护研究所，成都　610066；[2]四川省农业科学院作物研究所，成都　610066；[3]越西县农业农村局，凉山州　616650）

Screen the Tolerance of Potato Variety Resources Against Potato Cyst Nematode*

Yi Jun[1**]，Huang Run[1]，Li Bing[2]，Fu Huijuan[1]，Hu Jianjun[2]，Xu Bingzhi[3]，Li Xingyue[1***]
（[1] *Institute of Plant Protection*，*Sichuan Academy of Agricultural Sciences*，*Chengdu* 610066，*China*；[2] *Institute of Crop Research*，*Sichuan Academy of Agricultural Sciences*，*Chengdu* 610066，*China*；[3] *Agriculture and Rural Affairs Bureau of Yuexi County*，*Liangshan* 616650，*China*）

摘　要：马铃薯金线虫是我国重要的检疫性有害生物，通常会导致马铃薯减产25%~50%，目前全世界尚无有效的根除办法。我国马铃薯近一半的种植区域位于西南地区，四川、云南、贵州等地的马铃薯主栽区为马铃薯金线虫适生区，属于高风险地区。因此，开展不同的马铃薯品种资源对马铃薯金线虫的耐病性筛选，有助于优化马铃薯品种种植布局，并可以对马铃薯抗病品种轮换做资源储备。在四川省越西县的马铃薯金线虫发生田块，以当地主栽品种“青薯9号”作为对照品种，再选择适合当地气候条件的在遗传背景上具有线虫抗性的马铃薯品种（品系）7个（川芋85、川芋117、187 、185-4、C48、C20、F6）作为处理。经过比较不同品种马铃薯根系被马铃薯金线虫的田间侵染危害程度及马铃薯田间产量发现，与青薯9号相比，川芋85、川芋117、187这3个品种（系）的根系新生孢囊数量显著低于青薯9号，分别减少62.9%、45.8%、20.6%。187、C48、C20这3个品种（系）的产量和大薯率显著高于比其他品种（系）的马铃薯。

关键词：马铃薯金线虫；耐病性筛选；品种资源

---

* 基金项目：四川省科技计划项目（2021YFN0009，2022JDRC0110）

** 第一作者：易军，助理研究员，从事作物绿色栽培研究。E-mail：donnyj123@163.com
*** 通信作者：李星月，副研究员，从事线虫学与害虫生物防治研究。E-mail：michelle0919lee@126.com

马铃薯不同品种对马铃薯孢囊线虫的抗性鉴定

兰世超，甘秀海
（绿色农药国家重点实验室/绿色农药与农业生物工程教育部重点实验室/
贵州大学精细化工研发中心，贵阳　550025）

The Resistance of Different Potato Varieties to Potato Cyst Nematodes

Lan Shichao，Gan Xiuhai
（*National Key Laboratory of Green Pesticide*，*Key Laboratory of Green Pesticide and Agricultural Bioengineering*，*Ministry of Education*，*Center for R&D of Fine Chemicals of Guizhou University*，*Guiyang*　550025，*China*）

摘　要：我国是世界第一马铃薯生产和消费国，马铃薯产业的健康可持续发展关系到我国粮食安全，对国家乡村振兴战略顺利实施至关重要。马铃薯孢囊线虫（*Globodera rostochiensis*）是马铃薯上最重要的植物寄生线虫，在马铃薯规模化种植地区，马铃薯孢囊线虫发病严重时，产量损失高达80%~90%，甚至绝收。近年来，我国云南、四川等地发现了马铃薯孢囊线虫，给我国马铃薯产业带来了一定的影响。利用抗性种质选育抗的品种是马铃薯孢囊线虫病防治较为经济、有效的方法。

本研究以筛选抗马铃薯孢囊线虫病的马铃薯资源为目的，采用室内检测与田间试验相结合，在不同区域和季节，分别测定了24个马铃薯品种根围土壤虫口数、根系马铃薯孢囊线虫侵入率和根围土壤孢囊数3个指标参数。表明在22~25℃时较适宜马铃薯孢囊线虫的孵化，在马铃薯苗出土一周时根际土壤虫口数达最高，苗出土四周左右开始有黄色孢囊产生；同时，3个指标参数结果显示，24个马铃薯品种中，荷兰5号、沃土5号、云薯505和尤金为抗病品种，青薯9号、中华9号、黔芋8号和费乌瑞它为感病品种，其中抗性最强的是荷兰5号，抗性最弱的是青薯9号。

关键词：马铃薯；马铃薯孢囊线虫；抗性筛选

中国线虫学研究（第九卷）Nematology Research in China（Vol. 9）：84-85

杀线剂田间防控马铃薯金线虫（*Globodera rostochiensis*）效果*

黄立强[1**]，许　翀[2]，宋家雄[2]，江　如[1]，王晓亮[3]，杨毅娟[2]，陈　敏[2]，
李永青[2]，张汉学[2]，彭　焕[1]，黄文坤[1]，彭德良[1***]
（[1] 中国农业科学院植物保护研究所/植物病虫害综合治理全国重点实验室，北京　100193；
[2]云南省昭通市植保植检站，昭通　657000；[3] 全国农业技术推广服务中心，北京　100125）

Filed Evaluation of the Nematicides Against the Potato Golden Nematode（*Globodera rostochiensis*）in the Field*

Huang Liqiang[1**], Xu Chong[2], Song Jiaxiong[2], Jiang Ru[1], Wang Xiaoling[3], Yang Yijuan[2], Chen Ming[2], Li Yongqing[2], Zhang Hanxue[2], Peng Huan[1], Huang Wenkun[1], Peng Deliang[1***]
（[1]*The State Key Laboratory for Biology of Plant Disease and Insect Pests, Institute of Plant Protection, Chinese Academy of Agricultural Sciences, Beijing*　100193, *China*；[2]*Plant Protection and Quarantion Station of Zhaotong city, Yunnan Province, Zhaotong*　657000, *China*；
[3]*National Agro-Tech Extension and Service Center, Beijing*　100125, *China*）

摘　要：马铃薯金线虫是马铃薯上一种严重的病原线虫，严重时可减产 80%以上。中国贵州、云南和四川的部分马铃薯产地已检测出了马铃薯金线虫。杀线虫剂是防治金线虫的重要措施之一。本研究目的是通过田间药效试验对现有杀线剂进行马铃薯金线虫防治效果评价，进而筛选出可有效控制马铃薯金线虫发生的药剂，为马铃薯金线虫的防治提供技术支撑。

结果表明，7 种非熏蒸型药剂中 1%阿维菌素颗粒剂、41.7%氟吡菌酰胺悬浮剂、10%噻唑膦颗粒剂、6%寡糖·噻唑膦水乳剂、40%氟烯线砜乳油对马铃薯金线虫有较好的防治效果。防效分别为 1%阿维菌素颗粒剂 36.27%~66.04%、41.7%氟吡菌酰胺悬浮剂 41.09%~58.54%、10%噻唑膦颗粒剂 54.81%~66.73%、6%寡糖·噻唑膦水乳剂 33.94%~39.48%、40%氟烯线砜乳油 20.89%~31.34%。10%噻唑膦颗粒剂虽然防效也可以达到 50%以上，但是不稳定。而 10.5%阿维·噻唑膦颗粒剂和 2 亿孢子/g 淡紫拟青霉粉剂则防效不明显。熏蒸型杀线剂 98%棉隆微粒剂的 3 个剂量对马铃薯金线虫防效均超过 90%，其中

* 基金项目：国家自然科学基金（32072398）；政府购买服务项目（15190025）；中国农业科学院科技创新工程（ASTIP-02-IPP-15）

** 第一作者：黄立强，硕士研究生，从事植物线虫病害研究。E-mail：hlq292573628@163.com
*** 通信作者：彭德良，研究员，从事植物线虫和线虫病害综合治理技术研究研究。E-mail：pengdeliang@caas.cn

40kg/667m^{2}达 100. 00%。7 种非熏蒸型杀线剂对马铃薯株高有提高作用。6%寡糖・噻唑膦水乳剂处理组对株高的增长作用最大，与对照相比增长了 85. 33%。98%棉隆微粒剂处理后马铃薯株高显著高于对照组，20kg/667m^{2}处理组株高和鲜重值都是最大的，是对照组的 2. 28 倍。在产量方面，41. 7%氟吡菌酰胺悬浮剂处理组增产作用最显著，最高可增产 62. 06%；其他分别为 6%寡糖・噻唑膦水乳剂 22. 40% ~ 44. 22%、1%阿维菌素颗粒剂 17. 14%~42. 76%，40%氟烯线砜乳油、10%噻唑膦颗粒剂和 10. 5%阿维・噻唑膦颗粒剂不同剂量下最高也可增产 20%~30%，但是这三种药剂的增产效果不稳定。2 亿孢子/g 淡紫拟青霉粉剂的增产效果不明显。不同剂量棉隆处理后，产量也都显著高于对照组，增产率可达 62. 29%~113. 95%。

关键词： 马铃薯金线虫；防治；杀线剂；防效评价

中国线虫学研究（第九卷）Nematology Research in China（Vol. 9）：86-87

腐烂茎线虫 ISSR-PCR 反应体系的建立与优化*

韩　变[1**]，刘永刚[2]，倪春辉[1]，石明明[1]，张　敏[1]，李惠霞[1***]
（[1]甘肃农业大学植物保护学院/甘肃省农作物病虫害生物防治工程实验室，兰州　730070；
[2]甘肃省农业科学院植物保护研究所，兰州　730070）

Establishment and Optimization of ISSR-PCR Reaction System of *Ditylenchus destructor**

Han Bian[1**], Liu Yonggang[2], Ni Chunhui[1], Shi Mingming[1], Zhang Min[1], LI Huixia[1***]
([1] *College of Plant Protection, Gansu Agricultural University/Biocontrol Engineering Laboratory of Crop Diseases and Pests of Gansu Province, Lanzhou　730070, China*;
[2] *Institute of Plant Protection, Gansu Academy of Agricultural Sciences, Lanzhou　730070, China*)

摘　要：腐烂茎线虫（*Ditylenchus destructor*），又名马铃薯茎线虫、甘薯茎线虫和马铃薯腐烂茎线虫，主要危害马铃薯等寄主植物的地下部分，很少或不危害地上部分。目前，美国、加拿大、日本和中国等 25 个国家报道该线虫的危害。在我国，北京、天津、新疆、云南和福建等地均有发生。随着腐烂茎线虫的扩散蔓延，该线虫的种群分化日益严重，不同来源腐烂茎线虫群体在致病力、耐寒性、耐盐性和抗药性等方面均存在差异，在基因水平的差异更是复杂多样。

为明确不同来源腐烂茎线虫群体种内差异，本研究以分离自马铃薯的陕西腐烂茎线虫群体 SXP1 的基因组 DNA 为模板，通过 L_{16}（4^5）正交试验方法对腐烂茎线虫 ISSR-PCR 反应体系进行 5 因素（Taq 酶、Mg^{2+}、DNA 模板、dNTPs 和引物浓度）4 水平体系筛选，获得最优 25μL 反应体系为 14 号水平组合：Taq 酶 1.75 U、Mg^{2+} 0.25mmol/L、模板 DNA 50 ng、dNTPs 0.25mmol/L、引物 1.2μL 和 10×PCR buffer 2.5μL。以采自青海、甘肃和黑龙江的 DTA2、DTA3、HLJP1、DXP1 和 WYA13 5 个腐烂茎线虫群体基因组 DNA 为模板，对计算法和直观法最佳 ISSR-PCR 反应体系进行验证，结果显示，14 号水平组合反应体系具有更好的稳定性和重复性。进一步探索单因素不同水平对扩增反应的影响发现，5 个单因素对 ISSR-

* 基金项目：国家自然基金项目（31760507）；甘肃省现代农业产业体系（GARS-ZYC-4）
** 第一作者：韩变，硕士研究生，从事植物病原线虫研究。E-mail：3140186812@qq.com
*** 通信作者：李惠霞，教授，从事植物线虫学研究。E-mail：lihx@gsau.edu.cn

PCR 扩增反应结果影响程度为：Taq 酶 > dNTPs > Mg^{2+} > DNA 模板 > 引物，此结果与正交试验极差、方差分析结果一致。进一步确定引物 UBC862 的最佳退火温度为 58.5℃，最佳循环次数为 36 次。

关键词：腐烂茎线虫；正交试验；扩增反应；单因素；ISSR-PCR

腐烂茎线虫类毒液过敏原蛋白基因 *DdVAP2* 功能研究[*]

常　青[**]，杨艺炜，张　锋，李英梅[***]
（陕西省生物农业研究所/陕西省植物线虫学重点实验室，西安　710043）

Functional Analysis of *DdVAP2* from *Ditylenchus destructor*[*]

Chang Qing[**]，Yang Yiwei，Zhang Feng，Li Yingmei[***]
（*Bio-Agriculture Institure of Shaanxi*，*Shaanxi Key Laboratory of Plant Nematology*，*Xi'an*　710043，*China*）

摘　要：腐烂茎线虫（*Ditylenchus destructor*）严重威胁马铃薯、甘薯等作物生产安全。由于其危害巨大、防治困难，因此腐烂茎线虫被包括我国在内的世界多国列为重要的检疫性植物线虫。深入研究植物线虫效应蛋白功能有助于揭示植物线虫与植物的互作机理，从而开发新的病害防控策略。但是现有植物线虫效应蛋白功能研究主要集中于定居型植物线虫，而腐烂茎线虫作为一种重要的迁移型植物线虫，效应蛋白功能相关研究十分有限。研究发现，类毒液过敏原蛋白（venom allergen-like proteins，VAP）作为一类重要的植物线虫效应蛋白，普遍存在于定居型与迁移型植物线虫中，并在植物线虫与植物互作过程中发挥着重要作用。本研究从采自陕西省的腐烂茎线虫中克隆到一个腐烂茎线虫类毒液过敏原蛋白基因（*DdVAP2*）。实时荧光定量 PCR 分析结果显示，相比于取食真菌，*DdVAP2* 在腐烂茎线虫取食马铃薯或甘薯时表达量显著上调。在不同发育时期中，*DdVAP2* 基因除在卵期表达量较低之外，在二龄幼虫、雌虫、雄虫中表达量均明显升高，且在二龄幼虫中的表达量最高。原位杂交分析发现 *DdVAP2* 基因在腐烂茎线虫亚腹食道腺中特异性表达。亚细胞定位结果显示去除信号肽的 *Dd*VAP2 蛋白在植物中定位于细胞质与细胞核。通过 RNAi 沉默 *DdVAP2* 基因会显著抑制腐烂茎线虫对植物的侵染及在植物上的繁殖。综上所述，*DdVAP2* 基因在促进腐烂茎线虫对植物的侵染过程中发挥着重要作用。

关键词：腐烂茎线虫；类毒液过敏原蛋白；基因功能

* 基金项目：陕西省科学院科技计划研究项目（2020k-04）
** 第一作者：常青，副研究员，从事植物线虫致病机理研究。E-mail：changq@ xab. ac. cn
*** 通信作者：李英梅，研究员，从事植物线虫综合防控技术研究。E-mail：liym@ xab. ac. cn

中国线虫学研究（第九卷）Nematology Research in China（Vol. 9）：89-90

马铃薯腐烂茎线虫脂肪酸与视黄醇结合蛋白家族基因的全基因组水平鉴定及其功能分析*

王　喆**，马　娟，李秀花，王容燕，陈书龙，高　波***
（河北省农林科学院植物保护研究所/河北省农业有害生物综合防治工程技术研究中心/
农业农村部华北北部作物有害生物综合治理重点实验室，保定　071000）

Genome-wide Identification and Functional Analysis of Fatty Acid and Retinol Binding Protein Family Genes from the Sweetpotato Stem Nematode *Ditylenchus destructor**

Wang Zhe**, Ma Juan, Li Xiuhua, Wang Rongyan, Chen Shulong, Gao Bo***
(*Plant Protection Institute of Hebei Academy of Agricultural and Forestry Sciences, IPM Centre of Hebei Province, Key Laboratory of IPM on Crops in Northern Region of North China, Ministry of Agriculture and Rural Affairs, Baoding*　071000, *China*)

摘　要：马铃薯腐烂茎线虫（*Ditylenchus destructor*）是甘薯和马铃薯上的一种重要的植物内寄生线虫，也是国内外重要的检疫性线虫，严重制约着我国的甘薯和马铃薯产业的健康发展。深入开展马铃薯腐烂茎线虫的分子致病机制研究对于制定高效新型的病害防控策略具有重要的基础理论支撑作用。脂肪酸与视黄醇结合蛋白（fatty acid and retinoid binding protein, FAR）是目前仅在线虫中发现的一种具有脂肪酸和视黄醇结合活性的蛋白，在线虫的生长及寄生过程中具有重要作用。为深入了解该蛋白家族基因在线虫侵染过程中的功能，本研究通过生物信息学手段从马铃薯腐烂茎线虫的基因组中筛选获得了 3 个 FAR 家族基因（*Dd-far*-1、*Dd-far*-2、*Dd-far*-3），并扩增获得了 3 个基因的 cDNA 全长序列，长度分别为：801bp、740bp 和 860bp，开放阅读框的长度分别为：570bp、594bp 和 576bp，预测编码蛋白肽链长度分别为：189 aa、197 aa 和 191 aa。序列分析发现 3 个基因的 5′末端均含有 SL1 序列，且 3 个基因的蛋白序列中均含有 FAR 家族基因的保守结构域。通过对 FAR 家族基因在线虫不同发育时期的表达分析发现，*Dd-far*-1 基因的表达量随龄期的增加而增加，且在雌成虫中表达量最高；*Dd-far*-2 在卵中表达量最高，其次为成虫；*Dd-far*-3 在卵中表达量最低，而在其他龄期表达差异不显著。对 FAR 家族基因在初侵染过程中的表达情况研究发现，*Dd-far*-1 和

* 基金项目：国家甘薯产业技术体系（CARS-10）；河北省农林科学院科技创新专项（2022KJCXZX-ZBS5）

** 第一作者：王喆，科研助理，从事植物线虫学研究。E-mail：1315550572@ qq. com
*** 通信作者：高波，副研究员，从事植物线虫学研究。E-mail：gaobo89@ 163. com

Dd-far-2 在线虫侵染甘薯 3d 后的表达量显著上升，而 *Dd-far*-3 的表达量则显著下降，说明 FAR 家族基因在线虫侵染甘薯过程中发挥着重要作用。为验证 3 个 FAR 家族基因的脂肪酸与视黄醇的结合特性，本研究通过原核表达分别获得了 3 个大小约为 19.2kDa（FAR-1）、20.0kDa（FAR-2）和 17.7kDa（FAR-3）的高纯度蛋白溶液，并进行了荧光竞争性结合实验，结果显示 Dd-FAR-1、Dd-FAR-2、Dd-FAR-3 蛋白均与脂肪酸类似物 DAUDA 和视黄醇具有较强结合能力，而与油酸的结合能力较弱，说明这 3 个 FAR 家族基因在马铃薯腐烂茎线虫生长发育过程中也发挥着重要的作用。本研究的结果为进一步研究马铃薯腐烂茎线虫 FAR 家族基因的生物学功能及其应用奠定了基础。

关键词：马铃薯腐烂茎线虫；脂肪酸与视黄醇结合蛋白家族；原核表达；荧光竞争结合

马铃薯不同品种对腐烂茎线虫的室内抗病性评价*

霍宏丽[1,2**]，白松林[1,2**]，张冬梅[1]，陈　敏[1,2]，席先梅[1***]
（[1]内蒙古自治区农牧业科学院，呼和浩特　010031；[2]内蒙古农业大学园艺与植物保护学院，呼和浩特　010018）

Indoor resistance evaluation of potato cultivars to *Ditylenchus destructor**

Huo Hongli [1,2**]，Bai Songlin [1,2**]，Zhang Dongmei [1]，Chen Min [1,2]，Xi Xianmei [1***]
（[1] *Inner Mongolia Academy of Agricultural & Animal Husbandry Sciences*，*Hohhot* 010031，*China*；[2] *College of Horticulture and Plant Protection*，*Inner Mongolia Agricultural University*，*Hohhot*　010018，*China*）

摘　要：中国是马铃薯第一生产大国，全球马铃薯产量的23%都来自中国，确保马铃薯产业安全稳定生产对于促进农村经济发展以及保障粮食安全具有重要意义。马铃薯腐烂茎线虫（*Ditylenchus destructor* Thorne，1945）是一种迁移性内寄生线虫，是国际公认的检疫性线虫。该线虫已报道的寄主高达120多种，可危害甘薯（*Ipomoea batutas*）和马铃薯（*Solanum tuberosum*）的地下部。近年来，腐烂茎线虫对我国马铃薯危害日益严重，在北京、内蒙古、黑龙江、陕西和山东等10多个省（自治区、直辖市）均有发生，可造成作物产量减少20%~50%，严重时甚至绝收，已然成为制约内蒙古地区马铃薯产业发展的重要因素之一。选育对该病原线虫具有高抗性的品种，可为有效防治该线虫奠定基础。

在中国侵染马铃薯的腐烂茎线虫单倍型主要为B型和C型，侵染甘薯的单倍型主要为A型，且内蒙古地区的马铃薯腐烂茎线虫多为C型。本研究将培养的C型马铃薯腐烂茎线虫接种在22个马铃薯品种块茎中，进行品种抗病性评价。根据RF值进行抗性评价，发现8个抗性品种，14个感病品种；根据RS值进行抗性评价，发现8个抗性品种，10个中抗品种，以及4个感病品种。在2种评价方法中，Agria和Spunta品种接种180d后数值最低为0。而使用发病程度进行抗性评价，共发现5个抗性品种。综合3种评价方式，室内试验共筛选出5个抗性品种，分别为后旗红、228、旭丰2号、Agria和Spunta。

关键词：马铃薯；腐烂茎线虫；品种；抗病性；室内

---

* 基金项目：内蒙古农牧业创新基金项目（2023CXJJN15）；内蒙古自治区科技计划项目（2020GG0070）
** 第一作者：霍宏丽，博士研究生，从事植物保护研究。E-mail：huohongli0127@163. com
白松林，硕士研究生，从事植物保护研究。E-mail：1352908644@qq. com
*** 通信作者：席先梅，研究员，从事植物线虫研究。E-mail：xixianmei1975@163. com

中国线虫学研究（第九卷）Nematology Research in China（Vol. 9）：92-93

大豆孢囊线虫对酸碱盐化学信号响应的分子调控机制*

姜　野[1,2]**，李春杰[1]，黄铭慧[1]，秦瑞峰[1,2]，蒋　丹[1,2]，常豆豆[1,2]，谢倚帆[1,2]，王从丽[1]***
（[1]中国科学院东北地理与农业生态研究所，中国科学院大豆分子设计育种重点实验室，
哈尔滨　150081；[2]中国科学院大学，北京　100049）

Molecular Regulatory Mechanism Underlying the Response of Soybean Cyst Nematode to Acid, Base and Salt Chemical Signals*

Jiang Ye[1,2]**, Li Chunjie[1], Huang Minghui[1], Qin Ruifeng[1,2], Jiang Dan[1,2],
Chang Doudou[1,2], Xie Yifan[1,2], Wang Congli[1]***
([1] *Key Laboratory of Soybean Molecular Design Breeding, Northeast Institute of Geography and Agroecology, Chinese Academy of Sciences, Harbin* 150081, *China*; [2] *University of Chinese Academy of Sciences, Beijing* 100049, *China*)

摘　要：线虫能够感知外界信号的动态变化继而产生趋避行为。pH 和盐离子都是大豆孢囊线虫重要的趋化性信号。但在趋化过程中线虫行为学如何变化，相应的在整个基因组层面线虫基因表达如何变化尚不清楚；此外，关键基因如何调控线虫响应吸引性 pH 以及如何影响线虫侵染和寄生都是未知的。因此，本研究将线虫在最佳趋化酸性 pH、碱性 pH 和 NaCl（170mmol/L）盐溶液中处理后进行行为学和转录组测序分析，然后通过生物信息学分析获得潜在的候选基因，继而进行基因克隆、组织染色定位及生物学功能验证等。具体结果如下。

第一，利用显微镜结合实时追踪线虫移动的 Wormlab 分析软件对处于吸引性酸性 pH 值为 5.25、碱性 pH 值为 8.6、盐（170mmol/L NaCl）溶液以及极端酸碱（pH 值为 4.5 和 pH 值为 10.0）环境条件下的线虫进行了行为学分析。结果表明，面对极端 pH 环境，线虫的移动速度均高于对照（pH 值为 7.0），说明线虫始终处于胁迫的状态；面对吸引性 pH 环境，线虫的移动速度先升高后下降，说明线虫从开始的刺激转为平稳适应状态。

第二，为了从基因组层面解析线虫在不同 pH 和盐溶液中的基因表达变化，对上面 6 个处理共 18 个样品进行三代全长转录组测序，获得了平均每个样品 4.36 Gbp 的序列，全长转

* 基金项目：中国科学院战略性先导科技专项项目（XDA24010307）；国家自然科学基金项目（32272501）

** 第一作者：姜野，博士研究生，从事线虫与植物互作研究。E-mail：jiangye@iga.ac.cn
*** 通信作者：王从丽，研究员，从事线虫与植物互作研究。E-mail：wangcongli@iga.ac.cn

录本经参考基因组比对及功能注释后，共获得 3 972 个新基因和 29 529 条新转录本。转录本结构分析表明：可变剪接 AS、可变多聚腺苷酸化 APA 和融合转录本等结构与对照相比发生了显著变化，说明这些转录后修饰参与了线虫行为的调控。对差异表达基因和转录本进行 GO 和 KEGG 富集分析表明：当线虫受到酸碱盐刺激后，跨膜受体被激活，例如 G 蛋白偶联受体（GPCR）包括化学受体 SRSX、Wnt 受体 MOM-5、多巴胺受体 F59. D12. 1、肽类受体 NPR-18、乙酰胆碱受体 mAChR 和促甲状腺激素释放激素受体 TRHR 等，其他受体包括乙酰胆碱受体 nAChR、GABAB（γ-氨基丁酸受体 β 亚单元）和鸟苷酸环化酶受体 GCY-18 等能够调控离子通道。此外，离子转运蛋白 PMCA、离子通道 VGCC 和 TRP-1（TRPC4）都处于激活状态。然而氧化磷酸化代谢途径和核糖体代谢通路蛋白都被抑制，说明能量产生和蛋白合成受到抑制，这减少了能量的消耗。受体激活与生长发育受到抑制说明线虫在最佳趋化状态下通过调整代谢途径而维持自身能量的平衡。结合差异表达基因和蛋白与蛋白互作分析建立了大豆孢囊线虫响应 pH 和盐离子刺激的网络调控模型。

第三，根据测序和 qRT-PCR 验证结果，对 G 蛋白 α 亚基基因 *Hg-goa*-1 和瞬时受体电压 TRPV 通道中的关键基因 *Hg-osm*-9 和 *Hg-ocr*-2 进行了克隆、表达分析和功能验证。结果表明：这 3 个基因均在侵染前二龄幼虫期表达量最高；原位杂交结果表明基因在尾感器内表达；RNA 干扰某一基因，其他两个基因的表达量发生改变，说明三者存在互作关系；dsRNA-*Hg-osm*-9 和 dsRNA-*Hg-ocr*-2 分别干扰后，线虫对最佳酸性 pH 的趋化指数降低，对最佳碱性 pH 的趋化指数升高，说明两者正调控线虫趋化酸性 pH，负调控线虫趋化碱性 pH。3 个基因单独 RNA 干扰，导致线虫侵染率显著降低 17.7%~40.5%、寄主根表的雌虫或孢囊数量显著降低 24.7%~45%，说明 3 个基因能够调控对酸碱的趋化性，调控对寄主的侵染和繁殖。

综上，大豆孢囊线虫对最佳趋化酸碱盐环境和极端环境的反应有异同；转录后修饰、跨膜受体、离子通道、离子转运蛋白、氧化磷酸化代谢途径和核糖体代谢通路参与调控线虫对酸碱盐的响应行为；G 蛋白 α 亚基和 TRPV 通道相互作用共同调控大豆孢囊线虫对酸碱的趋化性、调控线虫对寄主的侵染和繁殖。本研究明确了大豆孢囊线虫在最佳趋化性和极端酸碱环境中的行为变化及其相应的基因调控过程，丰富了线虫趋化性知识。所鉴定的跨膜受体蛋白和 TRPV 通道关键基因可以作为开发新型杀线剂的分子靶标。

关键词： 大豆孢囊线虫；化学信号；三代全长转录组测序；跨膜受体；离子通道

大豆孢囊线虫新种群 X12 致病基因分析*

都文振[1]**，练　云[2]，张刘萍[1]，田　栓[1]，彭德良[1]，卢为国[2]，刘世名[1]***
（[1]中国农业科学院植物保护研究所/植物病虫害综合治理全国重点实验室，北京　100193；
[2]河南省作物分子育种研究院，郑州　450002）

Analyses of Pathogenic Genes of a New Soybean Cyst Nematode Population X12*

Du Wenzhen[1]**，Lian Yun[2]，Zhang Liuping[1]，Tian Shuan[1]，
Peng Deliang[1]，Lu Weiguo[2]，Liu Shiming[1]***
（[1] *Institute of Plant Protection*，*Chinese Academy of Agricultural Sciences*，*Beijing* 100193，*China*；[2] *Henan Institute of Crop Molecular Breeding*，*Zhengzhou*　450002，*China*）

摘　要：大豆孢囊线虫（soybean cyst nematode，SCN，*Heterodera glycines* Ichinohe）病是世界范围内大豆（*Glycine max*（L.）Merr.）生产上危害最严重的一种病害，每年造成巨大的产量与经济损失。在中国，已明确 SCN 在 21 个省、市发生和分布，每年有超过 3 000 万亩的大豆田受到 SCN 的危害。

种植抗性品种是大豆孢囊线虫最为经济、有效且环保的防治措施。然而，随着抗性品种的单一种植以及环境条件等的变化，SCN 容易克服大豆的抗性，并不断扩散，其致病力也表现增强趋势。2012 年，在中国山西省古交市邢家社首次发现并报道了 1 个新的大豆孢囊线虫种群（生理小种）X12。该小种可以侵染目前几乎所有的抗性大豆，包括广谱抗性的山西灰皮支黑豆和 PI437654，并表现出比 4 号生理小种（此前认为毒性最强）更强的毒力，构成大豆生产的一种严重威胁。

本研究前期对黄淮海地区采集的 54 个线虫种群（包括 X12）的基因组进行了全基因组重测序，经基因组比较分析，初步鉴定了 9 个 X12 致病候选基因，在此基础上，设计引物从 X12 中克隆了相应的基因。目前，正通过大豆发根遗传转化体系进行这些致病候选基因的功能分析，希望鉴定出 X12 的超强致病基因。
关键词：大豆孢囊线虫；X12；大豆；致病基因

---

* 基金项目：河南省优势学科培育联合基金；国家自然科学基金（31972248）
** 第一作者：都文振，硕士研究生，从事植物线虫病害研究。E-mail：duwenzhen2021@163. com
*** 通信作者：刘世名，研究员，主要从事植物线虫分子生物学和线虫病害管理的研究。E-mail：liushiming01@caas. cn

中国线虫学研究（第九卷）Nematology Research in China（Vol. 9）：95-96

大豆孢囊线虫（*Heterodera glycines*）HgUIM1的功能研究*

张刘萍**，刘　峙，黄文坤，孔令安，彭德良，刘世名***
（中国农业科学院植物保护研究所/植物病虫害综合治理全国重点实验室，北京　100193）

Function Analysis of Soybean Cyst Nematode *Heterodera glycines* Effector HgUIM1*

Zhang Liuping**，Liu Zhi，Huang Wenkun，Kong Ling'an，
Peng Deliang，Liu Shiming***
（*State Key Laboratory for Biology of Plant Diseases and Insect Pests*，*Institute of Plant Protection*，*Chinese Academy of Agricultural Sciences*，*Beijing*　100193，*China*）

摘　要：大豆孢囊线虫病作为大豆生产上最重要的病害之一，每年造成巨大的产量与经济损失，栽培抗性品种是目前最有效的防治该线虫措施，随着大豆孢囊线虫的不断进化，大豆的SCN抗性有被逐步克服的风险。因此，迫切需要不断深入大豆与大豆孢囊线虫的互作机制研究，为大豆生产提供科学依据。本研究以挖掘与SCN抗性蛋白GmSNAP18互作的线虫蛋白为切入点，通过筛选SCN cDNA酵母文库，获得可能与其互作的效应蛋白，进而对其功能与互作机制进行研究，以期进一步明确GmSNAP18的抗SCN机制。

首先利用酵母双杂交技术筛选到与SCN抗性蛋白GmSNAP18互作的一个效应蛋白HgUIM1，其编码区全长1 275bp，编码424个氨基酸，N-端含一段24个氨基酸的信号肽序列，含3个UIM结构域，且不含跨膜结构域。*HgUIM*1在侵染后二龄中表达量最高，在SCN亚腹食道腺中特异性表达，在烟草细胞中则定位于细胞膜上。经酵母双杂交、BiFC以及GST pull-down三种体内体外实验与分析，明确了HgUIM1与*rhg*1-*a* GmSNAP18存在互作关系且互作发生在细胞膜上，且功能结构域分析表明HgUIM1$^{223\text{-}237AA}$区段与C-端HgUIM1$^{296\text{-}424AA}$为发挥该相互作用的重要区段。随后利用大豆发根转化系统和拟南芥-BCN亲和互作系统验证了HgUIM1的功能，异源表达*HgUIM*1的大豆转基因发根显著降低了对SCN的敏感性，异源表达*HgUIM*1和*Hs*9131（*HgUIM*1的BCN同源基因）的转基因拟南芥植株均显著降低了对BCN的敏感性。异源表达*HgUIM*1$^{25\text{-}257AA}$（与*rhg*1-*a* GmSNAP18不互作区段）及*HgUIM*1$^{223\text{-}424AA}$（与*rhg*1-*a* GmSNAP18互作区段）的大豆转基因发根的SCN侵染表型实验结果表明，HgUIM1$^{25\text{-}257AA}$不影响大豆对SCN的抗性，而HgUIM1$^{223\text{-}424AA}$可显著提高大豆对SCN

---

* 基金项目：国家自然科学基金（31972248）

** 第一作者：张刘萍，博士研究生，主要从事大豆孢囊线虫的抗性研究。E-mail：liupingz2013@163.com
*** 通信作者：刘世名，研究员，主要从事大豆孢囊线虫的抗性与突变育种研究。E-mail：smliuhn@yahoo.com

的抗性。表明 HgUIM1 蛋白对 SCN 敏感性的影响和其与 *rhg*1-*a* GmSNAP18 互作与否存在密切关系。而通过烟草叶片的瞬时表达系统明确了在烟草中表达 HgUIM1 可引起细胞坏死并增强大豆 GmSNAP18 和 GmSHMT08 引起的植物免疫反应。

关键词： α-SNAPs；大豆孢囊线虫；泛素互作基序；相互作用

禾谷孢囊线虫效应子 Ha17370 靶向植物 CBSX*

张笑寒**，杨姗姗，简　恒，刘　倩***
（中国农业大学植物病理学系，北京　100193）

The Ha17370 Effector of *Heterodera avenae* Targets Plant CBSX*

Zhang Xiaohan**，Yang Shanshan，Jian Heng，Liu Qian***
（*Department of Plant Pathology*，*China Agricultural University*，*Beijing*　100193，*China*）

摘　要：禾谷孢囊线虫（*Heterodera avenae*）是我国禾本科作物的重要病原，目前缺乏高效的防治方法，急需解析其致病机理，从而制定防治新策略。课题组前期从禾谷孢囊线虫中鉴定到一个带有 RxLR 基序的新效应子 Ha17370，在此基础上，本研究构建了异源表达 *Ha*17370 的拟南芥，通过免疫共沉淀-质谱联用技术发现，Ha17370 的潜在靶标是胱硫醚 β-合成酶（AtCBSX3），萤光素酶互补（LUC）和免疫共沉淀（Co-IP）进一步验证了效应子 Ha17370 与 AtCBSX3 蛋白的互作。考虑到拟南芥不是禾谷孢囊线虫的寄主植物，笔者在禾谷孢囊线虫的寄主大麦、小麦上找到 AtCBSX3 的同源基因——*HvCBX*3 和 *TaCBSX*3，通过 LUC 和 CO-IP 实验验证了 Ha17370 与 HvCBX3、TaCBSX3 也均能互作，且共定位实验结果表明 Ha17370 分别与 HvCBX3、TaCBSX3 共定位于线粒体中。据报道，CBSX3 可以调控细胞内过氧化氢的水平，笔者推测，禾谷孢囊线虫效应子 Ha17370 可能通过与宿主细胞 CBSX3 相互作用，调节植物体内过氧化氢的含量，从而抑制植物的免疫反应，促进线虫寄生。

关键词：禾谷孢囊线虫；效应子 Ha17370；CBSX3；相互作用

---

* 基金项目：国家自然科学基金“禾谷孢囊线虫新效应子 Ha17370 其 RxLR 基序转运作用和基因功能研究”（31871940）

** 第一作者：张笑寒，博士研究生，从事植物寄生线虫致病机理研究。E-mail：xiaohz18@163.com
*** 通信作者：刘倩，副教授，从事植物寄生线虫致病机理和防控技术研究。E-mail：liuqian@cau.edu.cn

杂交小麦品种对小麦孢囊线虫的抗性鉴定

于敬文[1*]，余曦玥[1]，于　清[1]，赵津田[1]，贾建平[1]，张改平[2**]，仝允正[2]，黄文坤[1**]
（[1]中国农业科学院植物保护研究所/植物病虫害生物学国家重点实验室，北京　100193；
[2]河南省禹州市农业农村局，禹州　461670）

Resistance Evaluation of Hybrid Wheat Varieties to Cereal Cyst Nematode

Yu Jingwen[1*], Yu Xiyue[1], Yu Qing[1], Zhao Jintian[1], Jia Jianping[1], Zhang Gaiping[2**], Tong Yunzheng[2], Huang Wenkun[1**]
([1] *The State Key Laboratory for Biology of Plant Disease and Insect Pests, Institute of Plant Protection, Chinese Academy of Agricultural Sciences, Beijing*　100193, *China*;
[2] *Plant Protection and Inspection Station of Agricultural and Rural Bureau of Yuzhou, Yuzhou*　461670, *China*)

摘　要：小麦孢囊线虫（cereal cyst nematode，CCN）是发生危害面积大、寄主范围广泛、对作物产量具有严重影响的植物病原线虫。我国小麦孢囊线虫的种类主要包括禾谷孢囊线虫（*Heterodera avenae*）和菲利普孢囊线虫（*Heterodera filipjevi*）。利用抗性种质资源选育抗小麦孢囊线虫的品种是防治孢囊线虫较为经济、有效的方法，由于国内小麦抗孢囊线虫病种质资源相对较少，所以亟须通过品种杂交选育新型抗小麦孢囊线虫品种。

本研究以筛选抗小麦孢囊线虫的小麦种质资源为目的，通过对国内外 20 个不同小麦品种进行品种杂交，对选育的 69 份杂交小麦材料进行田间抗性评价。田间调查结果表明：在小麦返青期时，通过对根部进行酸性品红染色统计根内线虫数量发现，69 份杂交小麦材料均受小麦孢囊线虫危害；在小麦抽穗扬花期时，通过对小麦根部白雌虫计数发现，69 份杂交小麦材料对小麦孢囊线虫的抗感性不同，其中 7 份杂交品种对小麦孢囊线虫表现为高抗，单株白雌虫数均在 3.0 个以下，其余杂交品种对小麦孢囊线虫的抗病指标变化趋势均不一致；通过对 69 份杂交小麦材料进行田间小区繁种后测产发现，2 份杂交品种相较于其亲本表现为高产，其中单株穗粒数相较亲本增加 29.8%~41.6%，单株穗粒重相较亲本提高 1.9~2.4 倍，后期将进行进一步的产量测定。上述研究结果对抗小麦孢囊线虫品种选育和开发具有重要的指导意义。

关键词：小麦；杂交品种；小麦孢囊线虫；抗性鉴定

* 第一作者：于敬文，硕士研究生，从事植物线虫致病机制研究。E-mail：jingwenyu1996@163.com
** 通信作者：张改平，高级农艺师，从事植物病虫害防控技术研究。E-mail：13608430689@163.com
黄文坤，研究员，从事植物线虫致病机理及防控技术研究。E-mail：wkhuang2002@163.com

南方根结线虫分泌粒蛋白基因 *MiSCG5L* 功能研究

叶梦迪[1, 2 *]，迟元凯[1]，赵　伟[1]，戚仁德[1 **]
（[1]安徽省农业科学院植物保护与农产品质量安全研究所，合肥　230031；
[2]新疆维吾尔自治区植物保护站，乌鲁木齐　830049）

Function Analysis of *MiSCG5L* Gene from the Root-knot Nematode *Meloidogyne incognita*

Ye Mengdi[1, 2 *]，Chi Yuankai[1]，Zhao Wei[1]，Qi Rende[1 **]
（[1] *Institute of Plant Protection and Agro-products Safety*，*Anhui Academy of Agricultural Sciences*，*Hefei*　230031，*China*；[2] *Xinjiang Uygur Autonomous Region Plant Protection Station*，*Urumqi*　830049，*China*）

摘　要：根结线虫（*Meloidogyne* spp.）是农业生产上危害最为严重的植物寄生线虫之一，每年对全球农业造成巨大的经济损失。根结线虫在寄生过程中向寄主植物分泌大量的效应蛋白，帮助其寄生与定殖。南方根结线虫的分泌粒蛋白基因 *SCG5*（*Secretogranin V*）-*MiSCG5L* 在线虫背食道腺细胞特异性表达，其编码蛋白含有一段 19 个氨基酸的分泌信号肽和一段 7 个氨基酸的核定位信号。该基因在线虫侵入植物 3d 后大量表达，蛋白进入植物细胞后定位于植物细胞核。利用农杆菌介导 *MiSCG5L* 基因在本氏烟草叶片中瞬时表达，能显著抑制由 flg22 和 elf18 诱导的活性氧爆发，以及植物防卫相关基因 *WRKY*7、*WRKR*8、*PTI*5、*ACRE*31，*GRAS*2 的表达，而去除信号肽和核定位信号的 MiSCG5L$^{\Delta SP\Delta NLS}$不能抑制 flg22 和 elf18 引起的活性氧爆发和植物防卫基因的表达。利用 dsRNA 介导南方根结线虫 2 龄幼虫的 *MiSCG5L* 基因沉默后接种本氏烟草，发现 *MiSCG5L* 基因沉默后显著降低南方根结线虫（*M. incognita*）2 龄幼虫的寄生能力；在本氏烟草中过表达 *MiSCG5L* 基因并不能促进南方根结线虫的侵染，但可以恢复由于 *MiSCG5L* 基因沉默导致侵染能力下降。以上结果表明，*MiSCG5L* 能够在根结线虫侵染早期帮助线虫抑制植物免疫反应，该基因对于南方根结线虫的寄生能力起关键作用。进一步研究发现，MiSCG5L 与番茄的 Ca^{2+} 信号传感器 CBL-CIPK 途径中的蛋白激酶 SlCIPK12（CBL-interacting protein kinase 12）存在互作，推测该蛋白可能通过干扰植物 Ca^{2+} 信号传导调控植物免疫反应。同时，本研究还发现以 *MiSCG5L* 为基因沉默靶标，沉默速度快、效率高，可作为 RNAi 防治根结线虫的候选靶标基因。

关键词：南方根结线虫；效应蛋白；MiSCG5L；RNAi；植物过表达

---

* 第一作者：叶梦迪，硕士研究生，从事植物线虫学研究。E-mail：1548154218@ qq. com
** 通信作者：戚仁德，研究员，从事土传病害综合防控技术研究。Email：rende7@ 126. com

南方根结线虫转录因子 Mi_03370 影响其发育及侵染能力的功能研究

廖宇澄*，代大东，张书荣，周雅艺，彭东海，郑金水，孙　明**
（华中农业大学农业微生物资源发掘与利用全国重点实验室，武汉　430070）

Functional Study of Transcription Factor Mi_03370 Affecting the Development and Infection Ability of *Meloidogyne incognita*

Liao Yucheng*，Dai Dadong，Zhang Shurong，Zhou Yayi，Peng Donghai，Zheng Jinshui，Sun Ming**
（*National Key Laboratory of Agricultural Microbial Resources Exploration and Utilization*，*Huazhong Agricultural University*，*Wuhan*　430070，*China*）

摘　要：植物寄生线虫是一类严重危害植物健康的病害，对全世界的作物生产和粮食安全构成严重威胁。其中根结线虫作为植物寄生线虫中威胁最严重的一种，对我国农业生产造成了巨大的破坏和经济损失。然而对其的防治手段却十分有限，基于线虫生物学的研究能有效提高新防治手段的开发效率。

本研究发现体外敲低南方根结线虫（*Meloidogyne incognita*，Mi）早期胚胎的去甲基化酶 Mi_*NMAD*-1 基因转录水平后能显著抑制 Mi 虫卵到 J2 的孵化率，通过生信分析结合实验验证的方法发现其调控的下游转录因子 Mi_03370 为影响 Mi 孵化率的主效转录因子。通过构建表达 Mi_03370 dsRNA 的转基因烟草，使用 1 000 条 Mi J2 侵染转基因烟草，与对照相比，宿主诱导的基因沉默（HIGS）株系根结数量减少 56%，并且侵染 HIGS 株系后的 Mi 产生的子代虫卵孵化率仅为对照组的 19%。进一步地，通过对 Mi_03370-HIGS 的虫卵转录组测序，并且将差异表达基因与正常发育的 Mi 虫卵到 J2 的差异基因取交集得到候选靶标基因，再结合 Mi_03370 ChIP-Seq 得到的基因共同分析，鉴定到其下游关键的靶基因 Mi_08398 和 Mi_22703。同时发现该转录因子在花生及爪哇根结线虫中保守存在，暗示同一株转基因植物可能同时对 3 种危害最严重的根结线虫产生抗性。本研究为南方根结线虫的发育与寄生提供了新见解，为生物防治提供了新颖的靶标基因，同时为寻找新型防治靶标和制定有效的防控策略提供了理论基础。

关键词：植物寄生线虫；表观调控；龄期转换；转录因子

---

* 第一作者：廖宇澄，博士研究生，从事植物寄生线虫生物信息学研究
** 通信作者：孙明，教授，主要从事微生物农药开发

象耳豆根结线虫 MeMSP1 效应子与植物谷胱甘肽转移酶家族蛋白（GSTs）互作帮助线虫寄生*

陈永攀**，刘　倩，简　恒***
（中国农业大学植物病理学系，北京　100193）

Meloidogyne enterolobii Effector MeMSP1 Target Plant Glutathione Transferase Family Proteins（GSTs）to Promote Nematode Parasitism*

Chen Yongpan**，Liu Qian，Jian Heng***
（*Department of Plant Pathology*，*China Agricultural University*，*Beijing*　100193，*China*）

摘　要：象耳豆根结线虫（*Meloidogyne enterolobii*）是近年来全球关注的新发生线虫。其效应子基因 *MeMSP*1 在背食道腺特异表达；在侵染后表达量升高；该效应子被线虫分泌进入巨大细胞中参与与植物的互作；*MeMSP*1 的异位表达拟南芥株系对线虫更敏感，RNAi 干扰拟南芥株系对线虫的敏感性下降；表明 MeMSP1 在该线虫寄生致病过程中发挥重要作用。但该效应子在植物中的靶标以及如何帮助线虫寄生的机理并不清楚。本研究通过免疫共沉淀-质谱联用技术发现 MeMSP1 的潜在靶标是植物特有的一类谷胱甘肽转移酶 AtGSTFs，该蛋白家族共有 13 个成员，双分子荧光互补（BIFC）和免疫共沉淀（Co-IP）进一步验证发现 MeMSP1 可与该家族中所有成员互作，且 AtGSTF9 的 GSH 结合位点（G-site）对二者的互作是必需的。对 MeMSP1 异位表达拟南芥进行转录组及代谢组分析表明，MeMSP1 影响了拟南芥的代谢，导致有利于线虫寄生的物质如 GSH 上调积累，不利于线虫寄生的代谢物如漆黄素（Butin）下调积累。这些结果表明 MeMSP1 通过与拟南芥的 GSTFs 蛋白互作调节寄主代谢，继而帮助线虫的寄生。

关键词：象耳豆根结线虫；MeMSP1；谷胱甘肽转移酶；植物代谢

---

* 基金项目：国家自然科学基金“象耳豆根结线虫新效应子 MeSP1 的功能研究”（31772138）
** 第一作者：陈永攀，博士后，从事根结线虫效应子研究。E-mail：chenyongpan1@ 163. com
*** 通信作者：简恒，教授，从事植物线虫研究。E-mail：hengjian@ cau. edu. cn

象耳豆根结线虫克服 Mi-1 抗性基因相关效应子的筛选

曾媛玲*，吴文涛，闫曦蕊，段山全，王　扬**
（云南农业大学，昆明　650201）

Screening of Effectors Related to Overcoming Mi-1 Resistance Gene of *Meloidogyne enterolobii*

Zeng Yuanling*，Wu Wentao，Yan Xirui，Duan Shanquan，Wang yang**
（*Yunnan Agricultural University*，*Kunming*　650201，*China*）

摘　要：象耳豆根结线虫具有致病力强、分布范围广、危害严重、能克服多数栽培作物中的抗根结线虫基因抗性的特性，对多种作物造成了严重的危害。目前针对象耳豆根结线虫效应蛋白的研究逐渐成为热点，但是在目前的研究中，关于象耳豆根结线虫如何克服抗线虫基因抗性机制的研究并不多。

根据前期研究结果，接种后的第 2 天至第 6 天是象耳豆根结线虫克服 *Mi*-1 基因抗性的关键时期，在该时期内线虫分泌了某些效应蛋白来逃避 *Mi*-1 基因介导的寄主抗性。因此本研究以象耳豆根结线虫与携带 *Mi*-1 基因番茄植株的互作为研究对象，基于转录组和比较基因组学，并通过生物信息学软件进行效应蛋白基因的初筛。结果表明：通过生物信息学软件分析发现无论是侵染关键时期差异表达上调的基因还是象耳豆根结线虫特有基因中都具有可能与线虫侵染相关的功能基因，通过 Singnal P5. 0 等生物信息学在线网站，对寄生关键时期转录组上调的 647 个基因以及比较基因组筛选出的象耳豆根结线虫特有的 2 522 个基因进行信号肽和跨膜结构域的预测，结果发现结合两个组学数据可筛选到具信号肽（信号肽可能性高于 90%）且无跨膜结构域的候选基因共有 47 个。从初筛的 47 个基因中选取 18 个或具有特殊结构域或在转录组测序中差异表达上调的基因进行实时荧光定量 PCR 实验，在验证转录组数据准确性的同时也对比较基因组所筛基因进行进一步的筛选，最后得到了在寄生关键时期表达量上调最高的 2 个致病相关候选效应蛋白基因 CAD2205632、CAD2175494，后续可对筛选出的候选效应蛋白进行功能验证，这无论是对于解释植物寄生线虫的分子致病机理，还是对于将它们作为靶标进行相关病害的防治都具有重要的研究意义和实践价值。

关键词：番茄；象耳豆根结线虫；效应蛋白；基因组学

* 第一作者：曾媛玲，硕士研究生，从事植物线虫病害研究。E-mail：944014545@ qq. com
** 通信作者：王扬，教授，从事植物线虫病害研究。E-mail：wangyang626@ sina. com

中国线虫学研究（第九卷）Nematology Research in China（Vol. 9）：103-104

爪哇根结线虫危害白及

杨艳梅[1*]，余兴华[2]，李云霞[1]，姚汉央[1]，刘福祥[1]，李乾坤[1]，胡先奇[1**]
（[1]云南农业大学植物保护学院/云南生物资源保护与利用国家重点实验室，昆明　650201；
[2]云南农业大学学生处，昆明　650201）

Bletilla striates was Harmed by *Meloidogyne javanica*

Yang Yanmei[1*], Yu Xinghua[2], Li Yunxia[1], Yao Hanyang[1], Liu Fuxiang[1], Li Qiankun[1], Hu Xianqi[1**]
([1] *State Key Laboratory for Conservation and Utilization of Bio-Resources in Yunnan/ College of Plant Protection*, *Yunnan Agricultural University*, *Kunming*　650201, *China*;
[2] *Student Affairs Departments*, *Yunnan Agricultural University*, *Kunming*　650201, *China*)

摘　要：白及（*Bletilla striates*）属兰科多年生草本球根植物，是我国一种重要、濒危的传统中药材，具有多种药理作用，常用于止血和消肿。2020 年，在云南省一白及种植区发现白及植株长势弱、叶片黄化，根部肿大、腐烂坏死等症状。采集受害根样进行病原分离，分离到根结线虫（*Meloidogyne* sp.），通过形态学（雌成虫、2 龄幼虫、会阴花纹形态观察）与分子生物学方法（rDNA-ITS 区序列比对、mtDNA-coxI 区序列比对、SCAR-PCR）相结合鉴定根结线虫种类，纯化、接种健康白及植株测定其致病性。形态学鉴定结果表明：雌成虫会阴花纹圆形至卵圆形，线纹平滑到稍波浪状，背弓低平或中等高，背区和腹区具有明显的沟形侧线（侧沟），极少有线纹通过侧沟，与爪哇根结线虫（*Meloidogyne javanica*）典型会阴花纹形态相符。分子鉴定结果表明：该线虫 rDNA 的 ITS 区序列和 mtDNA 的 coxI 区序列与 NCBI 数据库中已登录的爪哇根结线虫相应区序列相似度较高，分别达 99.35%和 99.75%；该线虫 rDNA 的 ITS 序列以 99.00%的支持率与 GeneBank 中登录的爪哇根结线虫（登录号 KX646187，MW672262，KJ739710，KP901063，MK390613）聚为同一分支；该线虫 mtDNA 的 coxI 序列以 97.00%的支持率与 GeneBank 中登录的爪哇根结线虫（登录号 OP646645，MZ542457，KP202352，KU372169，KU372170）聚为同一分支；利用 SCAR 特异性引物 Fjav/Rjav（5′-GGTGCGCGATTGAACTGAGC-3′/5′-CAGGCCCTTCAGTGGAACTATAC-3′）鉴定，该线虫能扩增出大小约 670bp 的基因特异性条带，与已报道的爪哇根结线虫 SCAR 特异性条带大小相符。致病性测定结果表明：接种纯化的爪哇根结线虫约 90d 后取根观察，根部

---

* 第一作者：杨艳梅，讲师，从事植物线虫病害研究。E-mail：yym_yunnong@163.com
** 通信作者：胡先奇，教授，从事植物线虫病害研究。E-mail：xqhoo@126.com

产生明显的根结，并形成雌虫、产卵，用前述形态学和分子生物学方法鉴定结果符合爪哇根结线虫的特征。综合形态学、分子生物学鉴定和致病性试验结果，将危害白及的根结线虫确定为爪哇根结线虫。

关键词： 白及；根结线虫病；爪哇根结线虫；病原鉴定

西班牙根结线虫侵染危害柚子根部的首次鉴定报道*

王 丽**，冯推紫，陈 园，孙燕芳，龙海波***
（中国热带农业科学院环境与植物保护研究所/农业农村部热带作物有害生物综合治理实验室，海口 571101）

Detection of Root-knot Nematode *Meloidogyne hispanica* Infesting Pomelo Tree in China*

Wang li**，Feng Tuizi，Chen Yuan，Sun Yanfang，Long Haibo***
(*Key Laboratory of Pests Comprehensive Governance for Tropical Crops*，*Ministry of Agriculture and Rural affairs*，*Environment and Plant Protection Institute*，*Chinese Academy of Tropical Agricultural Sciences*，*Haikou* 571101，*China*)

摘 要： 柚子是常见的柑橘类水果作物，我国南方亚热带和热带地区广泛种植。近期随机调查发现，海南省澄迈县柚子树生长矮小，叶片黄化、生长衰退，幼嫩根尖位置明显突起肿大，扭曲呈畸形。将病根带回实验室内检查，解剖镜下观测到根系表面附着大量胶质状卵囊，剥开根表皮可见白色梨形状雌虫，为典型的根结线虫特征。利用形态学和分子生物学方法对该根结线虫进行鉴定，形态学特征表现为：雌虫会阴花纹卵圆形，背弓少数较高，花纹多波浪至粗糙，侧线明显多分叉；二龄幼虫尾端明显变细，口针尖细，尾部细长，端部钝圆，尾部透明区不明显；二龄幼虫（$n=20$）测量指标具体为平均体长 362.6μm，平均体宽 13.9μm，口针长 11.0μm，尾长 46.3μm，与现有西班牙根据线虫 *Meloidogyne hispanica* 描述相吻合。同时，柚子根结线虫 mtDNA、D2D3 和 ITS 序列测序结果与 NCBI 数据库中 *M. hispanica* 高度一致。此外，利用选择限制性内切酶 *Taq* I 对柚子根结线虫 IGS2 区的 PCR 扩增产物进行酶切，获得了 554bp 的特异性条带。综上形态学和分子特征，柚子根部分离的根结线虫鉴定为西班牙根结线虫 *M. hispanica*，这是国内首次报道 *M. hispanica* 侵染危害柚子根部。

关键词： 西班牙根结线虫；*Meloidogyne hispanica*；柚子；寄生；鉴定

---

* 基金项目：中国热带农业科学院基本科研业务费专项资金（1630042022008）
** 第一作者：王丽，硕士研究生，从事植物寄生线虫学研究。E-mail：wangli23107@163.com
*** 通信作者：龙海波，副研究员，从事植物寄生线虫学研究。E-mail：longhb@catas.cn

中国线虫学研究（第九卷）Nematology Research in China（Vol. 9）：106-107

陕西省丹参根结线虫种类鉴定及侵染能力测定*

潘　嵩**，魏佩瑶，刘　晨，陈志杰，张　锋，张淑莲，李英梅***

（陕西省生物农业研究所/陕西省植物线虫学重点实验室，西安　710043）

Identification of *Meloidogyne* Species and Their Aggressiveness to *Salvia miltiorrhiza* in Shaanxi Province*

Pan Song**，Wei Peiyao，Liu Chen，Chen Zhijie，Zhang Feng，Zhang Shulian，Li Yingmei***

（*Bio-Agriculture Institute of Shaanxi*，*Shaanxi Key Laboratory of Plant Nematology*，*Xi'an*　710043，*China*）

摘　要：丹参是一类重要的多年生草本中药材，在中国已有上百年的种植历史。陕西省秦岭地区的商洛市是丹参的重要产区，丹参种植面积和产量较大。近年来调查发现，商洛市多个丹参产区有根结线虫病害发生，严重影响当地丹参的产量和品质。2021—2022年，笔者对陕西省商洛市丹参根结线虫病害的发生进行调查，对丹参根结线虫种类进行鉴定并对根结线虫对丹参的侵染能力进行了测定。调查结果显示，商洛市丹参根结线虫病害发生严重，一些田块丹参的发病率达到70%左右。丹参被根结线虫侵染后，地上部位主要表现为叶片发黄和长势衰弱，根部出现明显的膨大根结，根结上有时可见明显的淡黄色卵囊。采集发病样品带回实验室进行根结线虫种类鉴定。从发病样品根结处分离根结线虫雌虫，进行形态学与分子生物学鉴定。丹参根结线虫雌虫会阴花纹主要有两类：一类会阴花纹背弓较低，具有较明显的侧线，肛门附近有刻点，和此前所报道的北方根结线虫形态学特点相似；另一类会阴花纹背弓较高，无明显的侧线。提取分离自不同样品的单条雌虫DNA，分别对rDNA ITS和D2D3区域，以及线粒体*CoII*和*Nad*5区域进行扩增和核酸序列分析。结果显示，分离自陕西省丹参上的根结线虫为北方根结线虫与南方根结线虫。进一步采用SCAR特异性引物对所分离根结线虫种类进行鉴定，所得结果与序列分析结果相一致，因此确定侵染陕西省丹参样品的主要根结线虫种类为北方根结线虫与南方根结线虫。值得注意的是，感染北方根结线虫与南方根结线虫的样品采集自商洛市不同丹参产区，未发现不同种类根结线虫混合侵染的现象。对分离得到的北方根结线虫与南方根结线虫种群，挑取单卵囊在感病番茄植株上进行纯化扩繁后，分别接种丹参植株，结果显示所分离得到的北方根结线虫与南方根结线虫均可以

* 基金项目：西安市科技计划项目（22NYYF019）；陕西省科学院科技项目（2022K-02）

** 第一作者：潘嵩，助理研究员，从事植物线虫生物学研究。E-mail：letusgo2007@163.com

*** 通信作者：李英梅，副研究员，从事植物线虫生物学研究。E-mail：liym@xab.ac.cn

侵染丹参，其中北方根结线虫在丹参上的繁殖系数为 5. 06±1. 32，南方根结线虫在丹参上的繁殖系数为 5. 49±1. 46。该研究结果表明，根结线虫病已成为陕西省丹参种植中的重要土传病害之一，严重影响丹参的产量与品质，需要对其防控技术进行进一步的研究。

关键词：丹参；根结线虫；侵染

国内首次发现南方根结线虫侵染黄花菜*

魏佩瑶**，刘　晨，潘　嵩，陈志杰，张　锋，张淑莲，李英梅***
（陕西省生物农业研究所/陕西省植物线虫学重点实验室，西安　710043）

First Report of *Meloidogyne incognita* on Daylily (*Hemerocallis citrina*) in China*

Wei Peiyao**, Liu Chen, Pan Song, Chen Zhijie, Zhang Feng, Zhang Shulian, Li Yingmei***
(*Bio-Agriculture Institute of Shaanxi, Shaanxi Key Laboratory of Plant Nematology, Xi'an　710043, China*)

摘　要：2022 年，在陕西省大荔县的黄花菜产区，一些黄花菜植株出现叶片发黄和长势衰弱等症状，其根部出现明显的根结症状，一些田块黄花菜发病率超过 90%。出现明显根结的黄花菜植株花序数下降 30%左右，导致黄花菜的产量和品质严重下降。采集发病的黄花菜样品及根系周围 0～30cm 土壤带回实验室进行根结线虫种类鉴定。所分离得到的雌虫会阴花纹具有较高的背弓，缺乏明显的侧线。提取单条雌虫 DNA，对基因组 rDNA 区域与线粒体 *Nad*5 基因进行扩增并对所得片段进行序列比对，结果显示所分离得到的根结线虫为南方根结线虫。采用 SCAR 特异性引物以所分离线虫 DNA 为模板进行扩增，结果同样显示所分离得到的为南方根结线虫。将所分离得到的根结线虫卵囊进行孵化后，接种培养了 5 周左右的黄花菜植株，60d 后发现所接种的黄花菜植株根部出现了明显的根结症状，分离根结线虫并采用 SCAR 特异性引物对所分离线虫 DNA 进行扩增，结果显示侵染黄花菜的根结线虫均为南方根结线虫。这是在我国首次发现南方根结线虫侵染黄花菜，需要进一步对该病害的防控技术进行研究。

关键词：黄花菜；南方根结线虫；种类鉴定

---

* 基金项目：陕西省科学院科技项目（2022K-02）
** 第一作者：魏佩瑶，助理研究员，从事植物线虫生物学研究。E-mail：letusgo2007@163.com
*** 通信作者：李英梅，副研究员，从事植物线虫生物学研究。E-mail：liym@xab.ac.cn

广西水稻产区稻菜轮作田水稻根结线虫病发生情况调查及鉴定*

黄金玲[1**]，覃丽萍[1]，刘志明[1]，刘峥嵘[2]，李红芳[1]，陆秀红[1***]

（[1]广西壮族自治区农业科学院植物保护研究所/农业农村部华南果蔬绿色防控重点实验室/广西作物病虫害生物学重点实验室，南宁　530007；[2]广西大学农学院，南宁　530005）

Survey Identification of Rice Root-knot Nematode in Rice and Vegetable Rotation Fields in Guangxi*

Huang Jinling[1**]，Qin Liping[1]，Liu Zhiming[1]，Liu Zhengrong[2]，Li Hongfang[1]，Lu Xiuhong[1***]

（[1] *Plant Protection Research Institute，Guangxi Academy of Agricultural Sciences/Key Laboratory of Green Prevention and Control on Fruits and Vegetables in South China Ministry of Agriculture and Rural Affairs/Guangxi Key Laboratory of Biology for Crop Diseases and Insect Pests，Nanning*　530007，*China*；[2] *College of Agricuture，Guangxi University，Nanning*　530005，*China*）

摘　要：水稻根结线虫病是对水稻影响最严重的土传病害，广西由于近几年大力发展秋、冬菜生产，水稻耕作模式由过去“稻-稻”调整为“稻-菜”模式，使根结线虫在水稻主产区发生危害和传播加快趋势。2021—2022 年，笔者对广西南宁、柳州、梧州、北海、防城港、钦州、贵港、玉林、百色、贺州、河池、来宾、崇左共 27 个乡镇 28 个调查点的水稻产区稻菜轮作田水稻根结线虫病发生情况进行了调查，结果表明调查的 171 个田块，共有 64 个田块发生根结线虫危害，水稻种植季根结线虫检出率为 37.4%，蔬菜轮作季根结线虫检出率为 25.7%，轮作的小白菜、香葱、辣椒、番茄、菜花、黄瓜、豇豆、生菜、大白菜、萝卜、油麦菜、芹菜、芥菜等均能检测出根结线虫。通过对根结线虫 2 龄幼虫形态、雌虫会阴花纹特征等进行形态学鉴定，并利用核糖体 ITS 区和 28S rDNA D2D3 区序列比对及系统发育树分析方法，确定广西稻菜轮作田水稻根结线虫病原为拟禾本科根结线虫 *Meloidogyne graminicola*。

关键词：水稻根结线虫；调查；鉴定

* 基金项目：广西科技重大专项（桂科 AA22036001）；广西农业科学院科技发展基金（桂农科 2021YT062，桂农科 2021ZX24）

** 第一作者：黄金玲，副研究员，研究方向为植物寄生线虫及其防治。E-mail：386584315@ qq. com

*** 通信作者：陆秀红，副研究员，研究方向为植物寄生线虫及其防治。E-mail：447597587@ qq. com

首次在安徽省发现玉米短体线虫侵染危害玉米

王　硕*，赵湘媛，徐菲菲，孙梦茹，李洪连，王　珂，李　宇**
（河南农业大学植物保护学院/省部共建小麦玉米作物学国家重点实验室，郑州　450046）

Occurrence of *Pratylenchus zeae* Infect Corn in Anhui Province of China

Wang Shuo*, Zhao Xiangyuan, Xu Feifei, Sun Mengru, Li Honglian, Wang Ke, Li Yu**
(*College of Plant Protection /National Key Laboratory of Wheat and Corn Crop Science*, *Henan Agricultural University*, *Zhengzhou*　450046, *China*)

摘　要：短体线虫（*Pratylenchus* spp.，又称根腐线虫）分布广泛、寄主众多，是一类重要的迁移性植物内寄生线虫。短体线虫、根结线虫和孢囊线虫是作物上危害最为严重的 3 种植物病原线虫。2018 年，笔者课题组从安徽省宿州的玉米种植田采集发病的玉米根系和根际土壤样品。采用改良的贝尔曼漏斗法，从采集的样品中分离到大量的短体线虫，通过形态和分子生物学鉴定明确了该短体线虫的种类。通过胡萝卜愈伤组织纯化培养了该短体线虫种群，并测定了其对玉米的寄生性和致病力。

研究结果表明，安徽省宿州玉米根际分离的短体线虫与玉米短体线虫（*P. zeae*）的形态特征值较一致。rDNA-ITS 和 rDNA 28S D2-D3 序列与 NCBI 数据库中的玉米短体线虫序列具有高度相似性。基于 rDNA-ITS 和 rDNA 28S D2-D3 序列构建的贝叶斯进化树显示，该短体线虫与其他玉米短体线虫种群位于同一高度支持的分支。因此，将从安徽省玉米田根际采集到的短体线虫鉴定为玉米短体线虫。采用单雌接种胡萝卜愈伤组织的方法纯化扩繁，成功培养了该玉米短体线虫种群，并对该种群的繁殖力进行了测定。该种群在胡萝卜愈伤组织上能够大量繁殖，胡萝卜培养基接种 30 条雌虫在 27. 5℃下培养 60d 后线虫的繁殖倍数为 6 108。采用温室盆栽接种的方法测定了该种群线虫对玉米的寄生性和致病性，结果表明该种群在玉米根际的繁殖倍数为 5. 74（Rf > 1），玉米是该种群的适合寄主，该种群线虫对玉米具有很强的致病力。这是首次在安徽省玉米上发现玉米短体线虫的侵染危害，也是首次研究该种群对玉米的致病性。本研究对玉米短体线虫病的检测、鉴定及其他相关的研究奠定了基础。
关键词：玉米短体线虫；玉米；种类鉴定；寄生性；致病性

* 第一作者：王硕，博士研究生，从事植物线虫病害研究。E-mail：wssxx1220@163. com
** 通信作者：李宇，副教授，主要从事植物线虫学研究。E-mail：liyuzhibao@henau. edu. cn

广西火龙果根结线虫病的病原鉴定

伍朝荣[1*]，陈　董[1*]，林珊宇[2]，王如胭[3]，张晓晓[1]，吴海燕[1**]，杨姗姗[1**]
（[1]广西大学农学院广西农业环境与农产品安全重点实验室，南宁　530004；[2]广西农业科学院植物保护研究所，南宁　530007；[3]广西阳信祥农业科技有限公司，南宁　530007）

First Report of *Meloidogyne enterolobii* on *Selenicereus costaricensis* in Guangxi，China

Wu Chaorong[1*]，Chen Dong[1*]，Lin Shanyu[2]，Wang Ruyan[3]，Zhang Xiaoxiao[1]，Wu Haiyan[1**]，Yang Shanshan[1**]
（[1] *Guangxi Key Laboratory of Agro-Environment and Agric-Products safety*，*College of Agriculture*，*Guangxi University*，*Nanning*　530004，*China*；[2] *Plant Protection Research Institute*，*Guangxi Academy of Agricultural Science*，*Nanning*　530007，*China*；[3] *Guangxi Yangxinxiang Agricultural Technology Co.*，*Ltd.*，*Nanning*　530007，*China*）

摘　要：火龙果是仙人掌科的热带和亚热带水果，近年来在水果消费市场中受到广泛欢迎。2022年在广西多个地方发现火龙果遭受根结线虫的严重侵染，为明确病原线虫种类，笔者运用形态学、rDNA-ITS序列及特异性引物PCR扩增的方法对分离获得的根结线虫进行种类鉴定。形态学结果表明，火龙果根结线虫与象耳豆根结线虫（*Meloidogyne enterolobii*）的形态特征一致。rDNA-ITS序列比对结果表明，火龙果根结线虫与NCBI已登记象耳豆根结线虫序列相似对比度为100%。特异性引物检测结果表明，火龙果根结线虫为象耳豆根结线虫。同时，致病性测定结果显示，所有接种植株均产生根结症状，发病症状与田间发病症状一致。结果表明，火龙果根结线虫病的病原为象耳豆根结线虫，这是广西地区首次发现象耳豆根结线虫侵染火龙果，这一发现对发现地的根结线虫控制具有重要意义。

关键词：火龙果；象耳豆根结线虫；形态鉴定；分子生物学鉴定

---

* 第一作者：伍朝荣，博士研究生，从事植物线虫病害研究。E-mail：wuchaorong2020@163.com
陈董，硕士研究生，从事植物线虫病害研究。E-mail：chendong2217392002@126.com
** 通信作者：吴海燕，教授，从事植物线虫病害研究。E-mail：wuhy@gxu.edu.cn
杨姗姗，讲师，从事植物线虫病害研究。E-mail：yangshanshan12@126.com

河南省白术根结线虫病的病原鉴定*

许相奎**，陈昆圆，李荣超，郑　潜，王　宽，刘晓凯，崔江宽***
（河南农业大学植物保护学院，郑州　450002）

Pathogen Identification of Root-knot Nematode Disease on *Atractylodes macrocephala* in Henan*

Xu Xiangkui**，Chen Kunyuan，Li Rongchao，Zheng Qian，Wang Kuan，
Liu Xiaokai，Cui Jiangkuan***
（*College of Plant Protection*，*Henan Agricultural University*，*Zhengzhou*　450002，*China*）

摘　要：近年来，随着白术的大面积单一种植，白术病虫害问题明显加重，白术根结线虫病的发生也逐渐加重。该病通过侵染白术根部，影响植株对矿物质和水分的吸收，进一步影响白术产量和品质。为明确河南地区白术根结线虫病病原种类，笔者采用形态学、rDNA-ITS序列分析和特异性引物扩增相结合的方法对罹病白术的根结线虫进行种类鉴定。形态学结果表明，白术根结线虫与南方根结线虫（*Meloidogyne incognita*）的形态特征一致。rDNA-ITS序列比对结果表明，白术根结线虫与GenBank中的南方根结线虫序列具有高度相似性，最高可达100%。特异性引物检测结果也表明河南白术根结线虫为南方根结线虫。通过形态鉴定和分子鉴定，明确了引起河南地区白术根结线虫病的病原线虫为南方根结线虫，这是河南地区首次发现南方根结线虫侵染白术，有关部门应采取相应措施防止该病害在白术种植区域进一步扩散蔓延。

关键词：白术；根结线虫病；形态学鉴定；分子鉴定；南方根结线虫

* 基金项目：河南省高等学校重点科研项目（22B210004）；河南省重大科技专项（221100110100）；河南农业大学作物学科合作研究基金项目（CCSR2022-1）

** 第一作者：许相奎，硕士研究生，从事植物病理学研究。E-mail：xuxiangkui2022@163.com
*** 通信作者：崔江宽，副教授，主要从事植物与线虫互作机制研究。E-mail：jk_cui@163.com

中国线虫学研究（第九卷）Nematology Research in China（Vol. 9）：113-114

甘肃省3种中药材根结线虫病病原鉴定*

石明明[1**]，刘永刚[2]，李文豪[1]，倪春辉[1]，韩　变[1]，李惠霞[1***]
（[1]甘肃农业大学植物保护学院，甘肃省农作物病虫害生物防治工程实验室，兰州　730070；
[2]甘肃省农业科学院植物保护研究所，兰州　730070）

Identification of the Root-knot Nematode in Three Chinese Traditional Medicine, Gansu Province*

Shi Mingming[1**], Liu Yonggang[2], Li Wenhao[1], Ni Chunhui[1], Han Bian[1], Li Huixia[1***]
([1] *College of Plant Protection, Gansu Agricultural University, Biocontrol Engineering Laboratory of Crop Diseases and Pests of Gansu Province, Lanzhou* 730070, *China*;
[2] *Institute of Plant Protection, Gansu Academy of Agricultural Sciences, Lanzhou* 730070, *China*)

摘　要：甘肃省地形地貌复杂，气候类型多样，中药材种类丰富，种植规模较大的有当归、党参、甘草和黄芩等30余种。近年来，随着农业产业结构的调整，中药材种植年限和种植面积不断扩大，病原菌逐年积累，中药材上病害愈加严重。根结线虫作为一类重要的植物寄生线虫，寄主范围广，常造成严重的经济损失。笔者课题组于2020年9—11月对甘肃省中药材主产区根结线虫病进行调查，在党参、当归和黄芩3种中药材根系上发现根结线虫病症状。运用形态学和分子生物学相结合的方法对分离到的线虫进行种类鉴定。结果表明，雌虫虫体为柠檬形或梨形，乳白色，虫体前端突出如颈，后端呈圆球形。雌虫会阴花纹整体呈卵圆形，背弓略平缓，有些群体刻线明显，部分群体不明显，肛门处有刻点，根据形态特征初步将危害3种中药材的线虫鉴定为北方根结线虫（*Meloidogyne hapla*）。

采用线虫ITS-rDNA通用引物TW81/AB28对二龄幼虫DNA进行扩增，均得到557bp的片段。经测序，3种中药材群体ITS-rDNA序列与北方根结线虫（GenBank登录号MT490918、MN752202等）相似度为99.46%~100.00%。利用28S-rDNA D2/D3区段通用引物D2A/D3B进行扩增，得到的产物均为762bp，其序列与北方根结线虫（GenBank登录号MK213348、KJ645233等）相似度为99.60%~100.00%。利用北方根结线虫特异性引物Mh-F/Mh-R进行扩增，得到大小为462bp的特异性片段，该结果与冯光泉等描述的一致。因

---

* 基金项目：国家自然基金项目（31760507，32260654）；甘肃省现代农业产业体系（GARS-ZYC-4）
** 第一作者：石明明，硕士研究生，从事植物线虫学研究。E-mail：2690249575@qq.com
*** 通信作者：李惠霞，教授，从事植物线虫学研究。E-mail：lihx@gsau.edu.cn

此，根据形态学结合分子生物学特征，将这 3 种中药材上的线虫鉴定为北方根结线虫。

这是甘肃省首次发现北方根结线虫危害当归和黄芩。北方根结线虫是冷凉地区植物重要的寄生线虫，在我国 17 个省份均有分布，可侵染烟草、花生和甜菜等多种经济作物。甘肃省定西地区气候凉爽，中草药的种植不仅种类多，而且面积大。该地区也是我国重要的马铃薯种植基地，而马铃薯是北方根结线虫的模式寄主，且常与中药材倒茬轮作。因此，在甘肃省中部，北方根结线虫在中药材及马铃薯上的危害情况需进一步调研分析，以及时采取措施阻断其传播蔓延。

关键词：中药材；根结线虫；病原鉴定；北方根结线虫

中国线虫学研究（第九卷）Nematology Research in China（Vol. 9）：115-116

基于构建腐烂茎线虫全基因组转录调控网络揭示植物线虫动态调控模型及龄期特异调控模块

丛子文，霍诗天，廖雪玲，王雪雨，陈阳阳，刘　念，郑金水，彭东海，孙　明
（华中农业大学/农业微生物资源挖掘与利用全国重点实验室，武汉　430070）

Construction of Plant Parasite Nematode Dynamic Regulation Model and Age Specific Regulation Module Based on the Genome-wide Transcription Regulation Network of *Ditylenchus destructor*

Cong Ziwen，Huo Shitian，Liao Xueling，Wang Xueyu，Chen Yangyang，Liu Nian，Zheng Jinshui，Peng Donghai，Sun Ming
（*Huazhong Agricultural University/State Key Laboratory of Agricultural Microbiology*, *Wuhan*　430070，*China*）

摘　要：植物寄生线虫（Plant parasite nematode）作为一类全球广泛分布的植物病原体，几乎能感染所有的栽培植物，每年可造成全球超过 12.3%的粮食减产，带来上千亿美元的经济损失。然而，目前针对植物寄生线虫的防治手段仍相对落后，传统防治手段难以满足当前的防治需求，亟须寻找更加新颖且高效的防控策略。由于植物寄生线虫的遗传背景不明确，且复杂多变的生活史导致对其关键致病机制的研究难以开展，少有可针对的靶标基因。因而，笔者团队着眼于植物寄生线虫本身，以甘薯腐烂茎线虫（*Ditylenchus destructor*，Dd）为代表，通过构建全基因组转录调控网络揭示植物寄生线虫的关键致病机制，并对其整个龄期转换过程中动态调控规律进行探究，从中发掘关键防控靶标及应对策略。

本研究基于传统型酵母单杂交技术构建了适用于大规模筛选 TF-DNA 相互作用的重组型酵母单杂交筛选系统，并结合多个龄期的转录组数据在 Dd 中构建了随时间序列变化的动态转录调控网络，克隆了 376 个 TF、2 245 个启动子元件，获得了共计 5 549 对互作对。笔者从中发现了植物寄生线虫生活史相关的四种龄期特异性调控模型。通过对这些模型深入挖掘，笔者获得了以下结论。①在植物寄生线虫入侵宿主时期，133 个 effectors 中 45%受到多个 TFs 的分级调控，其中 13 个 TF 作为上游转录因子间接调控多个 effectors 的表达，它们主要参与植物细胞壁降解及植物免疫信号沉默机制，其余 effectors 则为协同调控或单一调控，这说明植物寄生线虫中 effectors 的调控可能是以分级调控为主的多种调控方式并存的调控模式。②植物寄生线虫中调控性别发育基因的转录因子中除了序列上有保守的家族之外，还存在许多未报道的转录因子，它们很可能是与秀丽隐杆线虫性别分

化差异有关的关键靶标基因。综上，本研究能对推动植物寄生线虫基础生物学发展提供理论指导，有助于研究人员认识植物寄生线虫致病机制并进一步发掘优质防控靶标，为绿色精准防控提供资源。

关键词：腐烂茎线虫；转录调控；杂交筛选；动态调控

基于线粒体 *COI* 序列分析中国北方地区腐烂茎线虫群体遗传结构*

李云卿[1]**，彭　焕[2]，黄文坤[2]，彭德良[2]***
（[1]丽水学院生态学院，丽水　323000；[2]中国农业科学院植物保护研究所/植物病虫害综合治理全国重点实验室，北京　100193）

Genetic Structure of *Ditylenchus destructor* Populations in Northern China by Mitochondrial *COI* Sequences*

Li Yunqing[1]**, Peng Huan[2], Huang Wenkun[2], Peng Deliang[2]***
（[1] *College of Ecology, Lishui University, Lishui* 323000, *China*; [2] *State Key Laboratory for Biology of Plant Diseases and Insect Pests, Institute of Plant Protection, Chinese Academy of Agricultural Sciences, Beijing* 100193, *China*）

摘　要：腐烂茎线虫是世界上重要的植物病原线虫和我国全国农业植物检疫性有害生物，严重威胁我国甘薯、马铃薯产业安全生产。利用线粒体 *COI* 序列研究国内 8 个省份 33 个腐烂茎线虫群体的遗传结构，依据单倍型网络图将腐烂茎线虫群体划分为两个类群，类群Ⅰ和类群Ⅱ，两个类群间没有共享单倍型，通过 83 个碱基彻底分开。其中类群Ⅰ腐烂茎线虫群体来自甘薯，类群Ⅰ腐烂茎线虫群体主要来自马铃薯。类群Ⅰ内单倍型呈星状分布，不同单倍型通过一个未知的共同祖先相连接；而类群Ⅱ整体呈树状分布。两个类群间遗传分化系数 Fst 为 0.821 3，遗传距离为 0.192，基因流为 0.108 8，表明不同类群的腐烂茎线虫群体间遗传分化显著。同时，利用软件 STRUCTURE 估算腐烂茎线虫群体的遗传结构，其最佳 *K* 值亦为 2。类群Ⅰ和类群Ⅱ中性检验表明，腐烂茎线虫整体水平上 Tajima's D 和 Fu's Fs 均为正值，差异显著，核苷酸不匹配分析呈多峰分布，表明腐烂茎线虫群体经历了平衡选择，可能是由于长期饮食差异导致的自然选择结果。

关键词：腐烂茎线虫；线粒体 *COI*；群体结构；单倍型

---

* 基金项目：国家自然科学基金（31571988）；国家 973 计划（2013CB127502）；公益性行业（农业）科研专项（201503114，200903040）

** 第一作者：李云卿，讲师，从事植物线虫分子生物学研究。E-mail：1549935275@qq.com
*** 通信作者：彭德良，研究员，从事植物线虫研究。E-mail：pengdeliang@caas.cn

中国线虫学研究（第九卷）Nematology Research in China（Vol. 9）：118-119

接种松材线虫后红松转录组分析

曹业凡*，汪来发，王曦茁**，汪　祥
（中国林业科学研究院森林生态环境与自然保护研究所，国家林业和草原局森林保护学重点实验室，北京　100091）

Transcriptome Analysis of *Pinus koraiensis* Inoculated by *Bursaphelenchus xylophilus*

Cao Yefan*，Wang Laifa，Wang Xizhuo**，Wang Xiang
（*Key Laboratory of Forest Protection of National Forestry and Grassland Administration*，*Ecology and Nature Conservation Institute*，*Chinese Academy of Forestry*，*Beijing*　100091，*China*）

摘　要：松材线虫病是以松材线虫（*Bursaphelenchus xylophilus*）为主要病原的系统性病害，主要危害松属（*Pinus* spp.）树种，目前该病害已蔓延扩散至中国、日本和韩国等多个国家，对当地林业造成巨大经济损失与生态破坏。近年来随着松材线虫病不断向我国北方地区扩散，红松（*Pinus koraiensis*）成为我国松材线虫病新的寄主，关于松材线虫对红松的致病机制亟须研究。

以4年生红松为研究对象，通过接种不同地理来源株系的松材线虫，研究松材线虫对红松的致病性，并对接种松材线虫后的红松幼苗转录组进行差异分析。接种实验所用红松幼苗购自辽宁省抚顺市清原县，松材线虫株系号及其来源如下：①QH-1，采自辽宁省清原县的感病红松；②NM-1，采自江苏省南京市的感病马尾松（*P. massoniana*）；③CM-1，采自重庆市的感病马尾松。每株系接种红松幼苗9株，接种量为2 000条/株，以无菌水接种为对照。致病性结果表明，QH-1对4年生红松幼苗致病性最强，NM-1与CM-1致病性低于QH-1。转录组分析结果表明，在接种10d后，QH-1处理组差异基因为22 050，而CM-1与NM-1处理组接种10d后与CK组相比差异基因分别为3 940与5 929，因此QH-1处理组差异基因数量远大于其他线虫处理组，因此接种第10天为4年生红松幼苗产生防御反应的关键时间节点。KEGG注释结果表明，植物-病原菌互作、MAPK信号通路-植物与植物激素信号转导途径是差异基因主要富集的代谢通路，其中QH-1株系松材线虫在接种红松10d后，上述通路相关基因数量显著大于其他线虫处理组。

对4年生红松幼苗进行不同致病力株系接种处理与转录组测序，确认接种第10天为4

* 第一作者：曹业凡，博士研究生，从事林木线虫病害研究。E-mail：cyf1995@ caf. ac. cn
** 通信作者：王曦茁，副研究员，从事林木线虫病害研究。E-mail：ladydal@ 163. com

年生红松幼苗产生防御反应的关键时间节点，4 年生红松幼苗转录组对不同致病力株系松材线虫侵染响应存在差异，4 年生红松幼苗接种不同致病力松材线虫后，转录组差异较大，其中强致病力株系松材线虫能够导致寄主松树发病更快，产生更多差异基因。至于哪些差异基因参与反应，有待于进一步基因筛选与功能验证研究。

关键词：红松；松材线虫；致病性；转录组

烟草过表达根结线虫异分支酸水解酶基因抑制 SA 介导的免疫*

方辰杰**，秦　鑫，田海洋，于家荣，陈　聪，王　暄***
（南京农业大学/农作物生物灾害综合治理教育部重点实验室，南京　210095）

Ectopic Expression of a Root-knot Nematode Isochorismatase Gene in *Nicotiana benthamiana* Suppresses SA-mediated Plant Immunity*

Fang Chenjie**，Qin Xin，Tian Haiyang，Yu Jiarong，Chen Cong，Wang Xuan***
（*Key Laboratory of Integrated Management of Crop Diseases and Pests*，*Ministry of Education*，*Nanjing Agriculture University*，*Nanjing*　210095，*China*）

摘　要：南方根结线虫（*Meloidogyne incognita*）是一类主要在植物根部危害的固定内寄生线虫，寄主种类多且分布范围广，给全世界农业生产造成了极大的危害。南方根结线虫在寄生过程中，通过向植物细胞分泌效应子抑制寄主的免疫反应，促进自身的寄生。Mi-ISC-1 是南方根结线虫分泌的一种包含异分支酸水解酶结构域的效应子，原核表达及离体试验证实 Mi-ISC-1 具有异分支酸水解酶活性。

为了明确 Mi-ISC-1 在活体条件下对植物免疫的影响，本研究利用农杆菌介导法培育了过表达 *Mi-isc*-1 的转基因烟草，选取两个 T3 代转化株系进行表型测定，结果显示：在烟草叶片接种辣椒疫霉（*Phytophthora capsici*）48h 后，转 *Mi-isc*-1 基因的烟草株系病斑直径比 GFP 对照增大了 25.8%～26.7%，而 SA 水平降低了 69.3%～77.9%，同时 *PR*1*a*、*PR*1*b* 和 *PR*2 基因转录水平也分别降低了 57.3%～59.5%、62.3%～75.6%和 59.0%～77.2%；进一步接种其他病原菌的测试结果表明，转 *Mi-isc*-1 基因株系对丁香假单胞菌（*Pseudomonas syringae*）的侵染更为敏感，接种 3d 后菌落数对数值比 GFP 对照增加了 11.3%～11.4%，差异显著；此外，过表达 *Mi-isc*-1 对灰霉菌（*Botrytis cinerea*）的侵染无明显影响，转 *Mi-isc*-1 基因株系、GFP 对照株系及野生型株系烟草叶片在接种灰霉菌 48h 后形成的病斑大小无显著差异。研究结果证实 Mi-ISC-1 能够抑制 SA 介导的免疫、促进病原菌的侵染。

关键词：南方根结线虫；异分支酸水解酶；过表达；SA

---

* 基金项目：国家自然科学基金（32272497）

** 第一作者：方辰杰，硕士研究生，从事植物线虫病害研究。E-mail：943764550@qq.com
*** 通信作者：王暄，教授，从事植物线虫学研究。E-mail：xuanwang@njau.edu.cn

土壤样本中菲利普孢囊线虫 TaqMan 实时荧光 PCR 快速检测及定量分析*

坚晋卓**，张梦涵，赵雨璇，黄文坤，刘世名，孔令安，彭　焕***，彭德良***
（中国农业科学院植物保护研究所/植物病虫害综合治理全国重点实验室，北京　100193）

Molecular Diagnosis and Direct Quantification of Cereal Cyst Nematode（*Heterodera filipjevi*）from Field Soil Using TaqMan Real-time PCR*

Jian Jinzhuo**，Zhang Menghan，Zhao Yuxuan，Huang Wenkun，Liu Shiming，Kong Ling'an，Peng Huan***，Peng Deliang***
（*State Key Laboratory for Biology of Plant Diseases and Insect Pests*，*Institute of Plant Protection*，*Chinese Academy of Agricultural Sciences*，*Beijing*　100193，*China*）

摘　要：菲利普孢囊线虫（*Heterodera filipjevi*）又称小麦孢囊线虫，是全球小麦生产上的一种重要病原物，一般造成禾谷类作物产量重大损失 20%~30%，严重威胁粮食的安全生产。早期快速准确地检测对于小麦孢囊线虫病的防控至关重要，然而，目前尚无 qPCR 技术检测和定量 *H. filipjevi* 的报道。在本研究中，根据 *H. filipjevi* 的 RAPD-SCAR 序列（KC529338）设计 TaqMan-qPCR 特异性的引物和探针，首次开发了小麦菲利普孢囊线虫 TaqMan-MGB 探针的实时荧光定量 PCR（real-time fluorescent quantitative PCR with TaqMan probes，TaqMan-qPCR）检测技术。

该检测方法具有极高的特异性和灵敏度，能从 13 种 27 个线虫种群中特异地检测出菲利普孢囊线虫，检测阈值低至 4^{-3}个单条 J2 DNA、10^{-3}个单条雌虫 DNA 和 0.01μg/μL 基因组 DNA。田间土壤样品检测发现，实时荧光定量 PCR 定量检测方法与传统显微镜检测法呈现出较好的线性关系（$R^2 = 0.8259$）。本研究建立的 TaqMan-qPCR 检测 *H. filipjevi* 技术特异性强、灵敏度高，能广泛适用于田间土壤开展 *H. filipjevi* 快速检测和定量分析，从而显著降低菲利普孢囊线虫的检测难度，缩短检测时间。

关键词：禾谷孢囊线虫；菲利普孢囊线虫；分子诊断；量化；TaqMan 实时 PCR

---

* 基金项目：国家自然科学基金（31772142，31571988）；公益性行业科研专项（201503114）；中国农业科学院农业科技创新工程（ASTIP-02-IPP-15）

** 第一作者：坚晋卓，博士后，从事植物线虫分子生物学研究。E-mail：jianjinzhuo@163.com
*** 通信作者：彭焕，研究员，从事植物寄生线虫研究。E-mail：hpeng@ippcaas.cn
彭德良，研究员，从事植物线虫及线虫病害综合治理研究。E-mail：pengdeliang@caas.cn

中国线虫学研究（第九卷）Nematology Research in China（Vol. 9）：122-123

拟禾本科根结线虫 qPCR 检测研究*

朱　衎**，严　丽，方　圆，韦郭鹏，凌瑞琪，杨梦婕，鞠玉亮***
（安徽农业大学植物保护学院，植物病虫害生物学与绿色防控安徽普通高校重点实验室，合肥　230036）

Study on qPCR Detection of *Meloidogyne graminicola**

Zhu Kan**，Yan Li，Fang Yuan，Wei Guopeng，Ling Ruiqi，
Yang Mengjie，Ju Yuliang***
（*School of Plant Protection*，*Anhui Agricultural University*，*Key Laboratory of Biology and Sustainable Management of Plant Disease and Pests of Anhui Higher Education Institutes*，*Anhui Agricultural University*，*Hefei*　230036，*China*）

摘　要：拟禾本科根结线虫引起的水稻根结线虫病在我国水稻种植区迅速蔓延，在水稻苗期发病尤其严重，导致水稻发育不良、分蘖减少，甚至死棵，严重时不得不将稻田翻耕重新播种。水稻根结线虫病属典型土传病害，主要采取病害监测和农业措施进行防治。基于此，本研究拟基于转录组数据开发 qPCR 引物，建立针对拟禾本科根结线虫的快速定量检测体系，为水稻根结线虫病的流行和预测提供技术支持。主要结果如下。

（1）拟禾本科根结线虫 qPCR 引物的筛选及反应条件优化。基于转录组学数据开发拟禾本科根结线虫 qPCR 引物的策略，选取拟禾本科根结线虫 2 龄幼虫转录组中 100 个高表达基因，选取与其他植物线虫基因序列相似度低于 20%的基因为靶标，筛选获得 16 对可用于常规 PCR 检测的特异性引物。在此基础上进一步筛选优化，最终获得 5 对可用于拟禾本科根结线虫 qPCR 检测的特异性引物 M356、M1178、M2315、M4322、M4562。对 qPCR 反应条件进行优化，M356、M1178、M2315、M4322、M4652 的最佳退火温度分别为 53.4℃、53.4℃、55.1℃、56.8℃、55.1℃；其最佳引物浓度分别为 0.6μmol/L、0.8μmol/L、0.6μmol/L、0.5μmol/L、0.8μmol/L。

（2）拟禾本科根结线虫基因组 DNA 及单条雌虫的 qPCR 检测。qPCR 引物 M356、M1178、M2315、M4322、M4562 仅可特异性扩增拟禾本科根结线虫，而对其他植物线虫无扩增。以基因组 DNA 为检测模板的灵敏度测试中，M356 的检测极限为 10^{-2}ng/μL，M1178 和 M2315 的检测极限为 10^{-3}ng/μL，M4322 和 M4562 的检测极限为 10^{-4}ng/μL。以单条雌虫 DNA 为检测模板的灵敏度测试中，M356 的检测极限为 $1/20\times10^{-3}$条，M1178 和 M4652 的检测极限均为 $1/20\times10^{-1}$条，M2315 和 M4322 的检测极限均为 $1/20\times10^{-2}$条。

* 基金项目：安徽省重点研究与开发计划（202204c06020028）；国家自然科学基金（31801714）
** 第一作者：朱衎，硕士研究生，从事植物线虫研究。E-mail：1253239065@ qq. com
*** 通信作者：鞠玉亮，副教授，从事植物线虫研究。E-mail：juyull@ 163. com

(3) 根组织和土壤中拟禾本科根结线虫的 qPCR 检测。利用本研究建立的 qPCR 体系对根组织中拟禾本科根结线虫 2 龄幼虫进行定量检测，M356、M1178、M2315、M4322、M4562 的检测极限分别为每 0.1g 根组织含 10 条、100 条、50 条、10 条、50 条 2 龄幼虫；标准曲线分别为 $y=-2.8192x+31.682$（$R^2=0.9837$，$E=126\%$）、$y=-2.9333x+33.568$（$R^2=0.9921$，$E=119\%$）、$y=-3.1668x+32.745$（$R^2=0.9964$，$E=107\%$）、$y=-2.8336x+32.101$（$R^2=0.9843$，$E=125\%$）、$y=-3.0687x+32.439$（$R^2=0.9673$，$E=112\%$）。利用该体系对土壤中拟禾本科根结线虫 2 龄幼虫进行定量检测，M356、M1178、M2315、M4322、M4562 的检测极限分别为每 0.5g 土壤中含 5 条、100 条、100 条、50 条、50 条 2 龄幼虫；标准曲线分别为 $y=-3.4376x+31.528$（$R^2=0.9982$，$E=95\%$）、$y=-2.6031x+30.885$（$R^2=0.9766$，$E=142\%$）、$y=-3.4862x+32.359$（$R^2=0.9984$，$E=94\%$）、$y=-3.2166x+31.452$（$R^2=0.9957$，$E=100\%$）、$y=-3.6472x+34.154$（$R^2=0.9439$，$E=88\%$）。

关键词：拟禾本科根结线虫；转录组学；灵敏度测试；检测极限

水稻干尖线虫对不同水稻品种的趋性比较*

宛　宁[1**]，王宏宝[2]，史雨琪[1]，战　炜[1]，廖澳琳[1]，王　暄[1***]
（[1]南京农业大学/农作物生物灾害综合治理教育部重点实验室，南京　210095；
[2]江苏徐淮地区淮阴农业科学研究所，淮安　223001）

Chemotactic Responses of the *Aphelenchoides besseyi* to Different Rice Varieties*

Wan Ning[1**], Wang Hongbao[2], Shi Yuqi[1], Zhan Wei[1], Liao Aolin[1], Wang Xuan[1***]
([1] *Key Laboratory of Integrated Management of Crop Diseases and Pests, Ministry of Education, Nanjing Agriculture University, Nanjing*　210095, *China*; [2] *Huaiyin Research Institute of Agricultural Sciences, Xuhuai District, Huai'an*　223001, *China*)

摘　要：水稻干尖线虫（*Aphelenchoides besseyi*）是一种严重威胁水稻生产的迁移性植物病原线虫，引起水稻叶尖部位干枯、扭曲，并导致水稻后期出现“小穗头”等症状，进而影响水稻产量和品质。水稻干尖线虫主要隐藏在稻种中，随着水稻的播种和萌发，线虫停止休眠逐渐恢复活性，通过水稻芽缝侵入水稻危害。

本研究以 Pluronic F-127 凝胶为介质，比较了水稻干尖线虫对 20 个不同水稻品种芽尖的趋性，结果显示：在放入水稻芽尖 10min、30min、60min 3 个时间段，线虫在强两优 698、宁乡粳 9 号、新稻 89、宏科共 4 个水稻品种的芽尖无聚集，表明水稻干尖线虫对上述 4 个品种无明显趋性；而相同时间段内，在徽两优、丝 5、天隆优、苏秀 857 共 4 个水稻品种的芽尖均有少量线虫聚集成团，60min 时 4 个水稻品种芽尖周围线虫数量为 10~20 条；而水稻干尖线虫对 9 优 418 等 12 个品种的水稻芽尖趋性明显，60min 时在上述品种的芽尖有大量线虫聚集成团，数量为 40~60 条；不同水稻品种芽尖的对峙试验显示，当将对线虫有吸引活性的不同品种水稻芽尖置于同一胶体中时，线虫对水稻芽尖的趋性发生变化，即绝大部分线虫向 9 优 418 聚集，而原稻 108 芽尖周围几乎不再有线虫聚集。

关键词：水稻干尖线虫；水稻品种；趋性

* 基金项目：淮安市农业科学研究院科研发展基金项目（HNY202122）
** 第一作者：宛宁，硕士研究生，从事植物线虫病害研究。E-mail：wanning0422@126.com
*** 通信作者：王暄，教授，从事植物线虫学研究。E-mail：xuanwang@njau.edu.cn

水稻抗拟禾本科根结线虫基因筛选和鉴定

王东伟*，成飞雪**
（湖南省园艺作物病虫害综合治理重点实验室，湖南省农业科学院植物保护研究所，长沙　410125）

Selection and Identification of Rice Resistant Genes to Root-knot Nematode，*Meloidogyne graminicola*

Wang Dongwei*，Cheng Feixue**
（*Key Laboratory of Integrated Management of the Pests and Diseases on Horticultural Crops in Hunan Province*，*Institute of Plant Protection*，*Hunan Academy of Agriculture Sciences*，*Changsha*　410125，*China*）

摘　要：拟禾本科根结线虫（*Meloidogyne graminicola*）是一种营固着寄生的植物线虫，近年来发生日趋严重，严重危害我国水稻生产。挖掘抗线虫基因并进行水稻遗传育种是防治水稻根结线虫最直接有效的方法。实验室利用已报道的抗拟禾本科根结线虫水稻品种中花 11 和感病品种日本晴，在 0. 15%琼脂培养基培养水稻并接种拟禾本科根结线虫，以不接种线虫的两个品种作为对照，通过转录组分析和 qPCR 验证，挖掘和鉴定水稻抗拟禾本科根结线虫基因。转录组结果表明：同一品种水稻的处理组与对照组之间相似度较高，不同品种的水稻之间相似度较低；抗病品种中花 11 的处理组与对照组之间有 38 个差异表达基因，感病品种日本晴的处理组与对照组之间有 161 个差异表达基因，包括共有的差异表达基因有 5 个，中花 11 特有的差异表达基因有 33 个，日本晴品种特有的差异表达基因有 156 个，表明感病品种的基因更容易受到拟禾本科根结线虫的诱导。同时分析发现，接种线虫和未接种线虫 *RPM*1 基因表达差异极其显著，且在抗病品种中花 11 中的表达量显著高于感病品种日本晴。*RPM*1 基因为植物免疫受体蛋白，对水稻稻瘟病菌和白叶枯病菌具有广谱抗性，可正调控水稻的防卫反应。qPCR 验证结果表明，水稻 *RPM*1 基因（*OsRLR*1）在接种拟禾本科根结线虫水稻中显著上调表达，与高通量测序结果一致。特别是接种线虫后中花 11 根中 *RPM*1 基因表达上调超 4. 18 倍，而在感病品种日本晴中 *RPM*1 基因主要在叶片中表达，根中接种与未接种线虫表达差异并不显著，推测水稻 *RPM*1 基因可能与水稻抗拟禾本科根结线虫相关。研究克隆获得了水稻 *RPM*1 基因（*OsRLR*1）序列，包含一个 2 772 bp 的 ORF，目前正在进行抗性验证。此外，*RPM*1 基因如何参与水稻抗拟禾本科根结线虫也需进一步研究。

关键词：拟禾本科根结线虫；水稻；转录组分析；抗病基因

---

* 第一作者：王东伟，助理研究员，从事植物线虫病害研究。E-mail：wangdongwei@ hunaas. cn
** 通信作者：成飞雪，研究员，从事植物线虫病害研究。E-mail：cfx937207@ 126. com

中国线虫学研究（第九卷）Nematology Research in China（Vol. 9）：126-127

水稻类钙调素蛋白 OsCML 在调控抗拟禾本科根结线虫中的功能研究*

魏　英[1,2**]，彭　焕[2]，彭德良[2]，刘　敬[1***]
（[1]湖南农业大学植物保护学院，长沙　410128；[2]中国农业科学院植物植物保护研究所，北京　100193）

Functional Studies on the Rice Calmodulin-like Protein OsCML in the Regulation of Resistance to *Meloidogyne graminicola**

Wei Ying[1,2**]，Peng Huan[2]，Peng Deliang[2]，Liu Jing[1***]
（[1] *College of Plant Protection*，*Hunan Agriculture*，*Changsha*　410128，*China*；
[2] *Institute of Plant Protection*，*Chinese Academy of Agricultural Sciences*，
Beijing　100193，*China*）

摘　要：拟禾本科根结线虫是危害水稻的重要病原线虫，严重威胁了水稻的生产安全。通过揭示拟禾本科根结线虫的致病机制，解析在寄主与线虫互作中发挥重要作用的效应蛋白，将为开发防治新技术提供必要的理论依据。通过同源克隆获得了拟禾本科根结线虫效应蛋白 MgCRT1 编码基因，CDS 全长为 1 352 bp，其 N 端编码信号肽，C 端具有内质网滞留信号区域 HDEL，不存在跨膜结构域。发育表达分析也表明 MgCRT1 在侵染后 J2（PJ2）幼虫中表达量最高。原位杂交显示 MgCRT1 在线虫的亚腹食道腺中特异性表达，且在线虫体内的免疫组化结果也与原位杂交结果相一致，亚细胞定位发现 MgCRT1 在植物细胞表达部位在内质网上。免疫组化证明了 MgCRT1 是一种分泌蛋白。

采用酵母双杂交技术筛选水稻膜体系 cDNA 文库。得到了候选的互作类钙调素蛋白 OsCML，并通过 BIFC 体内和 GST-pull down 体外等实验，都证明了线虫效应蛋白 MgCRT1 可与水稻 OsCML 发生特异性互作。通过接种实验证明了 MgCRT1 促进了线虫的寄生，OsCML 在线虫-水稻相关防御机制中属于正调控作用，而 OsCML 的下游靶标蛋白 OsHMGB 在线虫-水稻相关防御机制中属于负调控作用。通过 BIFC 体内和 GST-pull down 体外验证了 OsCML 与 OsHMGB 的互作关系。

经 qPCR 检测，致病相关基因 *PR1b* 和 *PR10* 的转录水平结果也表明 OsCML 正调节水稻免疫防御反应，OsHMGB 是水稻免疫防御反应中属于负调控因子。采用 100nmol/L flg22 处

---

* Funding：National Natural Science Foundation of China（31801716）；Natural Science Foundation of Hunan（2019JJ50273）；Scientific Research Project of Hunan Provincial Department of Education（19B259）

** First author：Wei Ying，mainly engaged in plant nematode and host interaction

*** Corresponding Author：Liu Jing，mainly engaged in the research of plant nematode and host interaction. E-mail：liujing3878@ sina. com

理野生型及超表达 *MgCRT*1 水稻植株的离体叶片，通过实时荧光定量检测不同时间段由 flg22 诱导的 PTI 途径中抗病相关基因的表达变化。表明 MgCRT1 在水稻中的表达可以抑制由 flg22 引起的 *Ks*4，*PAL*1 转录水平的上调。以上结果表明，MgCRT1 参与了由 flg22 诱导的水稻 PTI 免疫反应。

通过染色质分离实验表明 OsHMGB 是一种 DNA 结合蛋白。体外 GST-pull down 和体内 LCI 分析表明，MgCRT1 以竞争方式阻碍了 OsCML 和 OsHMGB 的结合，使 OsHMGB 在植物体内的积累，从而抑制寄主的免疫防御反应。

关键词： 拟禾本科根结线虫；钙调素类蛋白；免疫组化；原位杂交；酵母双杂；钙网蛋白；DNA 结合蛋白

水稻生物钟基因 *CCA*1 在拟禾本科根结线虫与寄主互作中的功能研究*

朱诗斐[1]**，刘　敬[2]***，彭　焕[1]***
（[1]中国农业科学院植物保护研究所/植物病虫害综合治理全国重点实验室，北京　100193；
[2]湖南农业大学植物保护学院，长沙　410128）

Functional Analysis of Rice Circadian Clock Gene *CCA*1 in the Interaction between *Meloidogyne graminicola* and Host*

Zhu Shifei**，Liu Jing[2]***，Peng Huan[1]***
（[1] *The State Key Laboratory for Biology of Plant Disease and Insect Pests*，*Institute of Plant Protection*，*Chinese Academy of Agricultural Sciences*，*Beijing*　100193，*China*；[2] *College of Plant Protection*，*Hunan Agriculture*，*Changsha*　410128，*China*）

摘　要：根结线虫（*Meloidogyne* spp.）作为一种重要的植物病原物，严重危害农业经济作物。水稻是全世界重要的粮食作物之一，同时也是拟禾本科根结线虫（*Meloidogyne graminicola*）的重要寄主。生物在适应地球环境的过程中进化出了一种内源性的计时系统——生物钟。通过生物钟系统植物能预测外界环境的变化，从而协调自身的生长发育和抗逆功能，能有效调控植物生理过程的发生时间减少不必要的能量消耗。植物的免疫反应是生物钟调控的重要生理生化活动之一。

本研究以探究水稻生物钟基因 *CCA*1 对拟禾本科根结线虫的互作和功能为目的，通过构建水稻生物钟基因超表达材料并进行人工接种，测定了水稻在因为 *CCA*1 基因超表达时失去生物钟节律时对水稻的拟禾本科根结线虫抗性的影响。结果表明，野生型日本晴水稻对根结线虫抗性具有明显节律，早上抗病，晚上感病。而这样对拟禾本科根结线虫带有明显昼夜节律的抗性在超表达生物钟基因 *CCA*1 水稻中没有出现。

由此可以得出结论，生物钟基因在水稻对拟禾本科根结线虫的抗性中发挥作用，且在早上生物钟基因 *CCA*1 大量表达时，水稻对拟禾本科根结线虫相比于晚上表现出更强的抗性。而当水稻中的生物钟节律消失时，对拟禾本科根结线虫的抗性也失去昼夜节律。

关键词：生物钟基因 *CCA*1；昼夜节律；拟禾本科根结线虫；水稻

---

* 基金项目：公益性行业（农业）科研专项（201503114）；国家自然科学基金（31672012，31972247）
** 第一作者：朱诗斐，硕士研究生，从事植物线虫病害研究。E-mail：Zsf121355@163.com
*** 通信作者：刘敬，讲师，从事植物线虫病害研究。E-mail：liujing3878@sina.com
彭焕，研究员，从事植物线虫病害研究。E-mail：hpeng83@126.com

甜菜孢囊线虫调控寄主基因可变剪切的效应蛋白筛选及作用机制初探*

张梦涵**，彭　焕***
（中国农业科学院植物保护研究所/植物病虫害综合治理全国重点实验室，北京　100193）

Screening of Efficent Proteins and Their Mechanisms in Regulating Variable Shear of Host Genes in *Heterodera schachtii**

Zhang Menghan**, Peng Huan***
(*StateKey Laboratory for Biology of Plant Diseases and Insect Pests*, *Institute of Plant Protection*, *Chinese Academy of Agricultural Sciences*, *Beijing*　100193, *China*)

摘　要：甜菜孢囊线虫（*Heterodera schachtii*）是重要的植物病原线虫，严重影响甜菜生产的品质和产量。我国甜菜孢囊线虫的适生区在新疆和内蒙古地区，因此开展对甜菜孢囊线虫的致病机理及治理研究迫在眉睫。

近年来，可变剪切被认为是植物抵御病原菌侵染的重要调控机制。董莎萌团队于2020年根据转录组的可变剪切发生位置克隆出目的基因中发生可变剪切的区域构建到植物表达载体上，并在其后面连接一个萤光素酶报告基因，这些目的基因由于发生可变剪切会产生不同形式的转录本：一种是内含子保留，这会导致终止密码子提前出现无法检测到萤光素酶信号；另一种是内含子剪切，编码框读通由此可以检测到萤光素酶信号。基于转录组和基因组数据，利用以萤光素酶为基础的可变剪切报告系统对孢囊线虫候选效应子进行筛选，确定具有调节寄主基因可变剪切的效应子，同时筛选出植物中与效应子互作的剪接因子，了解可变剪切在植物与寄生线虫互作过程中的机理与功能，可以为防控作物孢囊线虫提供新的思路。

通过生物信息学分析，筛选出41个大豆孢囊线虫、小麦孢囊线虫和甜菜孢囊线虫具有核定位信号的效应蛋白基因序列，构建到了植物表达载体PYBA-1132中。通过对含萤光素酶可变剪切报告系统的转基因烟草进行筛选，初步筛选出了甜菜孢囊线虫 *Gene*008347。

关键词：甜菜孢囊线虫；可变剪切；效应蛋白

---

* 基金项目：公益性行业（农业）科研专项（201503114）；国家自然科学基金（31672012，31972247）
** 第一作者：张梦涵，硕士研究生，从事植物线虫病害研究。E-mail：ZhangMH6689@163.com
*** 通信作者：彭焕，研究员，主要从事植物与线虫互作机制研究。E-mail：hpeng83@126.com

番茄 *LeMYB*330 基因过表达植株广谱抗病性分子机理研究*

康志强，邓小大，王新荣，蔡书静，袁永强，Waqar Ahmed，
王　燕，叶雯华，MD Kamaruzzaman
（华南农业大学植物保护学院，广州　510642）

The Molecular Mechanism of Broad-spectrum Disease Resistance of *LeMYB*330 Overexpressed Tomato Plant*

Kang Zhiqiang，Deng Xiaoda，Wang Xinrong，Cai Shujing，Yuan Yongqiang，Waqar Ahmed，
Wang Yan，Ye Wenhua，MD Kamaruzzaman
（*College of Plant Protection*，*South China Agricultural University*，*Guangzhou*　510642，*China*）

摘　要：根结线虫（*Meloidogyne* spp.，RKN）是一类植物专性寄生病原线虫，在寄生植物过程中，通过口针将食道腺效应蛋白注入植物根部细胞，诱导根细胞形成多核巨型细胞，抑制植物防卫反应。MYB 转录因子是最大的转录因子家族之一，调控寄主植物对病原真菌和病毒的抗病性。番茄抗根结线虫病由单一显性 *Mi* 基因控制，较易丧失功能，需要挖掘新的抗根结线虫病番茄种质资源。

（1）*LeMYB*330 基因过表达（简称 OE-*LeMYB*330）番茄具有广谱抗病性。人工接种结果表明：OE-*LeMYB*330 番茄抗南方根结线虫、灰霉病和番茄黄化曲叶病毒。OE-*LeMYB*330 番茄接种南方根结线虫后，番茄根结数和卵囊数相比对照组分别显著降低 49.17%、44.44%；OE-*LeMYB*330 番茄离体叶片接种灰霉菌 96h 后，病斑直径相比对照显著减少 35.05%；OE-*LeMYB*330 番茄接种番茄黄化曲叶病毒 21d 后，未表现出番茄黄化曲叶病毒病症状。

（2）OE-*LeMYB*330 番茄广谱抗病性与 SA 信号通路基因的表达相关。OE-*LeMYB*330 番茄接种南方根结线虫或者番茄黄化曲叶病毒后，*LeMYB*330 基因表达量均显著高于接种前的表达水平，同时 OE-*LeMYB*330 番茄抗南方根结线虫和抗番茄黄化曲叶病毒的部分防御基因表达量发生变化。qRT-PCR 分析 SA 信号通路相关基因结果表明，接种南方根结线虫后，OE-*LeMYB*330 番茄中 *EDS*1、*PAL*1、*NPR*1、*PR*1 和 *PR*5 相对表达量相比对照发生显著上调表达；而接种番茄黄化曲叶病毒后，则是 *ICS*、*EDS*1、*PR*1 和 *PR*5 基因相对表达量相比对照组发生显著上调表达。该研究结果为创制广谱抗病性番茄种质资源打下基础。

关键词：番茄；南方根结线虫；灰霉菌；*Le*MYB330

* 基金项目：广东省基础与应用基础自然科学基金（2019A1515012080）

番茄候选感病基因 *T*106 的全长克隆和功能验证

闫曦蕊*，吴文涛，曾媛玲，王　扬**
（云南农业大学植物保护学院，昆明　650201）

Cloning and Functional Verification of Tomato Candidate Susceptibility Gene T106

Yan Xirui*，Wu Wentao，Zeng Yuanling，Wang Yang**
（*College of Plant Protection*，*Yunnan Agricultural University*，*Kunming*　650201，*China*）

摘　要： 象耳豆根结线虫（*Meloidogyne enterolobii*）致病性强，寄主范围广，扩散迅速，并能克服番茄上的 *Mi* 基因和辣椒上的 *Me* 基因，对作物造成毁灭性的危害。现有研究表明，对某些病原体敏感的部分植物基因被称为易感基因或易感因子，它们功能的丧失或突变会赋予植物抗性。

本研究以象耳豆根结线虫-番茄互作体系为对象，通过转录组数据筛选出在象耳豆根结线虫侵染产生的根结组织中特异上调表达的基因 *T*106，采用原位杂交技术确定此基因在番茄中的表达位置，并利用 RACE 技术克隆获得 *T*106 基因 cDNA 全长，通过 TRV 病毒诱导的基因沉默技术沉默番茄 *T*106 基因后，观察线虫在根系发育情况和线虫诱导的巨型细胞发育情况，测定线虫侵染率、巨细胞面积、根结百分率以及产卵量等病理指标。结果表明：*T*106 基因 cDNA 序列全长 423bp，在根结中巨型细胞和周围不对称分裂的邻近细胞中表达。*T*106 基因沉默不影响象耳豆根结线虫的侵染率，但象耳豆根结线虫在根内的生长发育受到抑制，经基因沉默的番茄根每克根卵量仅为对照番茄的 20. 69%，巨细胞面积在各个侵染时期也均小于对照，其中 22d 达到最大差异，面积仅为野生型植株的 67. 97%。

综上所述，*T*106 为被象耳豆根结线虫靶向的感病相关基因，对 *T*106 基因沉默可明显降低线虫繁殖能力。通过对此基因的进一步研究，期望为防治象耳豆根结线虫寻找到持久抗病的新途径。

关键词： 象耳豆根结线虫；感病基因；原位杂交；RACE；基因沉默

* 第一作者：闫曦蕊，硕士研究生，从事植物线虫病害研究。E-mail：Yan784534151@ 163. com
** 通信作者：王扬，教授，从事植物线虫病害研究。E-mail：wangyang626@ sina. com

中国线虫学研究（第九卷）Nematology Research in China（Vol. 9）：132-133

负调控因子 OsWD40-193 与 OseEF1A1 的相互作用抑制水稻对尖细潜根线虫（*Hirschmaniella muccronata*）的抗性*

单崇蕾**，张连虎，孙晓棠，崔汝强***
（江西农业大学农学院，南昌 330045）

Interaction of Negative Regulator OsWD40-193 with OseEF1A1 Inhibits *Oryza sativa* Resistance to *Hirschmanniella mucronata* Infection*

Shan Chonglei**, Zhang Lianhu, Sun Xiaotang, Cui Ruqiang***
(*College of Agronomy*, *Jiangxi Agricultural University*, *Nanchang* 330045, *China*)

摘　要：水稻（*Oryza sativa* L.）是世界范围内至关重要的粮食作物，但其对一种迁移性内寄生线虫较为敏感。尖细潜根线虫（*Hirschmaniella muccronata*）在水稻的生产过程中带来了严重的产量损失。尽管进行了广泛的努力，但是至今还没有发现能够有效抵抗尖细潜根线虫感染的水稻品种。因此，研究水稻与尖细潜根线虫之间的相互作用并鉴定抗病蛋白是至关重要的，能够为开发抗病水稻品种提供重要依据。

本研究通过对抗性品种和感病品种的转录组数据进行综合分析和 qRT-RCR 验证发现，当线虫侵染水稻后，OsWD40-193 的表达量会升高 4~5 倍。创建 *OsWD*40-193 基因的敲除转基因水稻和过表达转基因水稻进行接虫测试，结果表明过表达植株的线虫感染水平显著升高，而敲除转基因植株的线虫感染水平呈下降趋势。这些结果表明 *OsWD*40-193 负调控水稻对线虫的抗性。为进一步探索 *OsWD*40-193 基因的功能，采用 pull-down 法筛选其假定相互作用因子，酵母双杂交和 BiFC 试验验证发现只有 *OseEF*1*A*1 表现出与 *OsWD*40-193 的相互作用。验证发现 *OseEF*1*A*1 在 *OsWD*40-193 敲除转基因水稻中表达量减少，在 *OsWD*40-193 过表达转基因水稻中表达量升高，表明 *Ose*EF1A1 受 *OsWD*40-193 下游信号通路的调节。*OseEF*1*A*1 基因在线虫侵染水稻中的表达量以及线虫侵染两种转基因植株的结果与 *OsWD*40-193 一致。*OseEF*1*A*1 敲除水稻和野生型水稻的转录组分析表明，*OseEF*1*A*1 缺失改变了水杨酸、茉莉酸和脱落酸信号通路相关基因的表达，并增加了次生代谢产物的积累，以增强水稻的抗性反应。

---

* 基金项目：国家自然科学基金（32060607，31860494）
** 第一作者：单崇蕾，硕士研究生，从事植物线虫病害研究。E-mail：shancl1119@126.com
*** 通信作者：崔汝强，教授，从事植物线虫病害研究。E-mail：cuiruqiang@jxau.edu.cn

研究表明，负调控因子 *OsWD*40-193 与伸长因子 *OseEF*1*A*1 相互作用，抑制水稻对尖细潜根线虫的抗性，这一研究揭示了水稻与线虫的相互作用，也为培育抗性水稻品种提供了依据。

关键词：尖细潜根线虫；水稻；*OsWD*40-193；*OseEF*1*A*1；负调控因子

钾离子转运蛋白 OsHAK 在水稻抗拟禾本科根结线虫中的功能研究

张家芊[1*]，彭　焕[2]，彭德良[2]，刘　敬[1**]
（[1]湖南农业大学植物保护学院，长沙　410128；[2]中国农业科学院植物保护研究所，北京　100193）

Function of Potassium Transporter OsHAK in Rice Resistant to *Meloidogyne graminicola*

Zhang Jiaqian[1*]，Peng Huan[2]，Peng Deliang[2]，Liu Jing[1**]
（[1] *College of Plant Protection*，*Hunan Agriculture*，*Changsha*　410128，*China*；
[2] *Institute of Plant Protection*，*Chinese Academy of Agricultural Sciences*，
Beijing　100193，*China*）

摘　要：拟禾本科根结线虫（*Meloidogyne graminicola*）是水稻生产中重要的致病线虫，近几年来，该线虫在我国其他主要水稻种植区发生面积逐年扩大，造成重大的经济损失。钾离子作为植物体内最丰富的阳离子，在植物生长发育过程中发挥着不可替代的作用。

前期工作已经确定拟禾本科根结线虫侵染诱导表达钾离子转运蛋白基因 *OsHAK*，本研究以 *OsHAK* 的功能为切入点揭示由 *OsHAK* 调控的水稻抗拟禾本科根结线虫的作用机理。通过利用酵母双杂实验筛选到了一个与 *OsHAK* 互作的线虫蛋白 MgHIP。MgHIP 在寄生前二龄幼虫的亚腹食道腺中合成，并以效应蛋白的形式被分泌进了植物巨细胞中。线虫侵染实验结果表明，缺失 *OsHAK* 会导致水稻对根结线虫的易感性提高，而超表达该基因则会提高水稻对根结线虫的抗性，其发育表达时期与 MgHIP 的发育表达时期相对应。反之 Ox-MgHIP 转基因水稻比野生型水稻更容易被侵染，RNAi-MgHIP 则不能引起这种效应。利用 BiFC 和 CoIP 进一步证明了 *OsHAK* 与 MgHIP 之间的相互作用，此外利用亚细胞定位实验推断，钾离子转运蛋白 *OsHAK* 极可能定位在细胞膜上，并在细胞膜上与根结线虫 MgHIP 互作。这说明线虫通过分泌 MgHIP 作用于 *OsHAK* 来调控植物的防御反应，促进自身寄生。

关键词：水稻；拟禾本科根结线虫；*OsHAK*

* 第一作者：张家芊，硕士研究生，从事植物线虫相关研究
** 通信作者：刘敬，讲师，主要从事植物线虫与寄主相互作用的研究

不同密度玉米孢囊线虫对玉米产量损失的研究*

李荣超**，郑 潜，王颢杰，武 肖，张 宇，汤继华，崔江宽***
（河南农业大学植物保护学院，郑州 450046）

Study on the Loss of Corn Yield by Different Density of *Heterodera zeae**

Li Rongchao**, Zheng Qian, Wang Haojie, Wu Xiao, Zhang Yu, Tang Jihua, Cui Jiangkuan***
(*College of Plant Protection*, *Henan Agricultural University*, *Zhengzhou* 450046, *China*)

摘 要：玉米是世界第三大粮食作物，玉米孢囊线虫给玉米的产量和品质带来严重损失。玉米孢囊线虫通过侵染玉米根部，影响玉米对矿物质和水分的吸收，限制玉米产量，降低玉米经济价值。为探究接种不同密度玉米孢囊线虫对玉米的产量损失的影响。以正大 618 为供试品种，每株接种 1 个、2 个、4 个、8 个、16 个、32 个玉米孢囊，以不接玉米孢囊作为对照。接种孢囊后 7d 调查植株的农艺性状、14d 调查植株的农艺性状和根内线虫数、30d 调查植株的农艺性状和土壤孢囊数，计算线虫侵染造成的产量损失。结果表明，随着植株接种孢囊密度的上升，植株的根长、叶绿素含量和干物质量呈现降低趋势，于接种 16~32 个孢囊趋于平缓。7d 时，每株接种 16 个孢囊相比于对照处理根长、叶绿素含量、干物质量分别降低 44. 70%、15. 63%、81. 32%。14d 时每株接种 16 个孢囊相比于对照处理根长、叶绿素含量、干物质量分别降低 17. 96%、26. 25%、23. 13%。每株接种 16 个孢囊根内线虫数量最多，为 134 条/100mL 土壤。30d 时每株接种 16 个孢囊相比于对照组根长、叶绿素含量干物质量分别降低 62. 40%、19. 21%、65. 58%，产量损失率 45. 45%。每株接种 16 个孢囊时土壤孢囊数最多，为 124 个/100mL 土壤。综上所述，每株接种 16 个孢囊及以上对玉米生长发育抑制最强，引起的产量损失最大。

关键词：玉米；玉米孢囊线虫；产量损失

* 基金项目：河南省重大科技专项（221100110300）；河南农业大学作物学科合作研究基金项目（CCSR2022-1）

** 第一作者：李荣超，硕士研究生，从事植物病理学研究。E-mail：rchli0805@163. com
*** 通信作者：崔江宽，副教授，主要从事植物与线虫互作机制研究。E-mail：jk_cui@163. com

玉米孢囊线虫孵化特性及不同品种玉米对其抗性研究*

王　媛**，林静雯，高福坤，王冬亚，吴海燕***
（广西农业环境与农产品安全重点实验室，广西大学农学院，南宁　530004）

Hatching Characteristics of *Heterodera zeae* and Resistance Identification of Different Maize Varieties*

Wang Yuan**, Lin Jingwen, Gao Fukun, Wang Dongya, Wu Haiyan***
(*Guangxi Key Laboratory of Agric-Environment and Agric-Products Safety, Agricultural College of Guangxi University*, *Nanning*　530004, *China*)

摘　要：植物线虫是引起农作物病害的重要病原物之一，每年因植物线虫的危害可导致800亿美元的损失。我国分别于2017年和2020年在广西和河南的玉米田报道发现玉米孢囊线虫病害，但尚没有具体产量损失的评估研究，该线虫对我国玉米产业存在潜在的风险。目前线虫病害的防治措施主要有种植抗病品种和化学防治等，其中利用抗病品种是最经济有效的防治措施。本研究通过生测和盆栽试验，研究广西玉米孢囊线虫群体孵化特性、侵染规律和发育进程，以及广西主栽玉米品种对该线虫群体的抗性。结果表明，玉米孢囊线虫在25～33℃均可孵化，较高温度（35℃，40℃）和较低温度（15℃，20℃）均不利于玉米孢囊线虫孢囊卵的孵化，33℃为最佳孵化温度，其累积孵化率最高为63.88%；在15℃和40℃条件下孢囊不孵化；20℃和35℃条件下孢囊线虫的孵化率仅为4.6%。在30℃条件下，根据玉米孢囊线虫的侵染动态推断玉米孢囊线虫15～18d完成一个生活史。广西主栽的16个玉米材料中仅有庆农13和万川973两个品种表现为高抗，同时，单株玉米根系孢囊数与根系长度和重量没有相关性。本研究结果为开展玉米孢囊线虫基础生物特性研究奠定了基础，并为广西玉米孢囊线虫病害的防治提供理论依据。

关键词：玉米孢囊线虫；玉米；孵化；抗性

---

* 基金项目：国家自然科学基金（31660511，32160627）；公益性行业（农业）科研专项（201503114）
** 第一作者：王媛，硕士研究生，从事植物线虫学研究。E-mail：wangy980518@163.com
*** 通信作者：吴海燕，教授，从事植物线虫学研究。E-mail：wuhy@gxu.edu.cn

乙烯信号通路参与大豆孢囊线虫抗性基因 *GmAAT* 表达调控机制研究[*]

何　龙[**]，王　靓，陈璐莹，刘倩男，蔡译枭，郑经武，韩少杰[***]
（浙江大学农业与生物技术学院生物技术研究所，杭州　310058）

Involvement of Ethylene Signaling Pathway in Regulation of *GmAAT* Expression for Soybean Cyst Nematode Resistance[*]

He Long[**], Wang Liang, Chen Luying, Liu Qiannan, Cai Yixiao, Zheng Jingwu, Han Shaojie[***]
(*Institute of Biotechnology, College of Agriculture and Biotechnology, Zhejiang University, Hangzhou*　310058, *China*)

摘　要：大豆孢囊线虫是威胁世界大豆生产安全的一个重要的土传病原生物，而抗性基因的应用是防控该线虫最经济有效的方法，其中 *Rhg*1 是大豆抗性品种中应用最广泛的抗性位点之一。尽管如此，有关 *Rhg*1 编码抗性基因的表达调控机制尚不完全清楚。研究表明，大豆孢囊线虫侵染后，*GmAAT* 基因表达明显上调。本研究首次对该基因的启动子区域的顺式作用元件（ERE）进行预测，并发现存在多个与植物乙烯通路相关的 ERE。通过酵母单杂交筛选，发现转录因子 *GmERF*18 与该 ERE 元件互作，并通过双萤光素酶活性测定和凝胶阻滞实验验证了它们之间的互作。目前，本研究通过 CRISPR 技术在大豆毛根中对 *GmERF*18 进行编辑，发现其参与对大豆孢囊线虫的抗性功能。本研究发现了参与 *Rhg*1 抗性的全新顺式作用元件和能与其结合的转录因子 *GmERF*18，初步阐释了 *Rhg*1 编码抗性基因 *GmAAT* 表达调控机制，并揭示了植物激素乙烯在 *Rhg*1 对大豆孢囊线虫抗性中的作用。这些结果将为大豆主效抗性基因 *Rhg*1 的进一步利用及抗性新种质的创制提供重要理论依据。

关键词：大豆孢囊线虫；乙烯；转录因子；*Rhg*1

---

[*] 基金项目：国家自然科学基金（32272478）；博士后面上资助（2022M710128）

[**] 第一作者：何龙，博士研究生，从事大豆孢囊线虫抗性基因表达调控分析和大豆基因编辑研究。E-mail：hnndhelong2@ 163. com

[***] 通信作者：韩少杰，研究员，从事大豆孢囊线虫抗性机制和大豆新种质创制研究。E-mail：hanshaojie@ zju. edu. cn

植物寄生线虫取食管调控取食分子量大小机制初探*

赵　薇**，彭　焕***
（中国农业科学院植物保护研究所/植物病虫害综合治理全国重点实验室，北京　100193）

Study on the Mechanism of Regulating the Size of Uptake by Plant Parasitic Nematode Feeding Tube*

Zhao Wei**, Peng Huan***
(*The State Key Laboratory for Biology of Plant Disease and Insect Pests, Institute of Plant Protection, Chinese Academy of Agricultural Sciences, Beijing* 100193, *China*)

摘　要：植物寄生线虫是重要的植物病原物之一，严重威胁我国农作物的生产安全。植物寄生线虫通过口针从植物细胞中吸取营养物质，维持其生长发育。目前在根结线虫、孢囊线虫、肾型线虫和毛刺线虫的取食细胞中均发现了取食管结构，该结构起着分子筛的重要作用，可以排除大的蛋白质或者细胞器进入口针，从而防止线虫口针的堵塞，然而植物线虫取食管调控取食分子量大小机制尚不明确。

本研究通过大豆发根转化体系初步明确了大豆孢囊线虫、象耳豆根结线虫、短体线虫取食分子量大小；通过拟南芥植株转化，明确了甜菜孢囊线虫和南方根结线虫取食分子量大小；并测定了转 Bt 蛋白的大豆发根对植物线虫的控制效果。结果表明，大豆孢囊线虫和甜菜孢囊线虫可以摄取 27~53kDa 的荧光蛋白。在表达不同分子量大小荧光蛋白的拟南芥植株上接种南方根结线虫后，荧光观察结果表明，象耳豆根结线虫可以摄取 27~90kDa 的荧光蛋白。此外，迁移性的短体线虫可以摄取 27~90kDa 的荧光蛋白。

采用大豆发根体系构建了表达 Cry6Aa2、Cry13Aa1、Cry5Ba2 基因的大豆发根，结果表明 Cry6Aa2 基因显著延缓了大豆孢囊线虫的生长发育。而表达 Cry13Aa1、Cry5Ba2、Cry14Ab1 3 种 Bt 蛋白的大豆发根对大豆孢囊线虫的侵染和繁殖无显著影响。

关键词：孢囊线虫；根结线虫；取食分子量；苏云金芽孢杆菌

---

* 基金项目：国家自然科学基金（31672012）
** 第一作者：赵薇，硕士研究生，主要从事线虫与植物互作研究。E-mail：zw2460018926@126.com
*** 通信作者：彭焕，研究员，主要从事植物与线虫互作机制研究。E-mail：hpeng83@126.com

丁香酚抑杀南方根结线虫三种生测方法比较研究*

陈宗雄**，杨紫薇，惠仁杰，丁晓帆***
（海南大学植物保护学院/热带农林生物灾害绿色防控教育部重点实验室，海口 570228）

Comparison of Three Bioassay Methods of Eugenol for Inhibiting *Meloidogyne incognita**

Chen Zongxiong**，Yang Ziwei，Hui Renjie，Ding Xiaofan***
（*Key Laboratory of Green Prevention and Control of Tropical Plant Diseases and Pests*，*Ministry of Education*，*School of Plant Protection*，*Hainan University*，*Haikou* 570228，*China*）

摘 要：为充分发挥植物提取物丁香酚（$C_{10}H_{12}O_2$）对南方根结线虫（*Meloidogyne incognita*）的抑杀效果，本研究利用水琼脂平板熏蒸法、土壤熏蒸法和浸渍法 3 种生物测定方法分别处理南方根结线虫的二龄幼虫（J2）和卵，评价丁香酚 3 种处理方法的适用性。研究结果表明，水琼脂平板熏蒸法对二龄幼虫的 LC_{50} 值仅为 122.98μL/L，而土壤熏蒸法和浸渍法的 LC_{50} 值分别为 154.38μL/L 和 400.78μL/L；在丁香酚为 100μL/L 时，水琼脂平板熏蒸法的卵孵化抑制率高达 98.75%，而土壤熏蒸法和浸渍法的卵孵化抑制率仅分别为 31.30%、31.61%；同时，水琼脂平板熏蒸法线虫回收率为 92.74%，极显著高于土壤熏蒸法的线虫回收率（24.93%），且水琼脂平板可直接在体式显微镜下观察判别线虫的活性。综上，水琼脂平板熏蒸法方便且准确性好，更适宜于熏蒸类精油药剂的杀线生测。

关键词：丁香酚；南方根结线虫；生测方法

* 基金项目：热区植物保护专业课程实习整合研究与实践（hdjy2163）
** 第一作者：陈宗雄，硕士研究生，从事植物线虫研究。E-mail：chen583036114@163.com
*** 通信作者：丁晓帆，副教授，从事植物线虫研究。E-mail：dingxiaofan526@163.com

不同品种根系分泌物对马铃薯金线虫卵孵化的影响

余曦玥[1*]，陈　敏[2]，付启春[3]，于敬文[1]，于　清[1]，黄文坤[1**]
（[1] 中国农业科学院植物保护研究所/植物病虫害综合治理全国重点实验室，北京　100193；
[2]云南省昭通市植保植检站，昭通　657099；[3] 云南省昭通市大关县植保植检站，昭通　657400）

The Effect of Root Exudates on the Hatching of Potato Nematode（*Globodera rostochiensis*）Eggs

Yu Xiyue[1*]，Chen Min[2]，Fu Qichun[3]，Yu Jingwen[1]，Yu Qing[1]，Huang Wenkun[1**]
（[1]*The State Key Laboratory for Biology of Plant Disease and Insect Pests/Institute of Plant Protection*，*Chinese Academy of Agricultural Sciences*，*Beijing*　100193，*China*；
[2]*Plant Protection and Quarantine Station of Zhaotong City*，*Zhaotong*　657099，*China*；
[3]*Plant Protection and Quarantine Station of Daguan County*，
Zhaotong　657400，*China*）

摘　要：马铃薯金线虫（*Globodera rostochiensis*）是一种马铃薯生产上的毁灭性线虫病原物，是我国进境检疫性的有害生物，目前在我国云贵川三省已有发现。马铃薯金线虫卵需被宿主根系分泌物中的化学成分刺激才能孵化，继而侵染宿主。近年来，根系分泌物的作用研究逐渐成为各国学者的研究热点。本研究利用马铃薯金线虫抗感品种根系分泌物 GC-MS 检测，分别鉴定根系分泌物中组分和含量变化，以筛选刺激马铃薯金线虫孵化有效成分为目的，结合卵孵化刺激实验，明确有效物质对卵孵化的影响。结果表明，筛选有效成分有 Alpha-Solanine、Alpha-Chaconine 等，抗感品种中有效成分含量差异明显，Alpha-Solanine、Alpha-Chaconine 浓度为 5~7μmol/L 孵化刺激率最高达 48.02%，高感品种较高抗品种提早 7d 刺激卵孵化成 J2。旨在通过上述问题的研究探讨为马铃薯金线虫的绿色防控提供理论基础及依据。

关键词：马铃薯金线虫；根系分泌物；Alpha-Solanine；Alpha-Chaconine；作用机制

---

* 第一作者：余曦玥，硕士研究生，从事植物线虫病害研究。E-mail：yuxiyue22@163.com
** 通信作者：黄文坤，研究员，从事植物线虫病害致病机理及防控技术研究。E-mail：wkhuang2002@163.com

中国线虫学研究（第九卷）Nematology Research in China（Vol. 9）：141-142

和硕县加工辣椒根结线虫病病原鉴定及其防治药剂筛选

曹　铭[1*]，张晓静[2]，李克梅[2**]
（[1]铜川市中药材产业化办公室，铜川　727199；[2]新疆农业大学农学院，乌鲁木齐　830052）

Identification of Root-knot Nematode and Screening of Nematicides for Controlling it in Processing Peper in Heshuo County

Cao Ming[1*]，Zhang Xiaojing[2]，Li Kemei[2**]
（[1] *Tongchuan Chinese Herbal Medicine Industrialization Office*，*Tongchuan* 650201，*China*；[2] *Agricultural University*，*Xinjiang Agricultural University*，*Urumqi*　830052，*China*）

摘　要：和硕县是新疆维吾尔自治区加工辣椒的重要产区。2020 年 6 月，新疆农业大学植物线虫课题组在新疆和硕县发现有根结线虫侵染加工辣椒，主要表现为植株矮小、发育迟缓，部分植株整株死亡，严重影响当地加工辣椒产量和品质。经形态学特征结合分子生物学技术鉴定了根结线虫的种类，并针对该线虫开展防病药剂筛选。主要得到以下结果。

（1）依据根结线虫的成熟雌虫会阴花纹特征、2 龄幼虫形态测计和特异性引物（Mh-F/Mh-R）PCR 扩增结果，将新疆和硕县加工辣椒根结线虫病病原鉴定为北方根结线虫。

（2）通过室内毒力测定试验分别测定了氟吡菌酰胺、三氟嘧啶酰胺、阿维菌素 3 种单剂及 9 种复配制剂对北方根结线虫 2 龄幼虫及卵的生物活性影响。结果表明，单剂处理中，阿维菌素对北方根结线虫 2 龄幼虫毒力最高，LC_{50}为 0. 657 3 mg/L；氟吡菌酰胺对线虫卵的毒力最高，LC_{50}为 0. 221 3 mg/L。药剂配比筛选中，阿维菌素和三氟嘧啶酰胺混配比例为 1：1 的制剂对北方根结线虫 2 龄幼虫及卵的致死作用增效最为显著（$P<0.05$），共毒系数分别为 152. 54 和 134. 39。研究结果将为北方根结线虫病的药剂选用及混配药剂的开发提供理论依据。

（3）运用室内盆栽法测试了不同杀线剂对辣椒根结线虫病的防治效果。结果表明，苗前施用噻唑膦与苗后施用阿维菌素相对防效最佳，达 100%，其次在苗前施用阿维・噻唑膦、三氟嘧啶酰胺、氟吡菌酰胺、厚孢轮枝菌、寡糖・噻唑膦，以及苗后施用阿维・噻唑膦、三氟嘧啶酰胺等 7 个处理，相对防效均达 90%以上。

（4）通过田间小区试验测试了不同杀线剂对加工辣椒的生长性状、产量的影响及其对

---

* 第一作者：曹铭，硕士研究生，从事植物线虫病害研究。E-mail：779693752@ qq. com
** 通信作者：李克梅，教授，从事植物线虫病害研究。E-mail：likemei@ xjau. edu. cn

根结线虫病的防效。结果表明，阿维·噻唑膦、厚孢轮枝菌两种药剂处理防效最高，分别达到了96.1%、95.6%，对辣椒的生长有促进作用，产量相比空白对照均有提高，分别增产7.9%、7.6%。

上述研究结果明确了和硕县加工辣椒根结线虫病的病原物，并为该病害的药剂防治提供参考依据。

关键词：加工辣椒；根结线虫病；病原鉴定；毒力测定；防治药剂筛选

寄主挥发物 BHT 对小卷蛾斯氏线虫觅食策略的影响*

唐凡希**，张 忱，汤华涛，李意璇，侯有明，吴升晏***
（福建农林大学闽台作物有害生物生态防控国家重点实验室/福建农林大学植物保护学院，福州 350002）

Effects of BHT on *Steinermema carpocapsae* Foraging Behaviors*

Tang Fanxi**，Zhang Chen，Tang Huatao，Li Yixuan，Hou Youming，Wu Shengyan***
（*State Key Laboratory of Ecological Pest Control for Fujian and Taiwan Crops*，*Fujian Agriculture and Forestry University/College of Plant Protection*，*Fujian Agriculture and Forestry University*，*Fuzhou* 350002，*China*）

摘 要： 昆虫病原线虫是昆虫的专性寄生天敌，主要包括斯氏线虫属（*Steinernematid*）和异小杆线虫属（*Heterorhabditid*），可应用于害虫可持续治理。昆虫病原线虫的觅食策略（即搜寻寄主的方式）是影响其寄生昆虫的重要因素之一。不同觅食策略的昆虫病原线虫对寄主昆虫有不同的偏好。因此，选择适宜的昆虫病原线虫物种用于防治相关农业害虫具有十分重要的意义。前人研究指出，昆虫病原线虫采取的觅食策略可能会受到寄主产生的挥发性物质影响而有所改变。然而，是哪些相关的化合物，以及其背后影响的机制，这部分的答案尚不清楚。本研究旨在探究寄主产生的挥发性物质对小卷蛾斯氏线虫（*Steinermema carpocapsae*）觅食策略的影响。通过顶空固相微萃取（HS-SPME）与气相质谱联用（GC-MS）技术，发现红棕象甲幼虫能产生较高的2,6-二叔丁基-4-甲基苯酚（2,6-di-tert-butyl-4-methylphenol，BHT）。生物学测试表明，当移动受限制的大蜡螟（*Galleria mellonella*）在200ng、20ng、2ng、0.2ng、0.02ng BHT的处理下，与对照组相比，200ng的BHT处理的大蜡螟身上得到的线虫数量是对照组的2倍，且存在显著差异（$P=0.02$）。结果显示200ng的BHT能显著诱导*S. carpocapsae*发现该移动受限制的寄主，这暗指BHT可能具备调控*S. carpocapsae*觅食策略的潜力。未来，将进一步分析BHT对*S. carpocapsae*的扩散、移动、跳跃等方面能力的影响。同时，从基因层面探究导致*S. carpocapsae*觅食行为发生改变的原因。本研究成果将有助于扩大昆虫病原线虫的寄主范围，为害虫防治技术提供新的思路。

关键词： 昆虫病原线虫；BHT；小卷蛾斯氏线虫；觅食策略

* 基金项目：国家自然科学基金青年基金项目（32202374）；福建省科技厅对外合作项目（2021I0006）
** 第一作者：唐凡希，硕士研究生，从事昆虫病原线虫研究。E-mail：1097194722@qq.com
*** 通信作者：吴升晏，副教授，从事昆虫病原线虫觅食行为调控机制、植物-昆虫-昆虫病原线虫三营养级互作和入侵害虫生物防治研究。E-mail：sywu531@163.com

健康与根结线虫病柑橘根际土壤微生物群落对比分析

杨姗姗[1*]，赵 微[1*]，林宇明[1]，吴海燕[1]，黄晶晶[1]，单 彬[2**]，张晓晓[1**]
（[1]广西大学农学院广西农业环境与农产品安全重点实验室，南宁 530004；
[2]广西壮族自治区亚热带作物研究所，南宁 530001）

Comparative Analysis of Rhizosphere Soil Microbial Communities between Healthy and Root-knot Nematodes in Citrus

Yang Shanshan[1*], Zhao Wei[1*], Lin Yuming[1], Wu Haiyan[1], Huang Jingjing[1], Shan Bin[2**], Zhang Xiaoxiao[1**]
([1] *Guangxi Key Laboratory of Agro-Environment and Agric-Products safety*, *College of Agriculture*, *Guangxi University*, *Nanning* 530004, *China*; [2] *Guangxi Subtropical Crops Research Institute*, *Nanning* 530001, *China*)

摘 要：根结线虫（*Meloidogyne* spp.）是一种严重危害农作物的土传病原物，柑橘在生产过程中会受到根结线虫危害。本课题组发现广西柑橘根结线虫的病原为番禺根结线虫（*Meloidogyne panyuensis*），且它的侵染与根际土壤微生物群落相关。利用根际土壤微生物来防治根结线虫是较为绿色、经济的方法。

为了研究健康柑橘根际土壤与发病柑橘根际土壤微生物群落的关系，通过高通量测序研究了健康柑橘根际土壤与发病柑橘根际土壤微生物多样性、组成结构、组间比较和与环境因素相关性。结果表明：健康柑橘根际土壤微生物群落的 Alpha 和 Beta 多样性与发病柑橘根际土壤差异不显著。健康柑橘根际土壤与发病柑橘根际土壤真菌与细菌群落在门、科和属水平上组成相似，但物种丰度存在差异，门水平上，发病柑橘根际土壤 Basidiomycota 的丰度占比明显高于健康柑橘根际土壤；科水平上，发病柑橘根际土壤 Burkholderiales 的丰度占比明显高于健康柑橘根际土壤；属水平上，发病柑橘根际土壤的 *Lycoperdon*、*Burkholderia-Caballeronia-Paraburkholderia* 丰度占比与健康柑橘根际土壤存在显著差异。且通过根际土壤微生物与环境因素相关性分析发现，土壤有机质、总 N、总 P、总 K、有效 N、有效 P、有效 K 与发病柑橘根际土壤的 *Lycoperdon*、*Burkholderia-Caballeronia-Paraburkholderia* 属呈显著正相关。因此，柑橘根结线虫病的发生可能与物种丰度及环境因素密切相关。研究结果可为通过调控根际微生物群落结构来防治柑橘根结线虫病提供理论依据。

关键词：柑橘；番禺根结线虫；土壤微生物

---

* 第一作者：杨姗姗，讲师，从事植物线虫病害研究。E-mail：yangshanshan12@ 126. com
赵微，硕士研究生，从事植物线虫病害研究。E-mail：fungizw68@ 163. com
** 通信作者：单彬，高级实验师，从事农产品质量安全与绿色防控研究。E-mail：zzbj219@ 163. com
张晓晓，讲师，从事植物病原与寄主互作研究。E-mail：zhangxiao0719@ 126. com

咖啡短体线虫不同种群对模式植物本氏烟的寄生性和致病性测定

牛文龙*，孙梦茹，史　琳，徐菲菲，李洪连，王　珂，李　宇**
（河南农业大学植物保护学院，郑州　450046）

The Parasitism and Pathogenicity of Different Populations of *Pratylenchus coffeae* to the Model Plant *Nicotiana benthamiana*

Niu Wenlong*, Sun Mengru, Shi Lin, Xu Feifei, Li Honglian, Wang Ke, Li Yu**
(*College of Plant Protection*, *Henan Agricultural University*, *Zhengzhou*　450046, *China*)

摘　要：短体线虫（*Pratylenchus* spp.）又称根腐线虫，它是一类分布范围广、寄主种类多且破坏性强的迁移性内寄生线虫。咖啡短体线虫（*P. coffee*）是短体线虫中较为重要的种之一，分布广泛，严重危害小麦、玉米 、大豆、花生、芝麻、山药、苎麻等多种粮食作物和经济作物。近年来，本氏烟也被作为研究线虫病害的模式植物，但本氏烟是否为咖啡短体线虫的寄主尚不明确。本研究旨在明确采集到的咖啡短体线虫 5 个不同种群对本氏烟的寄生性和致病性及不同种群间的致病力差异。希望利用模式植物本氏烟的优良遗传背景，为咖啡短体线虫与寄主互作等相关的研究奠定基础。

温室盆栽接种 60d 后的结果表明：供试咖啡短体线虫 5 个不同种群均能侵染本氏烟根系，接虫本氏烟的植株表现为地上部长势弱和分蘖减少等症状，植株的地上部鲜重和根鲜重均显著低于（$P<0.05$）未接虫的对照组（CK），接虫本氏烟根际线虫的 Rf 值均大于 1，本氏烟是供试咖啡短体线虫 5 个不同种群的适合寄主植物。根系染色和病理切片结果显示：咖啡短体线虫可以在本氏烟根系内生长发育并完成生活史。供试咖啡短体线虫 5 个不同种群对本氏烟均具有较强的致病性，但寄主或地理来源不同的种群之间的致病力存在明显的差异。来自山东潍坊烟草 SD-YC-2 种群的致病力最强，来自安徽宿州小麦的 AH-015A2 种群的致病力最弱。本研究可为利用本氏烟作为植物线虫-寄主互作研究的平台，为咖啡短体线虫致病机理、防治靶标筛选及防治方法等研究提供科学依据。

关键词：咖啡短体线虫；模式植物；本氏烟；寄生性；致病性

---

* 第一作者：牛文龙，硕士研究生，从事植物线虫病害研究。E-mail：nwl17737664255@ 163. com
** 通信作者：李宇，副教授，主要从事植物线虫学研究。E-mail：liyuzhibao@ henau. edu. cn

喹唑啉类化合物作为新型杀线虫骨架的发现

王　盛*，宋宏义，蔡庆峰，陈吉祥**
（绿色农药全国重点实验室/教育部绿色农药与农业生物工程重点实验室，贵阳　550025）

Discovery of Quinazoline Compounds as a Novel Nematicidal Scaffold

Wang Sheng*, Song Hongyi, Cai Qingfeng, Chen Jixiang**
(*State Key Laboratory Breeding Base of Green Pesticide and Agricultural Bioengineering/Key Laboratory of Green Pesticide and Agricultural Bioengineering, Ministry of Education, Guizhou University, Huaxi District, Guiyang* 550025, *China*)

摘　要： 植物病原线虫每年给全球农业造成的损失达 1 570 亿美元。使用化学杀线虫剂仍然是当前最主要的线虫防控手段之一。然而，长期使用传统杀线虫剂，如噻唑膦、阿维菌素等，不仅会导致线虫的耐药性增加，而且会对环境和植物健康产生不利影响。因此，创制新的杀线虫剂替代品对线虫综合防治具有重要意义。为了寻找新型的杀线虫骨架结构，笔者测试了一系列喹唑啉类化合物的杀线虫活性，部分化合物表现出了很好的杀线虫活性。其中，化合物 B1 对松材线虫（*Bursaphelenchus xylophilus*）、水稻干尖线虫（*Aphelenchoides besseyi*）的 LC_{50}分别为 5. 06mg/L、22. 34mg/L；化合物 B2 对松材线虫、水稻干尖线虫的 LC_{50}分别为 14. 34mg/L、33. 97mg/L。当化合物浓度为 100mg/L 时，化合物 B1 和 B2 对根结线虫的杀线虫活性分别为 48. 95%、22. 96%。喹唑啉可作为一种新的杀线虫骨架结构，在今后笔者会以化合物 B1 为先导结构，进一步优化化合物结构，设计合成新的喹唑啉类衍生物，然后进行杀线虫活性及作用机制研究。

关键词： 喹唑啉；杀线虫活性

* 第一作者：王盛，硕士研究生，从事绿色杀线虫剂创制与防控研究。E-mail：Shy2637649950@ 163. com
** 通信作者：陈吉祥，副教授，从事绿色杀线虫剂创制与防控研究。E-mail：jxchen@ gzu. edu. cn

利用抗坏血酸过氧化物酶揭示生防放线菌 XFS-4 对大豆孢囊线虫的抗性研究*

项　鹏**
（黑龙江省农业科学院黑河分院，黑河　164300）

Study on the Resistance of Biocontrol Actinomycete XFS-4 Strain Against *Heterodera glycines* Revealed by Ascorbate Peroxidase*

Xiang Peng**
（*Heihe Branch of Heilongjiang Academy of Agricultural Sciences*，*Heihe*　164300，*China*）

摘　要：为了解生防放线菌 XFS-4 对大豆孢囊线虫的作用机理，本研究以合成关键酶抗坏血酸过氧化物酶基因序列设计引物，利用 RT-PCR 技术，分析该基因在菌株 XFS-4 包衣黑河 43 后抗大豆孢囊线虫过程中的表达差异，从基因转录表达水平上对 Gm-Apx 进行研究，以发现该基因与 SCN 抗性之间的关系。试验结果表明 Gm-Apx 在接种大豆孢囊线虫侵染后，其菌株 XFS-4 包衣黑河 43 处理组中相对表达量在接种后第 4 天和第 7 天表达上调，而在接种后第 11 天和第 14 天表达下调，说明该基因参与了大豆早期防御孢囊线虫的侵染过程，对植物抗性反应及解除胁迫诱导的氧化损害起到了很重要的作用。

关键词：放线菌 XFS-4；大豆孢囊线虫；抗坏血酸过氧化物酶；RT-PCR

---

* 基金项目：国家大豆产业技术体系专项（CARS-04）；农业基础性长期性工作植保中心爱辉试验点（NAES-PP-033）；黑龙江省“揭榜挂帅”项目第五积温区大豆极早熟高产品种重茬障碍消减增产技术研究与示范（2021ZXJ05B011）

** 作者简介：项鹏，助理研究员，主要从事植物线虫学研究。E-mail：xp_303@126.com

中国线虫学研究（第九卷）Nematology Research in China（Vol. 9）：148–149

Study on Endophytic Fungi Identification and Nematicidal Activity of *Chaetomium ascotrichoides* 1-24-2 from Five-year-old Pine Wood Nematode-resistant Masson Pine（*Pinus massoniana*）*

Zheng Lijun，Ye Wenhua，Wang Xinrong，Wang yan，Liu Songsong，
Waqar Ahmed，MD Kamaruzzaman，Zhou Ziyang，Chen Bingjia
（*College of Plant Protection*，*South China Agricultural University*，*Guangzhou* 510642，*China*）

Abstract：Pine wild disease（PWD） causes serious losses to the production of pine forests in the world. The control of PWD is difficult，and the selection and cultivation of disease-resistant pine species are the fundamental measure to control PWD. Endophytic fungi are ubiquitous in various pine plants and play a crucial role in improving plant disease resistance. Masson pine is widely distributed in South China and Central China. There are different types of endophytic fungi in different regions，and the natural diversity are rich. Tongmian Masson pine is well distributed in Guangxi，and the research on the taxonomic composition and function of endophytic fungi has not been reported yet. In this study，the endophytic fungi were isolated and identified from the five-year-old Tongmian Masson pine with different resistance levels to PWD，and their functions were studied. It will lay a foundation for using plant endophytic fungi to improve the resistance to the pinewood nematode of Tongmian Masson pine.

The key innovations were as follows：

（1）Isolation and identification of endophytic fungi from Tongmian Masson pine. The branches of five-year-old Tongmian Masson pine with different levels of resistance to PWN were selected for endophytic fungi isolation. Altogether 67 fungal strains were isolated from 159 Masson pine tissues with an isolation rate of 43.39%. After the purification，the isolated strains were identified as 10 genera in Ascomycota，including *Alternaria*，*Beltraniella*，*Chaetomium*，*Colletotrichum*，*Daldinia*，*Diaporthe*，*Fusarium*，*Neofusicoccum*，*Penicillium*，and *Pestalotiopsis*.

（2）It was found that the endophytic fungi isolated from PWD resistant and susceptible pine plants were highly diversified. Among them，four genera such as *Beltraniella*，*Chaetomium*，*Colletotrichum* and *Fusarium* were only found in the resistant Masson pine，and another two genera *Neofusicoccum* and *Pestalotiopsis* were only available in susceptible Masson pine. The endophytic fungi of the genus *Diaporth* and *Colletotrichum* were the most widely distributed in the five-year-old Tongmian Masson pine. Total isolation frequency of a *Diaporthe*，*Colletotrichum* and *Alternaria* were accounting for 31.34%，23.88% and 16.41%.

* 基金项目：国家林草局重点研发项目（GLM〔2021〕）

(3) An endophytic fungus numbered 1-24-2 was screened out from the endophytic fungi on the PWD resistant Masson pine plants, and the fermentation parameters of the strain1-24-2 to produce active substances were optimized. The live nematode of *Bursaphelenchus xylophilus* was treated by the filtrate of the fermentation broth of endophyte fungus 1-24-2 with immersion method for 48h and after recovery for 24h, the corrected nematode mortality rate of *B. xylophilus* reached more than 85%. The optimization of the fermentation conditions of endophytic fungi 1-24-2 showed that when the filling volume was 100/250, the fermentation temperature was 28℃, the pH was 6, and the fermentation broth was 3d, the supernatant of the fermentation broth of endophytic fungi 1-24-2 had the best nematicide activity against *B. xylophilus*.

Key words: Masson pine; Endophytic fungi; Pine wood nematode; Nematicidal activity; *Chaetomium ascotrichoides*

莓实假单胞 Sneb1990 通过多种途径增强番茄抗南方根结线虫*

王　帅[1**]，赵双玲[2]，谢　佳[1]，胡　展[1]，李　栋[1]，孙然锋[1]，陈立杰[2***]
（[1]海南大学植物保护学院，海口　570228；[2]沈阳农业大学植物保护学院，沈阳　110866）

Multifunctional Efficacy of the Endophyte *Pseudomonas fragi* in Stimulating Tomato Immune Response Against *Meloidogyne incognita**

Wang Shuai[1**], Zhao Shuangling[2], Xie Jia[1], Hu Zhan[1], Li Dong[1], Sun Ranfeng[1], Chen Lijie[2***]
([1] *College of Plant Protection, Hainan University, Haikou*　570228, *China*;
[2] *College of Plant Protection, Shenyang Agricultural University, Shenyang*　110866, *China*)

摘　要：根结线虫是世界农业上危害严重且非常难以防治的病原物，可以专性寄生 3 000 多种植物，每年造成数十亿美元损失。利用生防菌诱导作物防治根结线虫已经是现在绿色防治的热点。笔者前期通过种子处理从 271 株根瘤内生菌中筛选到对南方根结线虫防效显著的菌株——莓实假单胞菌 Sneb1990。通过对南方根结线虫毒力测定发现菌株 Sneb1990 发酵上清液处理 48h 时对根结线虫二龄幼虫致死率达 76.8%，其上清液处理 7d 时对线虫卵孵化抑制率为 71.9%。菌株 Sneb1990 具有固氮和溶磷能力，盆栽和田间试验表明，Sneb1990 菌悬液对番茄生长起到明显的促进作用，并显著减轻根结线虫发病情况。裂根实验显示，菌 Sneb1990 诱导番茄产生局部及系统免疫抑制南方根结线虫的侵入，并且菌株在不同时期激发番茄 PR1、PR2、PR3、PR5 和 PDF1.2 等抗性基因的表达。此外，浸根试验结果表明该菌株鞭毛蛋白多肽 $flg22_{pf}$ 能够激发番茄产生对南方根结线虫的抗性，其刺激番茄活性氧（ROS）和水杨酸（SA）途径抗性相关基因的表达。本研究结果明确了根瘤内生莓实假单胞菌 Sneb1990 防治南方根结线虫的潜力，为根结线虫病害的生物防治提供新思路。

关键词：莓实假单胞；番茄；根结线虫；鞭毛蛋白；生防潜力

---

* 基金项目：海南大学科研启动基金项目（RZ2200001238）；海南省植物病虫害防控重点实验室开放课题（KF2022HN03）

** 第一作者：王帅，讲师，从事根结线虫生物防治及线虫趋化性研究。E-mail：996018@hainanu.edu.cn
*** 通信作者：陈立杰，教授，从事植物线虫学研究。E-mail：chenlijie0210@163.com

中国线虫学研究（第九卷）Nematology Research in China（Vol. 9）：151–152

南方根结线虫对不同用途玉米苗期侵染观察*

高海英[1**]，丁晓琦[1**]，张红焱[1]，梁　晨[1,2]，史倩倩[1,2]，宋雯雯[1,2]，赵洪海[1,2***]
（[1]青岛农业大学植物医学学院/山东省植物病虫害绿色防控工程研究中心，青岛　266109；
[2]东营青农大盐碱地高效农业技术产业研究院，广饶　257347）

Observation on the Infection of *Meloidogyne incognita* on Corns for Different Purposes during Seedling Stage*

Gao Haiying[1**], Ding Xiaoqi[1**], Zhang Hongyan[1], Liang Chen[1,2], Shi Qianqian[1,2], Song Wenwen[1,2], Zhao Honghai[1,2***]
([1] *College of Plant Health and Medicine, Qingdao Agricultural University/Shandong Green Control of Plant Diseases and Pests Engineering Research Center, Qingdao* 266109, *China*; [2] *Academy of Dongying Efficient Agricultural Technology and Industry on Saline and Alkaline Land in Collaboration with Qingdao Agricultural University, Guangrao* 257347, *China*)

摘　要：南方根结线虫（*Meloidogyne incognita*）是一种寄主范围广、经济危害大的农业有害生物。笔者团队于2020年在我国山东省首次发现南方根结线虫在田间侵染玉米，造成部分幼株明显矮化。同时，还发现南方根结线虫对玉米的危害主要在苗期。明确南方根结线虫对玉米不同类型、品种侵染寄生性对于抗病品种利用具有重要意义。

本研究通过杯栽玉米进行人工接种试验，采用定期取样、根内线虫酸性品红染色等方法，测定了南方根结线虫对玉米2个普通品种（郑单958和登海圣丰168）、1个甜糯品种（中科糯3000）播种后30d内的侵染动态和27个普通品种（其中11个包衣）、11个甜糯品种、2个饲用品种播种后30d的侵染状况。侵染动态观察结果发现，普通和甜糯品种根内均出现J2侵入高峰、J3+J4发生高峰和雌虫发生高峰均分别出现播种后7d、14d和21d，但最早出现J3+J4和雌虫的时间普通品种均比甜糯品种晚。截至播种后30d，普通品种上未产生卵块，但甜糯品种上最早于播种后21d出现少量卵块。播种后30d，27个普通品种根内存在J2（9~387个/10g根）、J3+J4（6~210个/10g根）、雌虫（6~105个/10g根）和卵块（0~2个/条根）虫态，繁殖系数为0.12（0~0.20）；11个甜糯品种根内存在J2（69~

* 基金项目：黄三角国家农高区省级科技创新发展专项资金项目（2022SZX23）；山东省重点研发计划（重大科技创新工程）项目专题（2022CXGC020710-6）

** 第一作者：高海英，从事植物线虫病害研究。E-mail：754924633@qq.com
丁晓琦，从事植物线虫病害研究。E-mail：dingxqhzau@foxmail.com
*** 通信作者：赵洪海，教授，从事植物线虫病害研究。E-mail：hhzhao@qau.edu.cn

222 个/10g 根）、J3+J4（15~153 个/10g 根）、雌虫（8~137 个/10g 根）和卵块（1~5 个/条根），繁殖系数为 0.27（0.10~0.50）；2 个饲用品种根内各虫态均存在，繁殖系数为 0.1 和 0.5。结果表明南方根结线虫能够侵入玉米品种的根系并进一步发育，但侵染发育繁殖水平在品种间差异很大。

关键词：玉米；南方根结线虫；侵染；繁殖；寄主适合性

内蒙古地区腐烂茎线虫的适生区预测*

杨　帆[1**]，赵远征[2]，张晓明[3]，王　东[1***]，周洪友[1***]
（[1]内蒙古农业大学园艺与植物保护学院，呼和浩特　010018；[2]内蒙古自治区农牧业科学院植物保护研究所，呼和浩特　010031；[3]内蒙古农业大学草原与资源环境学院，呼和浩特　010018）

Potential Prediction of *Ditylenchus destructor* in Inner Mongolia*

Yang Fan[1**], Zhao Yuanzheng[2], Zhang Xiaoming[3], Wang Dong[1***], Zhou Hongyou[1***]
([1] *Collage of Horticulture and Plant Protection*, *Inner Mongolia Agricultural University*, *Hohhot* 010018, *China*; [2] *Institute of Plant Protection*, *Inner Mongolia Academy of Agricultural & Animal Sciences*, *Hohhot* 010031, *China*; [3] *College of Grassland*, *Resources and Environment*, *Inner Mongolia Agricultural University*, *Huhhot* 010018, *China*)

摘　要：为保障马铃薯产业健康发展，加强腐烂茎线虫的检疫防控工作，对腐烂茎线虫在内蒙古地区的适生区进行预测分析。根据腐烂茎线虫在我国最新的分布数据和环境数据，通过 MaxEnt 生态位模型和 ArcGIS 软件，预测了当前和未来气候条件下腐烂茎线虫在内蒙古地区的适生区。结果显示：MaxEnt 模型的 AUC 平均值为 0.929，预测结果准确可靠。当前气候条件下腐烂茎线虫在内蒙古自治区适生区面积 41.8 万 km^2，约占自治区面积的 35.3%。研究表明，在 2021—2040 年和 2041—2060 年，腐烂茎线虫在内蒙古自治区的总适生区面积将不断扩大；其中在 SSP1_2.6 气候情景下，在赤峰市、通辽市和兴安盟将出现腐烂茎线虫的高适生区；在 SSP3_7.0 气候情景下，在鄂尔多斯市、赤峰市、通辽市将出现腐烂茎线虫高适生区。利用 Jackknife 刀切法计算得到影响腐烂茎线虫分布的主导环境因子为温度季节性变化方差（911.94～1 121.32）、最湿月降水量（129.44～220.11mm）、最冷季平均温（-2.38～4.34℃）、昼夜温差月均温（9.31～11.59℃）、最湿季平均温（23.24～29.13℃），上述环境因子累计贡献率为 91.5%。腐烂茎线虫在内蒙古地区有进一步扩散的风险，建议在未来气候情景下出现腐烂茎线虫高度适生区的盟市，应加强调运检疫和疫情监测，防止腐烂茎线虫的进一步扩展，以保障马铃薯产业健康稳定发展。

关键词：腐烂茎线虫；MaxEnt；潜在适生区；气候模式

---

* 基金项目：内蒙古自治区重点研发项目（2022YFHH0036）；中央引导地方科技发展资金（2021ZY0005）；内蒙古自治区科技重大专项（2021ZD0005）

** 第一作者：杨帆，硕士研究生，从事植物线虫研究。E-mail：1330214926@qq.com
*** 通信作者：王东，副教授，从事植物线虫研究。E-mail：wangdong19852008@163.com
周洪友，教授，从事有害生物综合防治研究。E-mail：hongyouzhou2002@aliyun.com

山苍子精油对南方根结线虫的防效评价*

王　丽**，Hamza Shahid，刘　倩***
（中国农业大学植物病理学系，北京　100193）

Nematicidal Activity of *Litsea cubeba* Essential Oil Against *Meloidogyne incognita**

Wang Li**，Hamza Shahid，Liu Qian***
（*Department of Plant Pathology*，*China Agricultural University*，*Beijing*　100193，*China*）

摘　要：植物寄生线虫对全世界的粮食、经济作物安全构成了严重的威胁，在世界范围内造成了约 1 570 亿美元的损失，其中危害最严重的是根结线虫。植物源的活性成分及其衍生物可用作线虫的抑制剂，是目前对线虫病害进行防控的一种有效方法。山苍子（*Litsea cubeba*）是樟科木姜子属的一种植物，其果皮中所含多种芳香油，山苍子精油是从山苍子果实中提取的天然香精油，在食品行业、医药和化工等行业中都有广泛的应用。

室内生测试验发现，山苍子精油对南方根结线虫二龄幼虫具有较高的杀线活性，处理 24h 时 LC_{50}为 44.19μL/L。通过盆栽试验和田间试验发现，穴施 200mL 浓度为 50μL/L 山苍子精油后，番茄植株定植 60d 时的病情指数分别降低了 21.3%和 20%，防效分别达 44.64%和 45.5%，说明山苍子精油对番茄根结线虫病具有良好的防治效果。运用气相色谱-质谱联用（GC-MS）分析山苍子精油的化学成分，结果表明，从精油中共鉴定出 30 种挥发性化合物，主要成分为柠檬醛、2,6-辛二烯，3,7-二甲基、双环［5.2.0］壬烷、2-亚甲基-4，8-三甲基-4-乙烯基、异黄酮、α-松油醇、6-辛烯、7-甲基-3-亚甲基和石竹烯。其中，柠檬醛浓度为 0.4μL/L 时，对南方根结线虫二龄幼虫的致死率为 55%左右。综上所述，山苍子精油有成为植物源杀线剂的潜力。

关键词：南方根结线虫；山苍子精油；GS-MS

* 基金项目：三亚崖州湾科技城管理局引导资金项目（SYND-2021-11）
** 第一作者：王丽，硕士研究生，从事植物寄生线虫防控技术研究。E-mail：1910350892@qq.com
*** 通信作者：刘倩，博士，副教授，从事植物寄生线虫致病机理和防控技术研究。E-mail：liuqian@cau.edu.cn

中国线虫学研究（第九卷）Nematology Research in China（Vol. 9）：155-156

蔬菜根结线虫高效生物熏蒸植物筛选*

李秀花**，马　娟，高　波，王容燕，万艳争，赵　惟，陈书龙***
（河北省农林科学院植物保护研究所/农业农村部华北北部作物有害生物综合治理重点实验室/
河北省农业有害生物综合防治工程技术研究中心，保定　071000）

Screening of Plants with Efficient Biological Fumigation Effects on Vegetable Root-knot Nematodes*

Li Xiuhua**，Ma Juan，Gao Bo，Wang Rongyan，Wan Yanzheng，Zhao Wei，Chen Shulong***
(*Key Laboratory of Integrated Pest Management on Crops in Northern Region of North China*，*Ministry of Agriculture*，*IPM Center of Hebei Province*，*Institute of Plant Protection*，*Hebei Academy of Agricultural and Forestry Sciences*，*Baoding*　071000，*China*)

摘　要：根结线虫是一类全球性分布、寄主范围广泛、对重要经济作物造成巨大损失的重要植物寄生线虫。近年来，随着农业种植结构的调整，保护地蔬菜栽培面积逐年扩大，为根结线虫的发生、发展提供了适宜的环境，使土壤中的根结线虫数量逐年增加、危害逐年加重，危害严重时减产70%以上，甚至绝收。对根结线虫的防治方法主要包括轮作、种植抗病品种、土壤消毒、化学防治、物理方法、生物防治以及使用免疫和抗性砧木。但这些方法都有一定的局限性或对环境有害，因此需寻找环境友好的替代方法防治保护地蔬菜根结线虫病害，逐步趋向于应用非化学的控制方法进行综合治理。生物熏蒸是利用植物残体在土壤里降解过程中产生的挥发性气体抑制或杀死土壤中根结线虫而减轻对蔬菜的危害并促进蔬菜的生长，提高产量。我国植物资源丰富，寻找熏蒸效果较高的植物，对发展保护地蔬菜产业具有重要意义。

选用西兰花、卷心菜、菠菜、茼蒿、香麦、油菜、小白菜、白菜花、茴香、香菜、紫苏、万寿菊、紫甘蓝、芥菜（2个品种：百草园芥菜和德超芥菜）、芥蓝、青苤蓝、苋菜（2个品种：糯香苋菜、花红苋菜）、朝天椒（叶、茎）、菜心、胡萝卜（根茎）、空心菜、快菜、白萝卜（3个品种：玉丽秋、春晓、三尺白）共23种作物为供试植物。每种植物均用新鲜的组织。把30g新鲜植物，用刀切成1cm×2cm或2cm×2cm的小碎片，混到200mL灭菌土壤里，先把植物碎片和含水量在10%～11%的土壤混合均匀后，再加入4mL含有

* 基金项目：河北省农林科学院科技创新专项（2022KJCXZX-ZBS-5）
** 第一作者：李秀花，副研究员，从事线虫学研究。E-mail：lixiuhua727@163.com
*** 通信作者：陈书龙，研究员，从事线虫学研究。E-mail：chenshulong63@163.com

4 000 条南方根结线虫二龄幼虫悬浮液，滴加线虫悬浮液时要分散开滴加到已混合好的植物碎片和土壤中，再充分混合，混合好后再置于直径 6.5cm 、高 9.5cm 的罐头瓶中，用两层保鲜膜封口，置于（25±1）℃恒温培养箱里，分别放置 9d 和 15d 后，取 50mL 混合物用蔗糖离心法分离线虫，在显微镜下分别记录死线虫数和活线虫数，并计算线虫死亡率。以不加植物碎片只加线虫悬浮液为对照。重复 3 次。

实验结果表明，在供试的 23 种作物中，生物熏蒸效果最好的是白萝卜中的 2 个品种春晓和玉丽秋的秧、朝天椒的叶子、青苤蓝和芥蓝，在熏蒸 9d 时校正死亡率均达到 90%以上；其次是菜心、百草园芥菜，在熏蒸 9d 和 15d 时校正死亡率分别达到 77.8%～84.7%和 95%以上。除上述之外，在熏蒸 15d 时，校正死亡率达到 90%以上的有香菜、朝天椒的茎、胡萝卜（根茎）、快菜、白萝卜（三尺白根茎），校正死亡率在 80%～90%的有万寿菊、白菜花、菠菜、白萝卜（三尺白的秧），其他作物校正死亡率为 30%～80%。

关键词：南方根结线虫；生物熏蒸；校正死亡率

烟酰胺酶抑制剂在大豆对孢囊线虫的抗性中的作用机制及应用研究*

陈璐莹**，王　靓，刘倩男，蔡译枭，何　龙，郑经武，韩少杰***
（浙江大学农业与生物技术学院生物技术研究所，杭州　310058）

Mechanism and Application of Nicotinamidase Inhibitors in Soybean Resistance to Soybean Cyst Nematode*

Chen Luying**, Wang Liang, Liu Qiannan, Cai Yixiao, He Long, Zheng Jingwu, Han Shaojie***
(*Institute of Biotechnology, College of Agriculture and Biotechnology, Zhejiang University, Hangzhou* 310058, *China*)

摘　要：大豆孢囊线虫（SCN）是全球重要的植物病原线虫之一，寄生于大豆根部，可导致大豆产量显著下降，对大豆生产造成威胁。烟酰胺酶抑制剂类杀虫剂可过度激活 TRPV 离子通道，导致中毒症状并抑制烟酰胺酶功能，对包括线虫和果蝇在内的多种生物产生毒性。本研究发现，24h 内外源施加烟酰胺酶抑制剂并不会影响 SCN 二龄幼虫的生存率，但会影响其对寄主的识别。在 SCN 侵染 2 周内，通过培养基施加一定浓度的烟酰胺酶抑制剂，可显著提高大豆对 SCN 的抗性。进一步研究发现，通过沉默烟酰胺酶（Naam）合成基因可提高大豆对 SCN 的抗性，而过表达烟酰胺酶（Naam）合成基因则降低大豆对 SCN 的抗性。此外，线虫侵染的大豆根烟酰胺酶（Naam）合成基因在感病品种中受线虫侵染诱导而表达量显著上调。本研究首次发现了 SCN 调控寄主烟酰胺酶（Naam）的表达来影响大豆对孢囊线虫的感病性的分子机制，为烟酰胺酶抑制剂开发成新型杀线虫剂提供了潜在的应用前景。

关键词：大豆孢囊线虫；烟酰胺酶抑制剂；抗性机制

---

* 基金项目：国家自然科学基金（32102146）

** 第一作者：陈璐莹，硕士研究生，从事大豆孢囊线虫抗性基因挖掘和大豆转化技术研究。E-mail：2609972215@qq.com

*** 通信作者：韩少杰，研究员，从事大豆孢囊线虫抗性机制和大豆新种质创制研究。E-mail：hanshaojie@zju.edu.cn

中国线虫学研究（第九卷）Nematology Research in China（Vol. 9）：158-159

野艾蒿根系孢囊线虫种类鉴定及其孵化特性*

陈京环**，魏雪娟，石明明，张　洁，李惠霞***
（甘肃农业大学植物保护学院，甘肃省农作物病虫害生物防治工程实验室，兰州　730070）

Identification and Hatching Characteristics of the Root Cyst Nematode of *Artemisia lavandulaefolia* DC.*

Chen Jinghuan**, Wei Xuejuan, Shi Mingming, Zhang Jie, Li Huixia***
(*College of Plant Protection*, *Gansu Agricultural University*, *Biocontrol Engineering Laboratory of Crop Diseases and Pests of Gansu Province*, *Lanzhou*　730070, *China*)

摘　要： 孢囊线虫作为一类重要的植物寄生线虫，通过寄生根部引起病变，导致植物生长发育不良并造成严重减产。目前，世界上报道的孢囊线虫亚科内包含 8 个属 100 余种，其中孢囊线虫属（*Heterodera*）和球孢囊线虫属（*Globodera*）对植物造成的危害最大。Mulvey 和 Stone 建立了球孢囊线虫属，截至目前，该属已经报道了 10 多个种，其模式种是马铃薯金线虫（*G. rostochiensis*）。艾球孢囊线虫（*G. artemisiae*）最早在苦艾（*Artemisia absinihium*）上被发现，后在俄罗斯远东区的红足蒿（*A. rubripes*）上描述。在我国，该线虫仅在北京和甘肃天祝有发现，其寄主均为菊科蒿属（*Tanacetum*）植物。

本研究以野艾蒿（*A. lavandulifolia* DC.）根系上的孢囊线虫为材料，对其进行形态学和分子生物学鉴定，并研究了不同温度、pH 值、离子条件及寄主根系分泌物对孢囊线虫卵孵化的影响。形态鉴定结果显示，孢囊球形或卵圆形，颈略突出，无阴门锥，环膜孔，无下桥。二龄幼虫线形，头部稍缢缩，口针强壮，尾圆锥形，透明尾较长，根据上述特征将其鉴定为球孢囊属（*Globodera*）。采用 ITS-rDNA 通用引物 TW81/AB28 对该线虫 DNA 进行扩增，得到 943bp 的片段。经测序，该群体 ITS-rDNA 序列与艾球孢囊线虫（GenBank 登录号：AY519127、AF274415 和 MT233312）相似度为 99.4%～99.7%。利用 28S-rDNA D2-D3 区段通用引物 D2A/D3B 进行扩增，获得 788bp 的片段，其序列与艾球孢囊线虫（GenBank 登录号：KU845472 和 MT233316）相似度较高，均为 99.78%。因此，本研究结合形态学特征和分子生物学将该线虫鉴定为艾球孢囊线虫。生物学特性研究发现，20℃为该线虫卵孵化的最适温度，4℃或 30℃均不利于孵化。Mg^{2+}对卵的孵化有促进作用，但 Ca^{2+}、Zn^{2+}、Fe^{3+}、Cu^{2+} 均会抑制该线虫的卵孵化。pH 值为 9.2 时，艾球孢囊线虫卵的累积孵化率均最高，在根系

* 基金项目：国家自然科学基金（31760507，32260654）；甘肃省现代农业产业体系（GARS-ZYC-4）
** 第一作者：陈京环，硕士研究生，从事植物线虫学研究。E-mail：980112714@qq.com
*** 通信作者：李惠霞，教授，从事植物线虫学研究。E-mail：lihx@gsau.edu.cn

分泌物原液中，卵孵化率最高。

本研究通过形态描述和分子生物学鉴定，明确了野艾蒿根系上的线虫为艾球孢囊线虫。对孢囊线虫孵化特性的研究，有助于更好地了解该孢囊的孵化机制，并为进一步研究提供基础依据。此外，野艾蒿在农作物大田附近生长较为普遍，艾球孢囊线虫寄主范围是否广泛，能否危害种植的大田作物，还有待进一步调查和研究。

关键词：野艾蒿；艾球孢囊线虫；孵化特性；种类鉴定

野生豆孢囊线虫侵染对不同抗性大豆品种生理生化指标的影响研究*

郑刘春**，刘福祥，温亚娟，文艳华***
（华南农业大学植物保护学院植物线虫研究室，广州 510642）

Studies on the Physiological and Biochemical Indexes of Soybean Plants Infected by *Heterodera sojae* *

Zheng Liuchun**，Liu Fuxiang，Wen Yajuan，Wen Yanhua***
（*Lab of Plant Nematology*，*College of Plant Protection*，*South China Agricultural University*，*Guangzhou* 510642，*China*）

摘 要：为研究野生豆孢囊线虫（*Heterodera sojae*）胁迫下不同抗性大豆品种生理生化指标的变化与抗性的关系，以抗病品种中黄 27 和感病品种赣豆 4 号两个不同抗性大豆品种为材料，采用盆栽大豆并人工接种野生豆孢囊线虫，接虫后定期取样，采用分光光度计比色法测定大豆根内多酚氧化酶（PPO）、过氧化物酶（POD）、苯丙氨酸解氨酶（PAL）、超氧化物歧化酶（SOD）、过氧化氢酶（CAT）、丙二醛（MDA）及可溶性蛋白含量的变化，以明确几种重要生化酶的早期变化与大豆抗性之间的相关性，为抗性鉴定方法及进一步深入研究大豆对野生豆孢囊线虫侵染的抗性机制奠定基础。

在接虫的 3d 内，研究结果表明：①抗病品种接虫处理大豆根内 POD 活性迅速升高，抗病品种未接虫对照组中的 POD 活性几乎无变化，而感病品种接虫组和未接虫对照组中的 POD 活性均降低；②抗病品种接虫处理和未接虫处理的植株根内可溶性蛋白含量均上升，感病品种接虫处理植株根内中的可溶性蛋白含量几乎不变，而感病品种未接虫对照组中的可溶性蛋白含量则下降；③感病品种大豆根内的 MDA 含量迅速上升，而感病品种未接虫对照处理、抗病品种接虫处理和抗病品种未接虫对处理植株根内的 MDA 含量均呈下降趋势。研究结果表明，大豆在被野生豆孢囊线虫侵染后的 3d 内，可依据植株根内 POD、MDA 及可溶性蛋白含量的变化，初步判断大豆材料的抗性强弱，为大豆抗野生豆孢囊线虫的鉴定方法提供参考，同时为进一步深入探究大豆抗野生豆孢囊线虫机制奠定基础。

关键词：野生豆孢囊线虫；大豆；生理生化指标；抗性鉴定

---

* 基金项目：公益性行业（农业）科研专项经费项目（201503114）
** 第一作者：郑刘春，硕士研究生，从事植物线虫学研究。E-mail：541331569@qq.com
*** 通信作者：文艳华，副教授，从事植物线虫学研究。E-mail：yhwen@scau.edu.cn

西甜瓜根际微生态分析及其根结线虫生防菌株筛选*

豆浓笑[1,2]**，周　波[1]***，彭　焕[2]***
（[1]山东农业大学生命科学学院，泰安　271018；[2]中国农业科学院植物保护研究所，北京　100193）

Analysis of Microecology in the Rhizosphere of Watermelon and Screening of Biocontrol Strains Against Root-knot Nematodes*

Dou Nongxiao[1,2]**，Zhou Bo[1]***，Peng Huan[2]***
（[1] *School of Life Sciences*，*Shandong Agricultural University*，*Taian*　271018，*China*；
[2] *Institute of Plant Protection*，*Chinese Academy of Agricultural Sciences*，*Beijing*　100193，*China*）

摘　要：根结线虫危害严重，通常造成非常严重的经济损失，由于其本身传播范围广、宿主作物多，往往无法有效地防治根结线虫。西甜瓜是一种重要的经济作物，属于葫芦科，同时也是根结线虫的宿主之一。通常的防治方法是物理防治和化学防治，但是存在成本高、安全性差的缺点，逐渐被新型的生物防治所取代，因此需要进一步了解西甜瓜健康与发病根际土壤微生物群落结构变化，寻找健康根际土壤关键微生物用于防治根结线虫研究。

本研究以患病与健康西甜瓜根际土壤为材料，研究其菌群变化来筛选生防菌株。通过生物信息学的方法，了解其健康与患病根际土壤微生态变化，并采用生物标志物与网络图方法相结合，确定其差异物种与关键物种，最终结合物种丰度确定筛选目标菌株，并进行杀虫评价。结果表明：在获取的健康与发病的样品中，在门水平上，3个地区中，主要集中在变形菌门、拟杆菌门，在属水平上主要集中在鞘氨醇单胞菌属、假单胞菌属、芽孢杆菌属；通过多样性分析，健康组与发病组具有显著差异，且健康组的群落多样性高于发病组的群落多样性；最终将确定11个菌属作为筛选目标：*Bacillus*、*Achromobacter*、*Stenotrophomonas*、*Sphingobacterium*、*Pseudomonas*、*Acinetobacter*、*Rhodanobacter*、*Comamonas*、*Ralstonia*、*Clostridium*、*Flavobacterium*。目前，已经验证筛选出3株分离物，其中2株是芽孢杆菌属，另1株是无色杆菌属，12h触杀实验均达到80%以上；以生物信息学的方法筛选靶标微生物，可以更好地提高其筛选效率，而且可以更好地指导高效生防菌应用于大田。

关键词：根结线虫；西甜瓜；土壤微生态

---

* 基金项目：中国农业科学院农业科技创新工程（ASTIP-02-ipp-15）
** 第一作者：豆浓笑，硕士研究生，从事植物线虫病害研究。E-mail：2545730405@ qq. com
*** 通信作者：周波，教授，从事植物线虫病害研究。E-mail：zhoubo2798@ 163. com
彭焕，研究员，从事植物线虫病害研究。E-mail：hpeng@ ippcaas. cn

中国线虫学研究（第九卷）Nematology Research in China（Vol. 9）：162-163

淡紫紫孢菌 PLHN 与甲维盐减量复配防治象耳豆根结线虫的作用机理*

杨紫薇**，陈宗雄，惠仁杰，丁晓帆***
（海南大学植物保护学院/热带农林生物灾害绿色防控教育部重点实验室，海口 570228）

摘　要：象耳豆根结线虫（*Meloidogyne enterolobii*）严重威胁热带农业生产，了解药菌减量复配对象耳豆根结线虫的作用机理，有助于进一步开展防治研究。本研究通过测定线虫运动行为、总糖含量、可溶性蛋白含量、线虫解毒酶活和胚胎发育等，初步探讨淡紫紫孢菌（*Purpureocillium lilacinum*）（简称 PLHN）和 3%甲氨基阿维菌素苯甲酸盐微乳剂（简称甲维盐）减量复配对象耳豆根结线虫的作用机制。研究结果表明：经 0. 6mg/L 甲维盐（减量 25%）+50%PLHN 和 0. 4mg/L 甲维盐（减量 50%）+75%PLHN 复配组合处理后，象耳豆根结线虫二龄幼虫（J2）体长较 CK 对照略有缩短，而尾部透明区较 CK 对照均有所增长；线虫运动、呼吸受到显著抑制，且 J2 体内的总糖含量、可溶性蛋白含量和乙酰胆碱酯酶活性均降低；但线虫羧酸酯酶活性极显著增强，而且导致线虫体液渗漏；抑制象耳豆根结线虫卵胚胎发育，随着处理时间的增加而抑制增强，死亡卵的比例随之增加，卵胚胎发育多数停止在单胞期、原肠期和一龄幼虫期阶段，且孵育出的 J2 多数为死虫。

关键词：象耳豆根结线虫；淡紫紫孢菌；甲维盐；减量复配；作用机制

Mechanism of Decrement Compound of *Purpureocillium lilacinum* PLHN and Emamectin Benzoate Against *Meloidogyne enterolobii**

Yang Ziwei**, Chen Zongxiong, Hui Renjie, Ding Xiaofan***
(*Key Laboratory of Green Prevention and Control of Tropical Plant Diseases and Pests, Ministry of Education, School of Plant Protection, Hainan University, Haikou* 570228, *China*)

Abstract: *Meloidogyne enterolobii* is a serious threat to tropical agriculture. Understanding the mechanism of decrement compound of nematicide and biocontrol agent against *M. enterolobii* is helpful to carry out control research. In this study, the mechanism of decrement compound of

* 基金项目：淡紫紫孢菌 PLHN 和甲维盐复配防治象耳豆根结线虫及其作用方式研究（Qhys2021-55）；热区植物保护专业课程实习整合研究与实践（hdjy2163）
** 第一作者：杨紫薇，硕士研究生，从事植物线虫研究。E-mail：ziweiyzw@163. com
*** 通信作者：丁晓帆，副教授，从事植物线虫研究。E-mail：dingxiaofan526@163. com

Purpureocillium lilacinum PLHN and emamectin benzoate 3% ME against *M. enterolobii* was preliminarily investigated by determining the motional behavior, total sugar content, soluble protein content, detoxification enzyme activity and embryonic development of *M. enterolobii*. The results showed that the body length of the second-stage juveniles (J2) was slightly shorter than CK after being treated by 0.6mg/L emamectin (25% reduction) +50% PLHN and 0.4mg/L emamectin (50% reduction) +75%PLHN, while the tail transparent region of J2 was increased compared with CK. The movement and respiration of J2 were significantly inhibited, and the total sugar content, soluble protein content and Acetylcholinesterase activity of J2 were decreased. However, the Carboxylesterase activity was significantly enhanced and resulted in body fluid leakage. Both combination schemes had an inhibitory effect on the embryonic development of *M. enterolobii*. With the increase of treatment time, the proportion of dead eggs increased. Most of the development of eggs stopped at the single cell stage, gastrul stage and first instar larvae stage, and most of the hatched J2 were dead.

Key words: *Meloidogyne enterolobii*; *Purpureocillium lilacinum*; Emamectin benzoate; Decrement compound; Mechanism

中国线虫学研究（第九卷）Nematology Research in China（Vol. 9）：164-165

党参根结线虫病优势生防菌长枝木霉 TL16 作用方式探究*

张　洁[1**]，安丽婷[1]，刘永刚[2]，石明明[1]，陈京环[1]，李惠霞[1***]
（[1]甘肃农业大学植物保护学院，甘肃省农作物病虫害生物防治工程实验室，兰州　730070；
[2]甘肃省农业科学院植物保护研究所，兰州　730070）

Exploration on the Action Mode of the Dominant Biocontrol Bacteria *Trichoderma longibrachiatum* TL16 Against Root-knot Nematode Disease of *Codonopsis pilosula**

Zhang Jie[1**], An Liting[1], Liu Yonggang[2], Shi Mingming[1], Chen Jinghuan[1], Li Huixia[1***]
（[1] *College of Plant Protection, Gansu Agricultural University, Biocontrol Engineering Laboratory of Crop Diseases and Pests of Gansu Province, Lanzhou　730070, China*;
[2] *Institute of Plant Protection, Gansu Academy of Agricultural Sciences, Lanzhou　730070, China*）

摘　要：党参（*Codonopsis pilosula*）作为我国一种传统中药材，具有很高的经济价值。甘肃省是我国党参主产区，总产量和出口量均居全国前列。2020 年，本课题组在甘肃省发现北方根结线虫（*Meloidogyne hapla*）引起的党参根结线虫病，经调查，发现省内多个党参主产区均有该病发生，平均病田率为 40.7%，病株率为 13.2%，病情指数为 10.4。目前生产上主要通过农业防治及喷施广谱性杀线剂等方法来防治根结线虫病。鉴于化学防治的局限性和无公害中药材病虫害防治的要求，筛选安全高效的生防菌是防治党参根结线虫病病害及开发生防制剂的基础。

本研究测定了 11 株生防真菌孢子悬浮液和发酵滤液对北方根结线虫卵、卵囊和 J2 的抑制作用，发现综合作用效果最好的生防菌为长枝木霉 TL16。TL16 对根结线虫卵有寄生作用，其菌丝和孢子附着在卵表面，胚胎发育停止，内容物外渗，卵出现畸形、破裂和溶解等现象。同时，J2 虫体被菌丝缠绕，致使其生理活性受到影响。采用灌根法接种绿色荧光标记菌株 GFP-TL16 菌液 24h 后，在党参根表皮观测到 GFP-TL16 的分生孢子，3d 后发现菌丝沿着细胞间隙纵向生长，7d 后在党参根表皮和维管束细胞中，观察到 TL16 孢子和菌丝侵入，表明 TL16 能在党参根系定殖。室内盆栽试验结果显示，接种 1.5×10^{7} cfu/mL 的 TL16 分生孢子悬浮液后，党参根结指数较未接种显著降低，防治效果达 44.33%。综上所述，本

* 基金项目：国家自然科学基金（31760507，32260654）；甘肃省现代农业产业体系（GARS-ZYC-4）
** 第一作者：张洁，硕士研究生，从事植物线虫学研究。E-mail：1799314017@qq.com
*** 通信作者：李惠霞，教授，从事植物线虫学研究。E-mail：lihx@gsau.edu.cn

研究筛选出对党参根结线虫病有较好生防潜力的菌株 TL16，并初步明确了长枝木霉 TL16 孢子悬浮液对北方根结线虫的作用方式，为开发新型、高效的微生物杀线剂和合理应用提供理论依据。

关键词：党参；北方根结线虫；生防菌；长枝木霉

一种提高大豆气生根转化效率的技术及其在大豆孢囊线虫抗性分析中的应用研究*

王　靓**，陈璐莹，刘倩男，蔡译枭，何　龙，郑经武，韩少杰***
（浙江大学农业与生物技术学院生物技术研究所，杭州　310058）

A Technique to Improve Efficiency of Soybean Adventitious Root Transformation and Its Application in Analysis of Soybean Cyst Nematode Resistance*

Wang Liang**，Chen Luying，Liu Qiannan，Cai Yixiao，He Long，Zheng Jingwu，Han Shaojie***
（*Institute of Biotechnology*，*College of Agriculture and Biotechnology*，*Zhejiang University*，*Hangzhou*　310058，*China*）

摘　要：大豆是我国重要的粮食和油料作物之一，大豆优质基因资源的挖掘对于我国大豆育种至关重要。然而，大豆转化效率低、周期长一直是制约大豆种质基因挖掘的瓶颈。虽然大豆毛根转化技术已经得到广泛应用，但需要进行组培无菌操作、难以保留地上部分，不能检验转化根系在土壤中的真实生长状态。相比之下，大豆气生根转化技术可以避免无菌组织培养的操作，且具有保留地上部分和使根系在土壤中生长的优势，但其转化效率低，筛选困难一直是制约其应用的瓶颈。为此，本研究优化了大豆气生根转化流程，并添加使用助侵染剂和 *RUBY* 报告基因来提高转化效率。通过这一优化方案，首次获得了在侵染时添加助侵染剂和 *RUBY* 报告基因显著增加大豆气生根转化效率的结果。在此基础上，本研究利用气生根对大豆孢囊线虫抗性 QTL 位点编码基因进行了过表达和功能筛选，成功鉴定出了 3 个参与大豆孢囊线虫抗性的全新基因。本研究为大豆气生根转化技术的优化提供了一个实用的方案，并且成功地应用于大豆孢囊线虫抗性的研究中，为大豆基因功能研究和抗性育种提供了新的思路和方法。

关键词：大豆种质资源挖掘；大豆气生根转化；大豆孢囊线虫

---

* 基金项目：浙江省基础公益研究计划（LTGN23C130003）

** 第一作者：王靓，硕士研究生，从事大豆孢囊线虫抗性机制和大豆基因编辑技术研究。E-mail：liangwang105@ 163. com

*** 通信作者：韩少杰，研究员，从事大豆孢囊线虫抗性机制和大豆新种质创制研究。E-mail：hanshaojie@ zju. edu. cn

一株巨大芽孢杆菌对拟禾本科根结线虫防治效果研究*

叶　姗**，周思雨，阳祝红，丁　中***
（湖南农业大学植物保护学院，长沙　410128）

Study on the Control Effect of a *Bacillus megaterium* GC-5 Against *Meloidogyne graminicola**

Ye Shan**，Zhou Siyu，Yang Zhuhong，Ding Zhong***
（*College of Plant Protection*，*Hunan Agricultural University*，*Changsha*　410128，*China*）

摘　要：当今，植物线虫病害严重影响粮食作物的生产，化学农药的过度使用对生态环境与粮食安全造成不良影响。为减少农业生产对化学农药的依赖，保护生态环境，生物防治逐渐在防治植物线虫方面发挥越来越重要的作用。本研究从水稻根际土壤分离筛选出一株生防细菌 GC-5，经生理生化和 16S rRNA 序列分析鉴定为巨大芽孢杆菌（*Bacillus megaterium*）。室内直接触杀实验显示 GC-5 菌株发酵上清液对拟禾本科根结线虫（*Meloidogyne graminicola*）具有显著毒杀活性，其发酵液上清 10 倍稀释液处理 24h、48h 后，二龄幼虫的死亡率分别为 66.32%和 91.34%。温室盆栽和田间结果表明，菌株 GC-5 发酵液处理可显著降低水稻植株根结指数，对拟禾本科根结线虫防治效果分别达到 43.1%和 45.5%。同时 GC-5 发酵液处理可显著降低水稻根内线虫的侵染数量和田间根际土壤线虫群体密度并延缓线虫龄期发育。成熟期水稻产量调查结果显示，菌株 GC-5 可促进水稻生长发育，增产率达到 32.7%。进一步研究发现菌株 GC-5 可以激活水稻根部和叶片中抗病相关防御酶活性，并提高防御关键基因 *NPR*1、*PR*1*a*、*WRKY*45、*JaMYB* 和 *AOS*2 的表达量，从而提高水稻的系统抗性抵御根结线虫的侵染。结果显示 GC-5 可以有效防控拟禾本科根结线虫，具有开发为生防制剂的潜力。

关键词：生物防治；拟禾本科根结线虫；巨大芽孢杆菌

---

* 基金项目：国家自然科学基金（32001879）
** 第一作者：叶姗，讲师，从事植物线虫和线虫生防菌资源研究。E-mail：yeshan@hunau.edu.cn
*** 通信作者：丁中，教授，从事植物线虫研究。E-mail：dingzh@hunau.net

种子处理对东北地区水稻干尖线虫病防治作用研究

杨　芳[1*]，任化蓉[1]，杨行行[2]，吴晗霖[1]，彭云良[1]，姬红丽[1**]
（[1]四川省农业科学院植物保护研究所/农业农村部西南有害生物综合治理重点实验室，
成都　610066；[2]南京农业大学，南京　210095）

The Control Effect of Nematicides Treatment of Rice Seeds and Seedlings on Rice White Tip Nematode, *Aphelenchoides besseyi*

Yang Fang[1*], Ren Huarong[1], Yang Hanghang[2], Wu Hanlin[1], Peng Yunliang[1], Ji Hongli[1**]
([1] *MOA Key Laboratory of Integrated Management of Pests on Crops in Southwest China, Institute of Plant Protection, Sichuan Academy of Agricultural Sciences, Chengdu* 610066, *China*; [2] *Nanjing Agricultural University, Nanjing* 210095, *China*)

摘　要：水稻干尖线虫（*Aphelenchoides besseyi*）是一种种传病害，侵染水稻可使叶片产生“干尖”症状或使穗部产生“小穗头”症状，粳稻或偏粳型杂交稻品种感病症状尤为明显，产量损失一般在20%~30%，严重时可超过50%。前期研究发现，温汤浸种和干热处理均能显著降低收获种子中干尖线虫的数量，但温汤浸种和干热处理需要特定的设备，难以实现大面积应用。为寻找更适用于我国东北地区大面积生产上干尖线虫病的防控方法，本研究试验了3个在水稻上登记的新药剂处理带虫种子的防病效果。试验发现在15℃条件下，12%氟啶·戊·杀螟800倍稀释液浸种5d和3d的校正防效分别为98.3%和93.7%，1 000倍稀释液的校正防效分别为97.7%和92.2%；17%杀螟·乙蒜素200倍稀释液浸种5d和3d的校正防效分别为99.5%和71.7%，300倍、400倍稀释液浸种5d和3d的校正防效分别为71.0%、64.7%和64.4%、79.2%；16%咪鲜·杀螟丹400倍、600倍和800倍稀释液校正防效在各浸种时间均显著低于12%氟啶·戊·杀螟和17%杀螟·乙蒜素，最高为800倍稀释液浸种5d，校正防效为65.1%。各药剂及处理对种子发芽率均没有显著影响。

关键词：干尖线虫；东北地区；药剂浸种

---

* 第一作者：杨芳，助理研究员，从事水稻病害研究。E-mail：yfwe928@163.com
** 通信作者：姬红丽，研究员，从事作物病害流行与防控研究。E-mail：Hongli.Ji@Jihongli.com

中国线虫学研究（第九卷）Nematology Research in China（Vol. 9）：169-171

一个松材线虫病疫点拔除的成功案例

胡先奇*，王 扬，喻盛甫
（云南农业大学植物保护学院/云南生物资源保护与利用国家重点实验室，昆明 650201）

A Successful Case of Removing a Pine Wilt Disease Epidemic Site

Hu Xianqi*，Wang Yang，Yu Shengfu
（*State Key Laboratory for Conservation and Utilization of Bio-Resources in Yunnan/College of Plant Protection*，*Yunnan Agricultural University*，*Kunming* 650201，*China*）

摘 要：2004 年 5 月，随着检疫性病害松材线虫病（Pine wilt disease）在云南省瑞丽市部分林区的思茅松（*Pinu skesiya* Royle ex Gordon）上发生，云南省近 1.0 亿亩松林（占全国松林总面积的 20.0%）、德宏州逾 50.0 万亩便处于松材线虫病直接威胁之中，云南生态安全面临严峻的挑战。针对疫情，云南省科技厅立项“危险性病害松树萎蔫病的监测、预防和防治技术（2005NG03）”，资助云南农业大学、西南林业大学、云南大学、云南出入境检验检疫局、德宏州林业局森防站等单位的 23 名研究人员组成团队，并邀请植物线虫学家、松树萎蔫病研究与防控专家冯志新教授、杨宝君研究员、潘沧桑教授、廖金玲教授、胡学兵高级工程师及昆虫学家戴乐楷（N. S. Taleker）博士为攻关研究顾问，对松材线虫病除治相关技术进行联合攻关研究。至 2011 年成功拔除了疫点，为建设我国西南边境生态安全屏障，维护我国生态安全作出了应有的贡献。

关键词：松树萎蔫病；松材线虫；松墨天牛；疫点拔除；案例

1 工作实施主要成效

（1）制定了系统、有效的松材线虫病综合治理技术方案，经过 5 年实施，于 2011 年成功拔除了疫点，使云南成为我国用最短时间成功根除松材线虫病疫情的省份，保障了云南近 1.0 亿亩松林的安全，挽回了云南全省 725.0 亿元经济损失，取得了难以估量的经济、社会和生态效益。

（2）首创利用松材线虫病疫木种植茯苓进行疫木无害化处理技术应用效果显著。为减少资源浪费，改集中烧毁等疫木处理方法，探索创建了疫木处理的新方法和新技术——使用疫木种植茯苓菌（发明专利 CN200610010944.2，CN200910094134.3），挖沟埋入表面消毒的疫木木段，接种茯苓菌，6 个月后采收茯苓，疫木腐朽殆尽。种植茯苓过程中严格检测腐朽疫木与茯苓是否残存或带有松材线虫（*Bursaphelenchus xylophilus*）和松墨天牛（*Monochamus alternatus*）。在德宏州畹町、瑞丽疫区推广，创造直接经济效益 960.0 万元，

---

* 第一作者：胡先奇，教授，从事植物线虫病害研究。E-mail：xqhoo@126.com

经济效益显著。用这种方法处理疫木安全有效、投入少，同时能部分增加林权所有者的经济收入。这种疫木处理的新方法和新技术得到了国家和省林业部门和相关专家的认可。

（3）自主研发了一批具有较高推广应用价值的实用技术，并推广应用。松墨天牛综合防治技术（发明专利 CN200910094344.2，CN200910094346.1，CN200910094345.7）在瑞丽、新平、华宁、华坪等县（市）推广，有效控制了 5.1 万亩松林松墨天牛虫害的发生危害，累计挽回经济损失 718.0 万元；研发的松材线虫快速分离（发明专利 CN200510010920.2）、快速检测技术（发明专利 CN200510048722.5）在出入境检疫和森林检疫部门广泛应用，实践证明比常规的分离和检测技术更加快捷、简便，节约检测成本，提高检出率，为早发现、早除治松材线虫病提供了新的技术保障。

（4）针对松树萎蔫病进行广泛而深入的基础研究和技术储备。在全省范围内摸清了云南松树萎蔫病的病原线虫和媒介昆虫的种类及地理分布；对松材线虫病在云南定殖和扩散进行了风险评估；发现思茅松是自然条件下易感松材线虫病的新寄主松树，对思茅松感病症状、病害早期诊断进行了系统观测及研究；揭示了云南松材线虫病传播媒介昆虫松墨天牛的生物学和生态学特性；明确了云南主要松树品种对松材线虫和拟松材线虫（*B. mucronatus*）的抗性；进一步确认了拟松材线虫的致病性及内在因子；对松材线虫和松墨天牛生物防治的应用和作用机理进行了探索。上述研究成果丰富了松树萎蔫病的研究内容，既是云南松树萎蔫病研究获得的重要结果，又是云南松材线虫病疫点拔除的理论基础和实践支撑，为松材线虫病的防治提供了新的可资借鉴的资料和技术储备。

（5）项目实施期间，申报国家专利 17 项（授权 12 项）；主持、参与制定完成行业标准及国家标准各 1 项；培训松材线虫病防治基层技术人员 1 703人次；发表研究论文 24 篇，其中 SCI 源刊 5 篇。

2 主要经验

（1）项目攻关扣准林业生产实际，与林业生产部门紧密结合，解决生产实际问题为首要目标，做到 3 个结合：短期目标与长远目标结合，以短为主，尽快完成基础研究任务，形成实用技术为生产服务；预警机制与长效机制结合，预警为重，早发现早除治；创新与借鉴结合，结合疫点所在地林业生产实际，创建新的高效疫木处理技术，及时示范应用。

（2）联合攻关，发扬团队精神与奉献精神。组建了 5 家单位 23 人参与工作的科研团队，邀请 6 位相关专家为研究顾问；5 家单位大力支持、参加人员真心合作相互补充，所有参与者尽心竭力、甘于奉献。

（3）政府部门高度重视，并适时督查。2004 年 5 月发现疑似病树，至 2005 年 10 月立项正式开展工作，其间：云南省人民政府办公厅发文成立云南省松材线虫病预防和除治协调小组，云南农业大学作为松材线虫研究专家单位参加协调小组；省政府有关领导主持召开云南省松材线虫病预防和除治协调小组第一次成员会议；省科技厅召集申请参加攻关项目单位会议，决定设立攻关项目开展监测、预防与防治研究，由云南农业大学牵头协调申报攻关项目；云南农业大学承办召开申报攻关项目筹备会，拟参与攻关项目单位参加，并邀请了植物线虫学、昆虫学领域专家莅临指导等。项目实施过程中，各上级部门、相关领导适时督查、检查，考核工作成效。

（4）拔除疫点多措并举。从疫点外围、由外向内（疫点中心）实施皆伐；疫木就地处理，作为茯苓的种植材料，包括伐桩；及时进行林分更新改造，根据疫点所在地生态特点选择优质树种，并考虑经济效益、生态效益；严防疫木流出疫点，所有砍伐病木实施就地处理，一枝一叶不可流出。

（5）病害的研究和除治工作，宜立足当前着眼未来，持之以恒，不可取得成效而自喜、疏于恒久，亦不可劳而无功而自弃、废于短识。唯锚定目标，多维思虑，争取资助，坚持不懈，方得始终。

3　参加成员

项目“危险性病害松树萎蔫病的监测、预防和防治技术（2005NG03）”的主要参加人：胡先奇、喻盛甫、王扬、李正跃、肖春、陈斌、徐正会、司徒英贤、陈玉惠、陈太安、张克勤、乔敏、莫明和、周薇、叶辉、蒋小龙、白松、陈建涛、杜宇、周平阳、解芳、弄扎、李祖钦。

昆虫病原线虫外代谢物对觅食行为的调控研究*

张　忱**，唐凡希，汤华涛，李意璇，侯有明，吴升晏***
（福建农林大学闽台作物有害生物生态防控国家重点实验室/福建农林大学植物保护学院，福建　350002）

Study on the Regulatory Mechanism of Exometabolites in Entomopathogenic Nematode Foraging Behavior*

Zhang Chen**，Tang Fanxi，Tang Huatao，Li Yixuan，
Hou Youming，Wu Shengyan***
（*State Key Laboratory of Ecological Pest Control for Fujian and Taiwan Crops*，*Fujian Agriculture and Forestry University/College of Plant Protection*，*Fujian Agriculture and Forestry University*，*Fujian* 350002　*China*）

摘　要：昆虫病原线虫是一种在农林牧草、花卉和卫生等领域被广泛使用的害虫防治剂。为提高其在田间应用的效果，研究昆虫病原线虫的觅食行为及其调控机制十分必要。前期研究结果发现，混合昆虫病原线虫物种能提升其自身觅食行为。另外，线虫所分泌的外代谢物中含有一系列传递信号的重要物质，如蛔苷信息素，它们能诱导线虫的发育及社会化行为。而且这些化学物质也可能在觅食行为中起到关键作用。但是目前尚不清楚具体是哪些物质，以及其背后影响觅食行为的分子机制。本研究利用 2 种异小杆线虫属（*Heterorhabditis*）和 5 种斯氏线虫属（*Steinernema*）线虫的外代谢物，对小卷蛾斯氏线虫（*S. carpocapsae*）进行了刺激。结果发现，在斯氏线虫（*S. riobrave*）和异小杆线虫（*H. bacteriophora*）外代谢物的刺激下，*S. carpocapsae* 的寻找寄主效率提高了两倍。同时，*S. riobrave* 的外代谢物也显著提高了 *S. carpocapsae* 的扩散能力（20%）。通过靶向代谢组学，对 7 种昆虫病原线虫外代谢物中的 8 种蛔苷信息素进行定性和定量分析，结果发现 *S. riobrave* 和 *S. carpocapsae* 外代谢物中的 Ascr#11、Ascr#12、Ascr#14、Ascr#16 和 Ascr#18 的含量存在差异；而 *H. bacteriophora* 和 *S. carpocapsae* 外代谢物中的 Ascr#9、Ascr#12、Ascr#14 和 Ascr#16 的含量存在差异。根据该结果进行生物学测定，验证了不同浓度梯度的单一蛔苷信息素对 *S. carpocapsae* 觅食行为的影响，结果发现单一蛔苷信息素对 *S. carpocapsae* 的觅食行为无明显的作用。这些结果表明，不同种类的昆虫病原线虫能通过外代谢物进行交流，从而影响它们的扩散效率和觅食能力。但是单一蛔苷信息素没有明显效果，这可能是由于自然环境中蛔苷信息素要在混合的条件下才能发挥作用。

关键词：昆虫病原线虫；外代谢物；觅食行为；靶向代谢

---

* 基金项目：国家自然科学基金青年基金项目（32202374）；福建省科技厅对外合作项目（202110006）
** 第一作者：张忱，硕士研究生，从事昆虫研究。E-mail：zc18333789029@163. com
*** 通信作者：吴升晏，副教授，从事昆虫病原线虫觅食行为调控机制、植物-昆虫-昆虫病原线虫三营养级互作和入侵害虫生物防治研究。E-mail：sywu531@163. com

中国线虫学研究（第九卷）Nematology Research in China（Vol. 9）：173－174

昆虫病原线虫分泌蛋白对寄主激活的响应及致病功能分析*

常豆豆[1,2]**，王从丽[1]，黄铭慧[1]，姜　野[1]，谢倚帆[1,2]，秦瑞峰[1,2]，
蒋　丹[1,2]，赵亚男[1]，韦柳利[1]，李春杰[1]***
（[1]中国科学院东北地理与农业生态研究所，中国科学院大豆分子设计育种重点实验室，哈尔滨　150081；
[2]中国科学院大学，北京　100049）

Response of Secreted Proteins of Entomopathogenic Nematodes to Host Activation and Pathogenicity Analysis*

Chang Doudou[1,2]**, Wang Congli[1], Huang Minghui[1], Jiang Ye[1], Xie Yifan[1,2],
Qin Ruifeng[1,2], Jiang Dan[1,2], Zhao Ya'nan[1], Wei Liuli[1], Li Chunjie[1]***
（[1]*Key Laboratory of Soybean Molecular Design Breeding, Northeast Institute of Geography and Agroecology, Chinese Academy of Sciences, Harbin*　150081, *China*;
[2]*University of Chinese Academy of Sciences, Beijing*　100049, *China*）

摘　要：昆虫病原线虫（EPN）自然存在于土壤中，是一种专性寄生昆虫的线虫，已成功应用于害虫的生物防治。早期研究表明线虫体内的共生细菌是主要的致病因子，近年来研究发现线虫自身也能通过分泌毒性蛋白杀死寄主昆虫。但是线虫分泌蛋白的鉴定主要集中在斯氏属（*Steinernema*）线虫，对于寒区 EPN 还未见报道；并且分泌蛋白的致病性研究都是基于常温条件，寒区 EPN 分泌的蛋白是否对低温条件下害虫有很强的致病力，发挥致病作用的关键致病蛋白及组分均未知。因此，本研究采用从黑龙江省分离的昆虫病原线虫 *S. feltiae*-IGA（Sf-IGA），通过大蜡螟（*Galleria mellonella*）匀浆分别激活侵染期线虫 6h、12h、18h 作为处理，未激活（0h）作为对照，研究分泌蛋白在不同温度下对寄主昆虫的致病性，并对分泌蛋白进行鉴定及功能解析，主要结果如下。

（1）线虫 Sf-IGA 暴露在寄主匀浆 0h、6h、12h 和 18h 后显微镜下观察，通过咽球扩张判断线虫从休眠期恢复取食状态，即被激活，其激活率随着匀浆处理时间的延长而升高，同时对分泌物中的蛋白质初步分析表明，分泌蛋白总体上具有时间效应，即随着激活时间延长低分子蛋白量下降。

（2）将线虫体内共生细菌（*Xenorhabdus bovienii*）及线虫被激活前后分泌的蛋白粗提液

---

* 基金项目：中国科学院战略性先导科技专项（A 类）（XDA28070305）；国家自然科学基金（31601688）；吉林省与中国科学院科技合作高新技术产业化专项资金（2021STHZ0031）

** 第一作者：常豆豆，硕士研究生，从事农业害虫生物防治研究。E-mail：changdoudou@ iga. ac. cn

*** 通信作者：李春杰，研究员，从事农作物根部病虫害生物生态控制机理及应用研究。E-mail：lichunjie@ iga. ac. cn

通过血腔注射法注射到寄主昆虫大蜡螟（*Galleria mellonella*）体内，结果表明线虫 Sf-IGA 体内共生细菌在低温条件下（15℃）大蜡螟的致病力显著低于常温（25℃）；而在低温 15℃和常温 25℃下，激活 6h 和激活 12h 线虫分泌的蛋白表现出较强的致病力；随着激活时间的延长分泌蛋白的毒性呈现下降的趋势，激活 6h 和 12h 的分泌蛋白对寄主血细胞的裂解率显著高于激活 18h 和激活 0h 的分泌蛋白。可以看出，除了 EPN 体内共生细菌是对寄主的致病因子外，线虫分泌的蛋白也是其致病因子之一，但环境温度及激活时间都会影响分泌蛋白的致病性及免疫活性。

（3）运用 4D Label-free 定量蛋白质组学技术对分泌蛋白粗提液进行定性定量分析。共鉴定到 869 个蛋白，处理与对照每两组对比分析 CK：JH6h、CK：JH12h、CK：JH18h 共有 51（81）、75（76）、95（62）显著上调（下调）差异表达蛋白，说明随着激活时间的延长上调表达蛋白的数量升高，而下调表达蛋白的数量呈下降趋势。

（4）对差异表达蛋白进行 GO、KEGG 生物信息学分析，结果表明线虫随着激活时间的延长，先后出现参与机体的免疫调节、寿命延长和生长发育代谢过程的差异分泌蛋白，可见寄生的不同阶段线虫分泌的差异蛋白与侵染期线虫的生命活动密切相关。在寄生初期 0~12h，侵染期线虫主要通过分泌水解酶、肽酶、溶菌酶、凝集素等发挥应激反应、免疫等策略以完成侵染寄主和寄生的初级阶段，推测线虫在激活早期抑制寄主防御反应的同时也增强了线虫的信号传导以适应寄主环境；当被激活 JH18h 后通过分泌肽酶 S8 家族蛋白、脂肪酸视黄醇结合蛋白、假定蛋白、天冬氨酸型内肽酶、5-磷酸核糖异构酶、转醛醇酶等，还有血影蛋白、钙调蛋白等丰度较低的蛋白，获取寄主体内营养，同时调节自身细胞形成、开始发育、繁殖和代谢以完成完全寄生和进入生活史阶段。

总之，寒区线虫 Sf-IGA 分泌蛋白对低温（15℃）条件下的大蜡螟具有致病性，能削弱寄主免疫反应，与线虫体内共生细菌协同作用使寄主死亡，是寄主的主要致病因子之一。线虫寄生初期的分泌蛋白在调控或干扰寄主免疫应答中起着重要作用，侵入寄主昆虫后分泌蛋白对线虫开始新的发育周期起着积极作用。该研究结果对进一步筛选斯氏属夜蛾线虫 Sf-IGA 发挥致病作用的关键蛋白及组分提供非常有价值的信息，为寒区昆虫病原线虫这一生防资源高效防治低温下害虫提供理论依据。

关键词：昆虫病原线虫；分泌蛋白；蛋白组学分析；致病功能

新型1,2,4-噁二唑类杀线虫剂的设计与合成*

张　琪**，刘　丹，甘秀海***
（贵州大学绿色农药全国重点实验室/绿色农药与农业生物工程教育部重点实验室，贵阳　550025）

Design and Synthesis of New 1,2,4-Oxadiazole Nematocides*

Zhang Qi**, Liu Dan, Gan Xiuhai***
(*National Key Laboratory of Green Pesticide/Key Laboratory of Green Pesticide and Agricultural Bioengineering, Ministry of Education, Guizhou University, Guiyang* 550025, *China*)

摘　要：植物寄生线虫因种类繁多、不易被察觉且寄生在植物体内，导致植物线虫病害的防控极其困难，这严重影响着农业生产。当前防治线虫的商品药剂主要为有机磷和氨基甲酸酯类高毒杀线虫剂，长期大量的施用导致抗性严重，且对环境产生严重的污染，多数高毒药剂已被禁用或限用。因此研发高效、低毒、环境友好的杀线虫剂，仍然是线虫防治的关键科学问题。

本研究以 Tioxazafen 为先导结构，通过引入柔性基团的结构改造策略，设计并合成了含酯、酰胺、硫酯单元的新颖1,2,4-噁二唑类衍生物，并以南方根结线虫（*Meloidogyne incognita*）、水稻干尖线虫（*Aphelenchoides besseyi*）、松材线虫（*Bursaphelenchus xylophilus*）为测试对象，综合评价了衍生物的杀线虫活性。结果表明化合物 f1 对南方根结线虫、水稻干尖线虫、松材线虫均表现出了较好的杀线虫活性，LC_{50}值分别为 34.7μg/mL、19.0μg/mL 和 29.1μg/mL，优于先导化合物 Tioxazafen（LC_{50}值分别为 >200μg/mL、148.5μg/mL 和 >500μg/mL）。活体试验发现该化合物在 100μg/mL 的剂量下对水稻干尖线虫的抑制率为 55.1%，优于阳性对照 Tioxazafen（47.1%）和噻唑膦（51.9%）。

初步的作用机制发现化合物 f1 能一定程度上影响水稻干尖线虫的繁殖力并导致水稻干尖线虫会产生更显著的运动行为缺陷；同时，导致水稻干尖线虫体液渗透并抑制其呼吸作用；此外，化合物 f1 对乙酰胆碱酯酶（AchE）具有明显的抑制作用，揭示其潜在的靶标可能是 AchE。

关键词：植物寄生线虫；杀线虫剂；1,2,4-噁二唑类衍生物；乙酰胆碱酯酶

---

* 基金项目：国家自然科学基金（32060622）
** 第一作者：张琪，硕士研究生，从事植物线虫病害研究。E-mail：zhangqiqi0406@163.com
*** 通信作者：甘秀海，教授，从事植物线虫病害研究。E-mail：gxh200719@163.com

一种新病害——空心菜根腐病的病原鉴定

秦 玲[1*]，赵湘媛[1]，李志萌[2]，张洁[3]，田朝辉[2]，李洪连[1]，王 珂[1]，李 宇[1**]
（[1] 河南农业大学植物保护学院，郑州 450046；[2] 郑州市蔬菜研究所，郑州 450015；
[3]河南省农业科学院植物保护研究所，郑州 450008）

A New Disease of Water Spinach，Root Rot Caused by the Root-lesion Nematode

Qin Ling[1*]，Zhao Xiangyuan[1]，Li Zhimeng[2]，Zhang Jie[3]，
Tian Chaohui[2]，Li Honglian[1]，Wang Ke[1]，Li Yu[1**]
（[1]*College of Plant Protection*，*Henan Agricultural University*，*Zhengzhou* 450046，*China*；
[2]*Zhengzhou Vegetable Research Institute*，*Zhengzhou* 450015，*China*；[3]*Institute of Plant Protection*，
Henan Academy of Agricultural Sciences，*Zhengzhou* 450008，*China*）

摘 要：短体线虫（*Pratylenchus* spp.）又称根腐线虫，是一类发生普遍、寄主众多的迁移性内寄生线虫，可侵染危害多种粮食作物和经济作物，给农业生产造成了严重的经济损失。2021 年，笔者课题组在河北省保定市的空心菜种植地，发现部分空心菜出现黄化、长势弱，根系腐烂等症状。采用改良的贝尔曼漏斗法从采集样品中分离到一种短体线虫，通过形态和分子生物学鉴定明确了该短体线虫的种类。通过胡萝卜愈伤组织纯化、培养了该短体线虫种群，并测定了其对空心菜的寄生性和致病力。

研究结果表明，空心菜根际分离的短体线虫与玉米短体线虫（*P. zeae*）的形态特征值较一致，rDNA-ITS 区和 rDNA-28S D2-D3 区序列与 NCBI 数据库中的玉米短体线虫序列具有高度相似性。系统进化树显示，待鉴定短体线虫与其他玉米短体线虫种群聚在同一高度支持的分支。研究还发现该种群在胡萝卜愈伤组织上能够大量繁殖，接种 30 条雌虫在 27.5℃下培养 60d 后的繁殖率（Rf）为 6 993。盆栽接种该种群 60d，受侵染空心菜植株矮小、叶片黄化，地上部和根鲜重显著降低，根部出现褐色病斑甚至坏死腐烂，病根染色及组织切片发现接种空心菜的根系含有大量的玉米短体线虫及卵，该线虫在空心菜根际的 Rf 达到 6.1。

本研究明确了在河北省保定市发现的一种新病害——空心菜根腐病的病原为玉米短体线虫，空心菜是玉米短体线虫的适合寄主（Rf>1），该线虫种群对空心菜具有很强的致病性。本研究首次描述了采自空心菜根际的玉米短体线虫的形态特征，并证实该种群能够通过胡萝卜愈伤组织进行人工培养。本研究结果为空心菜短体线虫的检测、鉴定和其他相关的研究奠定了基础。

关键词：空心菜；新病害；病原鉴定；玉米短体线虫；寄生性；致病性

---

* 第一作者：秦玲，硕士研究生，从事植物线虫学研究。E-mail：qinling6912@163.com
** 通信作者：李宇，副教授，主要从事植物线虫学研究。E-mail：liyuzhibao@henau.edu.cn

新型不饱和酮类化合物的杀腐烂茎线虫活性、构效关系和作用机制*

周搏航**，王佳哲，张　锋，李英梅***
（陕西省生物农业研究所/陕西省植物线虫学重点实验室，西安　710043）

Nematocidal Activity Against Ditylenchus Destructor, Structure-activity Relationship and Action Mechanism of Novel Unsaturated Ketone Compounds*

Zhou Bohang**, Wang Jiazhe, Zhang Feng, Li Yingmei***
(*Bio-Agriculture Institute of Shaanxi, Shaanxi Key Laboratory of Plant Nematology, Xi' an* 710043, *China*)

摘　要：植物寄生线虫给世界范围内的农业造成了巨大的经济损失，严重威胁着现代农业的可持续发展。腐烂茎线虫（*Ditylenchus destructor*）又称马铃薯茎线虫，是一类多食性线虫，已报道的寄主超过120种，包括马铃薯、甘薯、洋葱、大蒜、当归、甜菜等，是国际公认的检疫性线虫。目前针对腐烂茎线虫的防治主要依靠化合物杀线剂，常见的杀线剂主要为有机磷类和氨基甲酸酯类，但是随着长时间单一且频繁使用这类杀线剂使线虫更容易产生抗药性，从而增加用药量。为了满足巨大的市场需求，减缓抗性的增长，需要新的杀线虫剂进入市场。目前，化学杀线剂开发的主要途径如下：①筛选商业农药的杀线虫活性（如氟普兰）；②中间化或先导化合物衍生化；③基于靶点的药物设计；④计算机辅助药物设计（如噻沙扎芬）。其中，中间体或先导化合物衍生化是公认的最快捷和最效率的药物开发途径。不饱和酮类化合物具有良好的反应活性，可以作为重要的前体化合物和有机合成中间体，广泛应用于药物合成、香料调味品生产中。近年来，科研工作者们致力于寻找简便高效的合成手段，并且对该类化合物的结构进行不断修饰，以期能够找到低毒、绿色的植物源药物。本研究通过设计合成一系列酮类化合物，制备了50个以上的目标化合物并完成其结构鉴定，其中7个目标化合物对腐烂茎线虫灭杀活性 EC_{50} 在20～11.4μg/mL，4号化合物在50mg/kg浓度下24h对腐烂茎线虫致死率达到74%，其72小时 EC_{50} 值为11.4μg/mL。通过对实验结果的分析，揭示目标化合物对马铃薯腐烂茎线虫及其虫卵的离体杀灭活性，阐明其构效关系。随后笔者针对施药虫体进行了活性氧（reactive oxygen species，ROS）和线粒体膜电位检测的杀虫机制进行初步探究，为目标化合物的进一步结构优化和新农药研发提供理论基础。

关键词：腐烂茎线虫；不饱和酮；构效关系

* 基金项目：陕西省科学院种子基金研究项目（2022K-19）
** 第一作者：周搏航，助理研究员，从事植物线虫致病机理研究。E-mail：zhoubh@ xab. ac. cn
*** 通信作者：李英梅，研究员，从事植物线虫综合防控技术研究。E-mail：liym@ xab. ac. cn

大蒜腐烂茎线虫对大蒜的损害阈值

成泽珺[1]，Toyota Koki[2]，Aoyama Rie[3]
（[1] 河南科技大学园艺与植物保护学院，洛阳　471003；
[2]东京农工大学生物应用与系统工程研究院，日本东京　184-8588；
[3]青森县产业技术研究中心蔬菜研究所，日本青森　033-0071）

Study on Damage Threshold of *Ditylenchus destructor* to Garlic Diseases

Cheng Zejun[1]，Toyota Koki[2]，Aoyama Rie[3]
（[1]*College of Horticulture and Plant Protection*，*Henan University of Science and Technology*，*Luoyang*　471003，*China*；[2]*Graduate School of Bio-Applications and Systems Engineering*，*Tokyo University of Agriculture and Technology*，*Tokyo*，*Japan*；[3]*Vegetable Research Institute*，*Aomori Prefectural Industrial Technology Research Center*，*Aomori*，*Japan*）

摘　要：大蒜腐烂茎线虫一直以来都是日本大蒜生产的主要病原，对其造成了严重的危害。本研究旨在探讨种植初期土壤、蒜种和蒜根中的大蒜腐烂茎线虫虫口密度在收获时对大蒜腐烂程度的影响，并制定与该线虫相关的损失阈值。

研究结果表明，大蒜腐烂茎线虫为 *Ditylenchus destructor*，并设计了特异性的实时 PCR 引物（Ddf 和 Ddr）。这对引物与 11 个不同国家的 *D. destructor* 序列中的 10 个，以及青森株 100%匹配，而与近缘属 *D. africanus* 和 *D. askenasyi* 序列则有 7 个和 16 个碱基不匹配。笔者采用 Real-time PCR 法制作了定量曲线，接种线虫数的对数转化数（x）与相应的 Ct 值（y）呈高度相关性（土壤：$y=-1.122\ 1x+35.225$，$R^2=0.997\ 3$；大蒜外皮：$y=-1.145x+35.295$，$R^2=0.988\ 3$）。

根据以上方法，研究探讨了土壤、蒜根、蒜皮、蒜种中虫口密度与大蒜腐烂程度之间的关系：①在 2016 年种植初期土壤和收获时蒜皮的虫口密度呈现高度正相关性，而在 2017 年却没有；②收获时，大蒜内虫口密度的损失阈值为蒜根 80 条/0.05g、蒜皮 300 条/0.05g；③蒜种皮中的虫口密度低于 285 条/0.05g 时，贮藏后的损害程度为 0%。这些结果表明，蒜根和外皮中的虫口密度可以作为预测大蒜腐败的一个良好指标。该技术为大蒜生产提供了一种有效的土壤线虫监测和防控手段，有望为大蒜病害防治提供理论支持和实际应用指导。

关键词：引物设计；定量 PCR；大蒜腐烂茎线虫；损失阈值

淡紫紫孢菌活性孢子储存技术研究

张雯欣*，杨利洁，练人杰，肖炎农，肖雪琼，王高峰**
（华中农业大学农业微生物资源发掘与利用全国重点实验室，武汉　430070）

摘　要：淡紫紫孢菌（*Purpureocillium lilacinum*）为一种商业化应用的植物线虫生防真菌。市场上销售的淡紫紫孢菌菌剂主要为固体发酵后获得的活性孢子，需施用大量的活性孢子以获得田间良好的线虫防效。淡紫紫孢菌菌剂主要采用干燥处理后进行常温常压储存运输。然而，实验室前期调查研究发现，以这种方式储存的淡紫紫孢菌菌剂货架期短，严重制约了淡紫紫孢菌的推广应用。为延长淡紫紫孢菌菌剂的货架期，本研究对淡紫紫孢菌活性孢子的储存技术进行了优化。

本研究评价了冷冻干燥、真空封存、添加干燥剂、添加活性炭及其组合处理对淡紫紫孢菌活性孢子储存期的影响。首先，将固体培养基（硅藻土、玉米粉和水）分装至 200 mL 广口瓶中，进行高压湿热灭菌。在接种适量淡紫紫孢菌孢子悬液后，于 28℃和 20%湿度条件下暗培养 8d。将获得的淡紫紫孢菌固体发酵物分别进行以下处理后室温保存：①冷冻干燥；②冷冻干燥 & 真空封存；③冷冻干燥 & 真空封存 & 干燥剂；④冷冻干燥 & 真空封存 & 活性炭；⑤冷冻干燥 & 真空封存 & 干燥剂 & 活性炭。以未冷冻干燥处理的淡紫紫孢菌固体发酵物为对照。分别在固体发酵物储存前后不同时间点测定各处理组淡紫紫孢菌活孢子数。试验结果显示，对照组、冷冻干燥组、冷冻干燥 & 真空封存组分别在储存 1 个、2 个、5 个月后未检测到淡紫紫孢菌活孢子。在冷冻干燥 & 真空封存组、冷冻干燥 & 真空封存 & 活性炭组、冷冻干燥 & 真空封存 & 干燥剂 & 活性炭组中活孢子数分别在储存 3 个、4 个、3 个月后显著减少。而以冷冻干燥 & 真空封存 & 干燥剂处理的活孢子存储时间最长，在保存 5 个月后活孢子数才开始显著减少，从最初的 3.04×10^5 个/g 减少至 0.68×10^5 个/g，在保存 11 个月后活性孢子数减少至 1.50×10^3 个/g。以上试验结果表明，在室温条件下，与未冷冻干燥处理的淡紫紫孢菌固体发酵物相比，淡紫紫孢菌固体发酵物冷冻干燥处理和冷冻干燥结合真空封存、添加干燥剂、添加活性炭的不同组合处理均可显著延长淡紫紫孢菌活孢子储存时间。其中，以冷冻干燥结合真空封存和添加干燥剂的储存方式最佳。

关键词：淡紫紫孢菌菌剂；货架期；冷冻干燥；真空封存；干燥剂；活性炭

---

* 第一作者：张雯欣，本科生，从事淡紫紫孢菌研究。E-mail：zhangwenxinzwx@mail. hzau. edu. cn
** 通信作者：王高峰，副教授，从事植物线虫致病机理及绿色防控技术研究。E-mail：jksgo@mail. hzau. edu. cn

拟禾本科根结线虫效应蛋白 MgCBP1 在线虫寄生中的作用

黄春晖*，黄秋玲，曹雨晴，林柏荣，廖金铃，卓　侃**
（华南农业大学植物保护学院，广州　510642）

Roles of the *Meloidogyne graminicola* Effector MgCBP1 during Nematode Parasitism

Huang Chunhui*，Huang Qiuling，Cao Yuqing，Lin Borong，Liao Jinling，Zhuo Kan**
（*College of Plant Protection*，*South China Agricultural University*，*Guangzhou*　510624，*China*）

摘　要： 由拟禾本科根结线虫（*Meloidogyne graminicola*，Mg）侵染水稻（*Oryza sativa*）导致的水稻根结线虫病是制约水稻产量的重要病害之一。Mg 在寄生过程中通过口针将食道腺分泌的效应蛋白注射至植物细胞中，促进线虫寄生。研究其效应蛋白不仅可了解线虫的寄生机理，也可为线虫防治提供理论依据。

本研究通过同源克隆获得 Mg 纤维素绑定蛋白基因 *MgCBP*1，并通过基因表达、纤维素绑定、纤维素水解及 RNAi 实验等研究 MgCBP1 在线虫寄生中的作用，结果如下：序列分析表明 MgCBP1 含分泌信号肽及纤维素绑定结构域，但不含跨膜结构域；原位杂交表明该基因在 Mg 侵染前二龄幼虫亚腹食道腺特异表达；qRT-PCR 检测发现 *MgCBP*1 在侵染前二龄幼虫中表达量最高，且在 Mg 寄生早期阶段还维持着较高表达水平；MgCBP1 具纤维素绑定活性，且该绑定活性与纤维素绑定结构域有关。将 MgCBP1 与纤维素酶 C0057 混合水解滤纸每分钟可产生 3. 5 ng 还原糖，显著高于纤维素酶 C0057 单独水解滤纸每分钟产生的 3 ng 还原糖，而 MgCBP1 本身并无纤维素水解活性。进一步通过克隆获得 Mg 的β-1，4 内切葡聚糖酶基因 *MgENG*2，MgENG2 具强的纤维素水解活性。纤维素水解实验表明 MgCBP1 可提高 MgENG2 的纤维素水解活性，并且该活性与纤维素绑定结构域相关。离体 RNAi 实验表明浸泡了 *MgCBP*1 双链 RNA 的线虫与浸泡了 *GFP* 双链 RNA 及 ddH_2O 的线虫相比，侵染水稻的总虫数分别下降 76. 3%和 74. 7%。

综上，MgCBP1 可能在线虫寄生早期通过绑定植物细胞壁纤维素，促进 MgENG2 对纤维素的水解，从而提高线虫的寄生能力。

关键词： 拟禾本科根结线虫；效应蛋白；纤维素绑定蛋白；纤维素水解；β-1,4 内切葡聚糖酶

---

* 第一作者：黄春晖，博士研究生，从事植物线虫学研究。E-mail：huang_ chunhui@ stu. scau. edu. cn
** 通信作者：卓侃，教授，从事植物线虫学研究。E-mail：zhuokan@ scau. edu. cn

OsBet v1 蛋白通过木聚糖酶抑制蛋白介导水稻对拟禾本科根结线虫的抗性

王　婧*，方天懿，曹雨晴，卓　侃，林柏荣**
（华南农业大学植物线虫研究室，广州　510642）

OsBet v1 Mediates Rice Resistance to *Meloidogyne graminicola* through Xylanase Inhibitor Protein

Wang Jing*, Fang Tianyi, Cao Yuqing, Zhuo Kan, Lin Borong**
(*College of Plant Protection*, *South China Agricultural University*, *Guangzhou*　510624, *China*)

摘　要：拟禾本科根结线虫（*Meloidogyne graminicola*）是水稻上的重要病原物，严重影响水稻产量和质量。课题组前期研究表明 *OsBet v*1 过表达水稻（OsBet v1^{OE}）能提高水稻对根结线虫抗性，*OsBet v*1 突变水稻（OsBet v1^{del}）能降低水稻对根结线虫的抗性。

本研究通过转录组测序对 OsBet v1^{OE}、OsBet v1^{del}以及野生型水稻（WT）的差异表达基因进行分析，并采用 RT-qPCR、亚细胞定位、RNAi 和酶活测定等方法发现 *OsBet v*1 可能通过促进水稻木聚糖酶抑制蛋白 1 基因（*OsXip*1）的表达提高植株对拟禾本科根结线虫的抗性。具体结果如下：转录组分析发现与 WT 相比，OsBet v1^{OE}和 OsBet v1^{del}水稻的差异基因均主要富集在糖基水解酶等通路，其中木聚糖酶抑制蛋白基因 *OsXip*1 在 OsBet v1^{OE}水稻中上调表达且在 OsBet v1^{del}水稻中下调表达。在 OsBet v1^{OE}水稻中拟禾本科根结线虫侵染后的植株木聚糖酶抑制蛋白 1 基因 *OsXip*1 的表达量与未侵染的植株相比提高 10～60 倍，而在 OsBet v1^{del}水稻和野生型水稻中线虫侵染不能显著改变 *OsXip*1 的表达。酶活测定表明 OsXip1 能降低拟禾本科根结线虫效应蛋白木聚糖酶 MgXyl1、MgXyl2 及 MgXyl3 对木聚糖的降解能力。

综上，水稻 OsBet v1 蛋白可能会促进木聚糖酶抑制蛋白的表达，抑制线虫木聚糖酶活性，降低线虫降解水稻细胞壁的能力，从而提高水稻对拟禾本科根结线虫的抗性。

关键词：拟禾本科根结线虫；木聚糖酶；水稻；木聚糖酶抑制蛋白；*OsBet v*1

---

* 第一作者：王婧，硕士研究生，从事植物线虫学研究。E-mail：1486823609@qq.com
** 通信作者：林柏荣，副教授，从事植物线虫学研究。E-mail：boronglin@scau.edu.cn

香芹酚防治象耳豆根结线虫及缓解噻唑膦药害的作用研究

龙昌文*，陈又琳，卓　侃，廖金铃，林柏荣**
（华南农业大学植物保护学院，广州　510642）

Study of Carvavol on Controlling *Meloidogyne enterolobii* and Alleviating the Phytotoxicity of Fosthiazate

Long Changwen*，Chen Youlin，Zhuo Kan，Liao Jinling，Lin Borong**
（*College of Plant Protection*，*South China Agricultural University*，*Guangzhou*　510624，*China*）

摘　要：噻唑膦是一种常用于防治植物线虫病害的药剂，对根结线虫（*Meloidogyne*）有较好的防治效果，但若使用不当，易对植物产生药害。香芹酚作为一种植物源的萜类化合物，对根结线虫等植物病原物具有一定的防控效果。

本研究通过浸虫法和盆栽法，研究香芹酚和噻唑膦复配后对根结线虫的防治效果，发现两者复配可提高防效，且香芹酚可减轻噻唑膦引起的药害，主要研究结果如下：浸虫法显示，单独使用噻唑膦或香芹酚处理象耳豆根结线虫（*Meloidogyne enterolobii*）二龄幼虫48h后，LC_{50}分别为62.7mg/L和1 695.0mg/L；两种药剂复配处理象耳豆根结线虫二龄幼虫48h后，LC_{50}为38.9mg/L，显著低于噻唑膦或香芹酚的LC_{50}，共毒系数为453。盆栽法显示，单独使用1 050g/亩的噻唑膦或15g/亩的香芹酚60d后，辣椒（*Capsicum annuum*）上的根结指数平均值分别为2.50和4.38，番茄（*Solanum lycopersicum*）上的根结指数平均值分别为0.85和1.88；使用香芹酚和噻唑膦复配药剂60d后，辣椒和番茄上的根结指数平均值分别为0.38和0.15，显著低于两种单剂处理后的根结指数；此外，与清水处理植株相比，单独使用1 050g/亩噻唑膦处理植株显著降低了辣椒和番茄的地上部重量和根重；单独使用15g/亩的香芹酚或两种药剂的复配剂处理植株，辣椒和番茄的地上部重量和根重与清水处理组相比均无显著差异，表明香芹酚能减轻噻唑膦引起的药害。

综上所述，香芹酚不仅能增强噻唑膦对象耳豆根结线虫的防效，还能减轻噻唑膦对辣椒和番茄的药害。

关键词：香芹酚；噻唑膦；复配；药害；象耳豆根结线虫

---

* 第一作者：龙昌文，硕士研究生，从事植物线虫病害研究。E-mail：changwenlong111@163.com
** 通信作者：林柏荣，副教授，从事植物线虫学研究。E-mail：boronglin@scau.edu.cn

中国线虫学研究（第九卷）Nematology Research in China（Vol. 9）：183-184

Challenges of Research in Plant Nematology in Colombia*

Lizzete Dayana Romero Moya**, Peng Deliang***
(*State Key Laboratory for Biology of Plant Diseases and Insect Pests. Institute of Plant Protection, Chinese Academy of Agricultural Sciences, Beijing, P. R.* 100193, *China*)

Abstract: Colombia is located on the Northwestern corner of South America between the Pacific and Atlantic Oceans. It is the 26th largest nation in the world, comprises 1.14 million km^2, 5 percent of which are under cultivation (pastures excluded). The country is entirely tropical, altitude ranges between 0 to 5.700 meters above sea level which combined with other climatic factors allows/provides all climate conditions for the production of a wide range of both tropical and temperate-zone crops and several harvests a year are obtained.

Colombia is predominantly an agrarian country and apart from few exceptions it is self-sufficient. Coffee, cut flowers and banana productions range between second to third largest on the world (FAO, 2022). Other important crops yielded are plantain, sugarcane, rice, palm oil, maize and potatoes. Agriculture production is affected by many factors. Although plant-parasitic nematodes are often not as important as some other biotic and non-biotic constrains on crop production in the tropics, they can nevertheless cause extensive damage and substantial yield losses (Jaraba, *et al.*, 2022; Hoyos and Moya; 2010; Múnera, 2000).

Many investigations have been carried out on the subject throughout the world. However, nematology is still an incipient science in Colombia. Colombian nematological research has been focused mainly on root-knot nematodes or the genus *Meloidogyne*. About 50% of the Colombian nematode literature has been produced by means of thesis research at level of bachelor. Despite this, some studies have been conducted on the prevalence of nematodes in different crops, such as coffee, banana, potato, and tomato (Leguizamon 2005; Riascos *et al.*, 2019; Vallejo *et al.*, 2021; Padilla *et al.*, 2022). These studies have helped to identify the species present in the country and the damage they cause to crops.

Topics most frequently dealt with are: identification to genus level of nematodes associated with different crops mainly banana, plantain, potatoes, fruit crops and cut flowers (Guzman *et al.*,

---

* 基金项目：国家自然科学基金（32072398）；政府购买服务项目（15190025）；中国农业科学院科技创新工程（ASTIP-02-IPP-15）

** 第一作者：Lizzete Dayana Romero Moya，博士研究生。E-mail：2020y90100067@caas.cn

*** 通信作者：彭德良，研究员，从事植物线虫致病机制和线虫病害综合控制技术研究。E-mail：pengdeliang@caas.cn

2005; Mùnera *et al.*, 2008; Ortiz *et al.*, 2012). There is also research being conducted on the use of biological control agents (Betancourth *et al.*, 2012), chemical control (Cardona *et al.*, 2014) and resistant plant varieties to manage nematode populations (Hurtado *et al.*, 2022; Polanco-Puerta *et al.*, 2018), genetic diversity (Vallejo *et al.*, 2021; Riascos *et al.*, 2019). Some success has been achieved with the use of biocontrol agents such as fungi and bacteria that parasitize nematodes or produce compounds toxic to them (Ortiz *et al.*, 2015), yield losses, interaction with other organisms, and entomopathogenic nematodes has received considerable attention in recent years (Tabima *et al.*, 2023; Melo *et al.*, 2007; Saenz *et al.*, 2020).

However, more investment and research are needed to develop effective and sustainable strategies for nematode management in Colombia. This will require collaboration between researchers, farmers, and policymakers to address the challenges posed by nematodes and protect the country's agricultural production. This neglect is partly due to the complexity of nematode management and the difficulty in achieving a scientific understanding of how smallholder cropping systems function. Additionally, the cost of acquiring and using technology to manage nematodes is often prohibitive for smallholder farmers. To address the challenge of nematode management in smallholder systems, there is a need to develop affordable and sustainable techniques that are adapted to the realities of smallholder farming. Collaborative research between scientists and smallholder farmers is essential to develop and promote innovative solutions to the problem of nematode infestation. This approach should involve inter-disciplinary collaboration to design and evaluate various management options, including the use of biopesticides, botanicals, crop rotation, and good agronomic practices, with a view to developing integrated nematode management systems. The promotion of sustainable agricultural practices that are based on ecological principles can also contribute significantly to nematode management. Ultimately, the successful management of nematodes in smallholder systems requires a concerted effort from all stakeholders, including policy-makers, researchers, extension workers, farmers, and the private sector.

Key words: *Meloidogyne* spp.; Developing country; Banana; Entomopathogenic nematodes

中国线虫学研究（第九卷）Nematology Research in China（Vol. 9）：185-193

生物源农药防治根结线虫病研究进展*

包玲凤**，苏银玲，木万福，方海东，杨子祥***
（云南省农业科学院热区生态农业研究所/元谋干热河谷植物园，元谋　651300）

摘　要：根结线虫是严重危害作物生长的植物病原线虫，且随着保护地栽培面积扩大，危害越来越严重。由于传统的防治方法效果不理想、破坏生态环境而不适合农业的可持续发展。近年来，利用生物源农药对根结线虫进行生物防治成为国内外研究热点。本文对生物源农药（微生物源农药和植物源农药）防治根结线虫机制及其应用展开综述，并对该领域进行了展望，希望能为进一步开发生防资源和防治根结线虫提供帮助。
关键词：根结线虫；生物源农药；机制

Research Progress in Biological Pesticides of *Meloidogyne**

Bao Lingfeng**，Su Yinling，Mu Wanfu，Fang Haidong，Yang Zixiang***
（*Instituteof Tropical Eco-agricultural Sciences*，*Yunnan Academy of Agricultural Sciences*，
Yuanmou Dry-hot Valley Botanical Garden，*Yuanmou*　651300，*China*）

Abstract：Root-knot nematode（*Meloidogyne* spp.）is a kind of plant pathogenic nematode，which is harmful to the growth of crops. The traditional control methods are not suitable for the sustainable development of agriculture because of their unsatisfactory effect and destruction of ecological environment. In recent years，biological control of root-knot nematodes with biological pesticides has become a research hotspot at home and abroad. In this paper，the mechanism and application of biological pesticides（microbial pesticides and botanical pesticides）to control root-knot nematodes were reviewed，and the prospect in this field was also discussed，it is hoped that it can be helpful for the further development of biocontrol resources and the control of root-knot nematodes.
Key words：*Meloidogyne*；Biological pesticide；Mechanism

根结线虫（*Meloidogyne* spp.）是危害性最为严重的植物病原体之一，被列为全球十大植物寄生线虫之首，属专性寄生病原物，寄主植物 3 000 余种[1]。目前已报道的植物根结线虫种类高达90 种，其中，引起植物根结线虫病的主要为南方根结线虫（*M. incognita*）、北方根结线虫（*M. hapila*）、爪哇根结线虫（*M. javanica*）和花生根结线虫（*M. arenaria*）[2]。寄主

---

* 基金项目：省部共建云南生物资源保护与利用国家重点实验室开放项目“元谋蔬菜根结线虫病绿色防控技术的开发与应用”（2022KF003）
** 第一作者：包玲凤，实习研究员，从事植物保护研究工作。E-mail：2938891074@ qq. com
*** 通信作者：杨子祥，副研究员，从事作物病虫害生态防控理论与技术研究工作。E-mail：ynymyzx@ sina. com

植物被根结线虫侵染后，根部呈瘤状凸起，形成大小不等的根结，地上部发育不良、生长缓慢，严重时植株黄萎死亡[3]。我国根结线虫主要危害水稻、马铃薯、烟草、番茄等作物。

目前，防治根结线虫病的传统方法主要包括抗病品种选育、农业防治和化学防治[4]。抗病品种选育是最为经济有效的防治措施，但由于根结线虫寄主广泛，品种抗性鉴定结果差异较大，导致抗病品种的使用存在一定的局限性[5]。农业防治缺乏健全的防治体系，部分种植户防治水平低，防治效果不理想[6]。化学农药是防治田间根结线虫病的常用手段之一，但化学农药的长期使用带来了农残、周边生态环境遭到破坏等一系列副作用[7]。因此，需要寻找一种安全有效、环境友好型的防治措施。在这方面，生物防治有望成为替代传统方法和化学杀线剂的新方法。生物防治是指利用生物及其产品或其他天然材料达到防治线虫的目的，对环境和人体没有或只有很小的威胁性[8,9]。其中，生物源农药以绿色、零残留和不易产生抗药性等优点成为当前国内外生物防治根结线虫病的研究热点，具有较大的发展潜力。生物源农药（biological pesticide）狭义是指生物活体或生物代谢过程中产生的具有生物活性的物质，或者从生物体中提取的物质制成的制剂，包括微生物源农药、植物源农药和动物源农药；广义则指人工合成的天然活性结构及类似物[10]。本文着重阐述了近年来生物源（微生物源和植物源）农药防治根结线虫的机制及其应用情况，旨在为根结线虫病的防治提供参考。

1 微生物源农药对根结线虫作用方式的研究

微生物农药（microbial pesticides）是指以微生物或基因修饰的微生物等活体为有效成分的农药[11]，一般具有较高的特异性，只对靶标生物具有致病性，对人畜低毒，也不能渗透到植物体内[12]。微生物是土壤中最丰富的有机体，对土壤健康、植物生长发育起关键作用，对植物病害管理也至关重要[13]，其中一些已对控制根结线虫表现出巨大潜力，例如，淡紫紫孢菌（*Purpureocillium lilacinum*）[14]、哈茨木霉菌（*Trichoderma harzianum*）[15]、芽孢杆菌（*Bacillus subtilis*）[16]。它们通过营养竞争、寄生、捕食、产生毒素或抗生素、干扰线虫宿主识别和诱导植物系统抗性等多种方式影响线虫[17]。

微生物与线虫竞争铁载体、营养物质和生态位是拮抗根结线虫的重要机制[18]。根结线虫定殖在根系后，会诱导宿主形成巨细胞提供生长发育的营养需求，利用植物内生真菌与线虫的竞争作用能阻碍巨细胞正常发育，削弱线虫繁殖能力[19]。Martinuz 等[20]研究发现内生菌尖孢镰刀菌 Fo162（*Fusarium oxysporum* Fo162）和根瘤菌 G12（*Rhizobium etli* G12）限制了南方根结线虫对营养物质的吸收，干扰了 J2 期向 J3 期的发育。

微生物寄生和捕食线虫是防治根结线虫的重要手段。其中，内寄生细菌和真菌产生的可移动孢子，可通过口腔或穿透角质层进入线虫体内[13,21]。例如，木霉菌丝能穿透线虫卵壳或角质层在卵和成虫体内定殖，将线虫作为潜在的营养来源，最终导致线虫死亡[22]。捕食真菌是借助捕食器官（收缩环）或黏附性捕获并杀死线虫，Cui 等[23]从番茄根际土壤中分离到一株泡盛曲霉 BS05（*Aspergillus awamori* BS05）捕食真菌，在添加线虫 18h 后产生菌丝环，菌丝环末端形成的收缩环能在 60h 后迅速捕获线虫并消化线虫，在盆栽试验中对南方根结线虫的防治效果为 44.9%。

微生物次生代谢产物能影响线虫生殖或吸引和杀死根结线虫。产毒真菌产生某些毒素、抗生素或酶，通过阻止寄生线虫孵化或杀死幼虫来控制线虫病害[24]，如糙皮侧耳菌（*Pleurotus ostreatus*）[25]。Park 等[26]从变黑链霉菌（*Streptomyces nigrescens*）发酵液中分离到的抗虫活性物质曲古抑菌素和去羟基曲古抑菌素对南方根结线虫表现出很强的抗线虫活性，杀线活性达 91.3%。Jang 等[27]发现黑曲霉 F22（*Aspergillus niger* F22）发酵液对南方根结线虫有较强的抑制作用，对 J2 期线虫、卵孵化有明显的致死和抑制作用，经鉴定其杀线成分为草酸。Cao 等[28]从地衣中分离到一株放线菌 130935，其代谢产物乙酸乙酯在 8h 和 24h 的杀线活性分别达到 89%和 96%，同时从乙酸乙酯相中鉴定出的 4 个化合物对南方根结 J2 表现出一定的毒害作用。Hu 等[29]以南方根结线虫为实验材料，分析了球毛壳菌 NK102 代谢产物球毛壳菌素 A 对线虫卵孵化产生负面影响，发现能抑制 J2 侵染宿主。此外，尖孢镰刀菌（*Fusarium oxysporum*）产生的许多挥发性有机化合物，包括 2-甲基丁基乙酸酯、3-甲基丁基乙酸酯、乙酸乙酯和 2-乙酸异丁酯，对植物寄生线虫具有杀线虫活性[30]。抗生链霉菌 M7（*Streptomyces antibioticus* M7）纯化的放线菌素 V、X_2 和 D 引起 90%以上的 J2 死亡，并抑制卵孵化[31]。韦氏芽孢杆菌（*Bacillus weihenstephanensis*）抗生素粗提物被发现是抑制卵孵化和引起南方根结线虫 J2 死亡的有效抗生素[32]。

很多根际细菌也被用于植物寄生线虫的生物防治研究，它与寄主、线虫和环境之间形成一个复杂的相互作用网络，在自然条件下控制植物寄生线虫的种群[33]。大多数根际细菌通过次生代谢产物、酶和毒素对根结线虫起作用，影响线虫的繁殖、卵孵化和幼虫存活，或直接杀死线虫[34]。例如，AbdelRazek 等[35]从感病根际土壤中分离到 6 株抗根结线虫活性较高的菌株，实验条件下对 J2 表现出 100%致死率。Nguyen 等[36]从黑胡椒根际土壤分离得到的 20 株根际细菌均对根结线虫有较强抑制作用，其中 14 株为根瘤菌，菌株代谢产物几丁质酶和蛋白酶对根结线虫卵孵化起抑制作用。其他一些根际细菌通过产生抗生素等化合物，减少有害生物，创造更有利于植物生长的环境[37]。

添加外源微生物能提高植物养分吸收率，促进植株发育，营养状况越好的植物根系对植物寄生线虫的耐受性越高，可有效减轻或抑制病害。其中，内生细菌存在于植物根组织内部，研究显示它们能促进植物生长并抑制根结线虫病，如 Osman 等[38]从番茄和茄子根际分离得到两株细菌蜡样芽孢杆菌（*Bacillus cereus*）Nem 212 和 Nem 213，在田间条件下南方根结线虫在马铃薯上的定殖受到抑制，马铃薯产量显著提升；Banihashemian 等[39]从猕猴桃根系和叶片分离出 31 株对南方根结线虫表现出较强拮抗能力的内生细菌，以芽孢杆菌属（*Bacillus*）和假单胞菌属（*Pseudomonas*）为优势菌属，对猕猴桃幼苗的根鲜重和地上部鲜重等生长参数均有显著影响；越南伯克霍尔德菌 B418（*Burkholderia vietnamiensis* B418）是一种多功能植物根际促生菌，具有固氮解磷能力，可用于多种作物和蔬菜的根结线虫管理[40]。另外，许多细菌产物在植物中产生系统信号，可以保护整个植物免受各种病原体诱导的疾病，或帮助植物对各种致病生物产生抗性。例如，根瘤菌定殖在马铃薯根系能提高对线虫的全身抗性[18]。Vu 等[41]从香蕉皮层组织中分离得到 2 株尖孢镰刀菌和 1 株分生孢子镰刀菌（*F.* cf. *diversisporum*）以及从番茄根系中分离到一株尖孢镰刀菌，研究发现 4 株镰刀菌能诱导香蕉对线虫产生全株抗性。

目前，微生物农药防治根结线虫病的研究主要集中在单一菌株的防病促生作用，但生防菌株效用受环境影响较大，生产中可选用多种生物制剂复配使用[5]。将两种及以上的生物源农药复配进行生物防治能显著提升对根结线虫病的防治效果，利用它们的协同作用提高对根结线虫病的防治效率，适应于新型药剂的开发趋势，一方面使生物防治效果更加稳定，另一方面可显著减少农药用量，具有较强的应用潜力和广阔的发展前景。Seenivasan 等[42]将荧光假单胞菌（*Pseudomonas fluorescens*）和淡紫拟青霉（*Purpureocillium lilacinum*）在感染根结线虫胡萝卜田中应用，土壤中 J2 数量减少 70%，根中雌虫数和卵数分别下降 71%、74%，同时胡萝卜产量得到提升；舒洁等[5]将发酵 48h 的 3 种生防菌（耐冷假单胞菌（*Pseudomonas psychrotolerans*）、粘金黄杆菌（*Chryseobacterium gleum*）和枯草芽孢杆菌（*Bacillus subtilis*））发酵液与1%苦参印楝素乳油剂稀释液（1 000 倍）复配制剂处理组对根结线虫病的防效达 67.75%，且番茄的生物量均显著性增加。

2 植物源农药对根结线虫作用方式的研究

截至目前已报道对植物寄生线虫有活性的植物达 102 多科 300 余种，其中菊科和豆科是研究最多的杀线虫植物[43]。植物防治根结线虫主要是利用杀线虫植物与农作物间作或轮作、植物提取物或有效成分、植物粉碎后还田等方式来防治[44]。例如，烤烟/蓖麻间作对根结线虫的防效为 41.67%，并且能够显著提升烟叶产量和质量[45]；施用万寿菊提取物降低了烤烟根结线虫病发病率，同时增加烤烟的产量[46]；山杏壳木醋液对南方根结线虫 J2 有较好的毒杀作用[2]；利用辣椒残体作绿肥能起到生物熏蒸剂的效果，有效减少南方根结线虫数和虫瘿数[47]。植物源农药（botanical pesticides）是指从植物中提取出的有效成分或活性物质，或植物产生的分泌物及次生代谢产物[12]。植物源农药杀线虫机理主要包括作用于寄主或线虫产生驱避和拒食效果或起到杀虫效果[44]。

趋化性是线虫定位寄主植物的主要手段，引诱剂在土壤中扩散，线虫可通过最短的路线到达引诱剂位置[48]。线虫的化感行为、触觉和光敏感等行为由不同的神经元调控，能感知不同的化学信号进而产生驱避性或影响线虫活力及休眠等[49]。植物根系分泌物对线虫的趋化性产生影响，起吸引或排斥行为。比如，从番茄、萝卜、黄瓜、白菜和甜椒中鉴定出 1-二十二烯对南方根结线虫有吸引作用[50]；某些植物内酰胺类抗生素反式茴香脑、（E,E）-2,4-癸二烯醛、（E）-2-十二烷、噻唑烷二酮和 2-十一烷酮对根结线虫具有引诱或驱避性能，可破坏线虫对寄主根系的趋化性[51]；雏菊特定的根系分泌物成分月桂酸能调节根结线虫的趋化性，干扰线虫感染寄主，番茄-雏菊间作减少线虫数量，缓解线虫危害[52]；蓖麻根系分泌物中含有的棕榈酸和亚油酸对南方根结线虫趋化性具有排斥作用，并抑制 Mi-flp-18 和 Mi-mpk-1 的表达，减少了土壤中的线虫数量[53]。

植物提取物对线虫的毒性作用效果表现为降低线虫活性、引起线虫死亡、抑制线虫卵的孵化、减少虫口密度和植物根的组织变化妨碍线虫发育繁殖等[43]。已知化感物质会伤害植物寄生线虫，能否成功控制线虫取决的他们的植株毒性程度[54]。从植物中分离获取的杀线活性物质根据化学结构的差异，可分为生物碱类、醇（酚）类、萜类、脂类和糖苷类等类别[55]。生物碱类指从植物体中提取出的碱类物质，是一类含氮有机化合物[44]，具有杀线活性的主要有苦参碱[56]、莨菪碱[57]、藜芦碱[58]等。刘晓宇等[59]研究表明使用 0.3%苦参碱水剂番茄根结线虫病的防效达到 64%，增产率为 90%以上。苦参碱还具有很好的协同作用，复配辣根素[60]或氨基寡糖素[61]能提高对线虫病害的防治效果。漆永红等[57]研究发现随着莨菪碱浓度和处理时间的增加，线虫卵囊、卵孵化的抑制率和J2的死亡率显著提升。从印楝中分离得到的萜类化合物印楝素具广谱杀虫活性、低毒和环境相容性好等优点，是公认活性最强的植物源杀虫剂，已被开发为植物源农药[62]。Pelinganga等[54]从非洲南瓜和野生黄瓜果实中分离得到的三萜类化合物葫芦素是黄瓜属类植物杀南方根结线虫的主要成分。Kumar等[63,64]收集的高粱、万寿菊植株根系分泌物经过适当的分级分离后，得到5种不同极性范围的化感物质，室内生测试验表明其中4种对爪哇根结线虫J2毒性为100%，对卵孵化也有98%的抑制作用。

3 问题与展望

综合近年来看，虽然生物源农药已成为国内外研究热点，但目前仍面临诸多问题：①越来越多的微生物和植物已显示出对根结线虫的抑制作用，然而，只有少数具有杀线虫潜力的被开发成商业产品，并用于农业系统；②大多数生物源农药在实验室条件下对根结线虫防效良好，但用于田间易受到环境因子影响，生防效果并不稳定；③生物源农药在各阶段防效不同，且药效缓慢，不能及时解决大面积发病或危害严重的病害；④生物源农药成分复杂，有些存在一定的风险性；⑤生物源农药推广宣传不到位，加之价格高昂假药多，农民不易接受。

针对上述问题，可以实行以下对策：①继续发现更多的杀线生防资源，准确了解生物源农药与寄主、线虫和环境之间的互作机制，开发作用性强的生物源农药；②采用新技术研发新型生物源农药，增加其稳定性和速效性；③结合低毒、高效、无残留农药及其他措施使用，以便更有效地控制根结线虫；④降低成本，使农民能够买到实惠药；⑤有关部门加大宣传普及力度，建立相应政策，确保生物源农药规范化管理。

参考文献

[1] 翟明娟，李登辉，马玉琴，等．绿色木霉菌株Tvir-6对黄瓜根结线虫的防治效果研究［J］．中国蔬菜，2017（10）：67-72.

[2] 李运朝，李洪涛，及华，等．不同生物制剂对南方根结线虫杀虫活性与防治效果［J］．农业环境科学学报，2022，41（12）：2810-2816.

[3] 雷敬超，黄惠琴．南方根结线虫生物防治研究进展［J］．中国生物防治，2007（S1）：76-81.

[4] 崔江宽，任豪豪，孟颢光，等．我国烟草根结线虫病发生与防治研究进展［J］．植物病理学报，2021，51（5）：663-682.

[5] 舒洁，张仁军，梁应冲，等．植物源与微生物源生物制剂复配防治根结线虫病［J］．生物技术通报，2021，37（7）：164-174.

[6] 于美钰．农业病虫害防治现状与方法［J］．南方农业，2021，15（35）：30-32.

[7] 梁兵，黄坤，李宏光，等．肥料和农药协同作用防治烟草根结线虫病研究［J］．西南农业学报，2016，29（8）：1894-1898.

[8] NAZ I，KHAN R A A，MASOOD T，*et al*. Biological control of root knot nematode，*Meloidogyne incognita*，in vitro，greenhouse and field in cucumber［J］. Biological Control，2021，152：104429.

[9] CETINTAS R，KUSEK M，FATEH S A. Effect of some plant growth-promoting rhizobacteria strains on root-knot nematode，*Meloidogyne incognita*，on tomatoes［J］. Egyptian Journal of Biological Pest

Control，2018，28（1）：1-5.

[10] 郭明程，王晓军，苍涛，等．我国生物源农药发展现状及对策建议［J］．中国生物防治学报，2019，35（5）：755-758.

[11] 中华人民共和国农业部公告第2569号《农药登记资料要求》［Z］．北京：中华人民共和国农业部，2017.

[12] 谭海军．中国生物农药的概述与展望［J］．世界农药，2022，44（4）：16-27，54.

[13] TIAN B Y，YANG J K，ZHANG K Q. Bacteria used in the biological control of plant-parasitic nematodes：populations，mechanisms of action，and future prospects［J］．FEMS microbiology ecology，2007，61（2）：197-213.

[14] DAS N，WAQUAR T. Bio-Efficacy of *Purpureocillium lilacinum* on management of root-knot nematode，*Meloidogyne incognita* in tomato［J］．Indian Journal of Nematology，2021，51（2）：129-136.

[15] KHAN M R，AHMAD I，AHAMAD F. Effect of pure culture and culture filtrates of *Trichoderma* species on root-knot nematode，*Meloidogyne incognita* infesting tomato［J］．Indian Phytopathology，2018，71：265-274.

[16] BAVARESCO L G，GUABERTO L M，ARAUJO F F. Interaction of *Bacillus subtilis* with resistant and susceptible tomato（*Solanum lycopersicum* L.）in the control of *Meloidogyne incognita*［J］．Archives of Phytopathology and Plant Protection，2021，54（7-8）：359-374.

[17] BHAT A A，SHAKEEL A，WAQAR S，*et al.* Microbes vs. Nematodes：Insights into Biocontrol through Antagonistic Organisms to Control Root-Knot Nematodes［J］．Plants，2023，12（3）：451.

[18] AFRIYIE BOAKYE T，KWADWO ANNING D，LI H X，*et al.* Mechanism of Antagonistic Bioagents in Controlling Root-Knot Nematodes（*Meloidogyne* sp.）：A Review［J］．Asian Research Journal of Agriculture，2022，15（2）：27-44.

[19] 易希，廖红东，郑井元．植物内生真菌防治根结线虫研究进展［J］．生物技术通报，2023，39（3）：43-51.

[20] MARTINUZ A，SCHOUTEN A，SIKORA R A. Post-infection development of *Meloidogyne incognita* on tomato treated with the endophytes *Fusarium oxysporum* strain Fo162 and *Rhizobium etli* strain G12［J］．BioControl，2013，58：95-104.

[21] MANKAU R，IMBRIANI J L，BELL A H. SEM observations on nematode cuticle penetration by *Bacillus penetrans*［J］．Journal of Nematology，1976，8（2）：179.

[22] KUBICEK C P，HERRERA-ESTRELLA A，SEIDL-SEIBOTH V，*et al.* Comparative genome sequence analysis underscores mycoparasitism as the ancestral life style of *Trichoderma*［J］．Genome biology，2011，12：1-15.

[23] CUI R，FAN C，SUN X T. Isolation and characterisation of *Aspergillus awamori* BS05，a root-knot nematode-trapping fungus［J］．Biocontrol Science and Technology，2015，25（11）：1233-1240.

[24] SIDDIQUI Z A，MAHMOOD I. Role of bacteria in the management of plant parasitic nematodes：a review［J］．Bioresource technology，1999，69（2）：167-179.

[25] SATOU T，KANEKO K，LI W，*et al.* The toxin produced by *Pleurotus ostreatus* reduces the head size of nematodes［J］．Biological and Pharmaceutical Bulletin，2008，31（4）：574-576.

[26] PARK E J，JANG H J，PARK J Y，*et al.* Efficacy evaluation of *Streptomyces nigrescens* KA-1 against the root-knot nematode［J］．Biological Control，2023（179）：105150.

[27] JANG J Y，CHOI Y H，SHIN T S，*et al.* Biological control of *Meloidogyne incognita* by *Aspergillus niger* F22 producing oxalic acid［J］．PloS one，2016，11（6）：e0156230.

[28] CAO X M, ZHANG R P, MENG S, *et al*. Biocontrol potential of *Agromyces allii* 130935 and its metabolites against root-knot nematode *Meloidogyne incognita* [J]. Rhizosphere, 2021, 19: 100378.

[29] HU Y, ZHANG W P, ZHANG P, *et al*. Nematicidal activity of chaetoglobosin A poduced by *Chaetomium globosum* NK102 against *Meloidogyne incognita* [J]. Journal of agricultural and food chemistry, 2013, 61 (1): 41-46.

[30] TERRA W C, CAMPOS V P, MARTINS S J, *et al*. Volatile organic molecules from *Fusarium oxysporum* strain 21 with nematicidal activity against *Meloidogyne incognita* [J]. Crop Protection, 2018, 106: 125-131.

[31] SHARMA M, JASROTIA S, OHRI P, *et al*. Nematicidal potential of *Streptomyces antibioticus* strain M7 against *Meloidogyne incognita* [J]. AMB Express, 2019, 9 (1): 1-8.

[32] SARANGI T, RAMAKRISHNAN S, NAKKEERAN S, *et al*. Effect of solvents to maximize the extraction of Crude antibiotics of *Bacillus* spp. and their influence on *Meloidogyne incognita* and *Fusarium oxysporum* f. sp. *lycopersici* [J]. Annals of Plant Protection Sciences, 2018, 26 (1): 181-186.

[33] KERRY B R. Rhizosphere interactions and the exploitation of microbial agents for the biological control of plant-parasitic nematodes [J]. Annual review of phytopathology, 2000, 38 (1): 423-441.

[34] GOWDA M T, PRASANNA R, KUNDU A, *et al*. Differential effects of rhizobacteria from uninfected and infected tomato on *Meloidogyne incognita* under protected cultivation [J]. Journal of Basic Microbiology, 2023.

[35] ABDELRAZEK G M, YASEEN R. Effect of some rhizosphere bacteria on root-knot nematodes [J]. Egyptian Journal of Biological Pest Control, 2020, 30 (1): 1-11.

[36] NGUYEN V B, WANG S L, NGUYEN T H, *et al*. Reclamation of rhizobacteria newly isolated from black pepper plant roots as potential biocontrol agents of root-knot nematodes [J]. Research on Chemical Intermediates, 2019, 45: 5293-5307.

[37] ZUCKERMAN B M, JANSSON H B. Nematode chemotaxis and possible mechanisms of host/prey recognition [J]. Annual Review of Phytopathology, 1984, 22 (1): 95-113.

[38] OSMAN H A I, AMEEN H H, HAMMAM M M A, *et al*. Antagonistic potential of an Egyptian entomopathogenic nematode, compost and two native endophytic bacteria isolates against the root-knot nematode (*Meloidogyne incognita*) infecting potato under field conditions [J]. Egyptian Journal of Biological Pest Control, 2022, 32 (1): 137.

[39] BANIHASHEMIAN S N, JAMALI S, GOLMOHAMMADI M, *et al*. Isolation and identification of endophytic bacteria associated with kiwifruit and their biocontrol potential against *Meloidogyne incognita* [J]. Egyptian Journal of Biological Pest Control, 2022, 32 (1): 1-12.

[40] LIU M, PHILP J, WANG Y L, *et al*. Plant growth-promoting rhizobacteria *Burkholderia vietnamiensis* B418 inhibits root-knot nematode on watermelon by modifying the rhizosphere microbial community [J]. Scientific Reports, 2022, 12 (1): 1-13.

[41] VU T, HAUSCHILD R, SIKORA R A. *Fusarium oxysporum* endophytes induced systemic resistance against *Radopholus similis* on banana [J]. Nematology, 2006, 8 (6): 847-852.

[42] SEENIVASAN N. Effect of concomitant application of *Pseudomonas fluorescens* and *Purpureocillium lilacinum* in carrot fields infested with *Meloidogyne hapla* [J]. Archives of Phytopathology and Plant Protection, 2018, 51 (1-2): 30-40.

[43] 李继平, 漆永红, 陈书龙, 等. 利用杀线植物资源防治植物寄生线虫的研究进展 [J]. 草业学

报，2013，22（3）：297-305.

［44］ 王佳，曾广智，汪哲，等．杀线虫植物以及植物源杀线虫活性化合物研究与应用进展［J］．中国生物防治学报，2018，34（3）：469-479.

［45］ 张宗锦，闫芳芳，孔垂旭，等．烤烟菽麻间作对烟草根结线虫防效及烟叶产质量的影响［J］．中国烟草科学，2019，40（2）：52-56.

［46］ 徐天养，桑应华，徐俊驹，等．万寿菊提取物对烤烟根结线虫发病率及产质量的影响［J］．湖南农业科学，2022（12）：35-39.

［47］ BUENA A P，GARCÍA-ÁLVAREZ A，DÍEZ-ROJO M A，*et al.* Use of pepper crop residues for the control of root-knot nematodes［J］．Bioresource Technology，2007，98（15）：2846-2851.

［48］ REYNOLDS A M，DUTTA T K，CURTIS R H C，*et al.* Chemotaxis can take plant-parasitic nematodes to the source of a chemo-attractant via the shortest possible routes［J］．Journal of the Royal Society Interface，2011，8（57）：568-577.

［49］ 李春杰，王从丽．植物寄生线虫对化感信号的识别及机制［J］．生物技术通报，2021，37（7）：35-44.

［50］ 丁正蛟．七种植物根系分泌物对南方根结线虫趋化性的作用［D］．昆明：云南大学，2018.

［51］ SOBKOWIAK R，BOJARSKA N，KRZYANIAK E，*et al.* Chemoreception of botanical nematicides by *Meloidogyne incognita* and *Caenorhabditis elegans*［J］．Journal of Environmental Science and Health，Part B，2018，53（8）：493-502.

［52］ DONG L，LI X L，HUANG L，*et al.* Lauric acid in crown daisy root exudate potently regulates root-knot nematode chemotaxis and disrupts *Mi-flp*-18 expression to block infection［J］．Journal of Experimental Botany，2014，65（1）：131-141.

［53］ DONG L L，LI X L，HUANG C D，*et al.* Reduced *Meloidogyne incognita* infection of tomato in the presence of castor and the involvement of fatty acids［J］．Scientia Horticulturae，2018，237：169-175.

［54］ PELINGANGA O，MASHELA P. Mean dosage stimulation range of allelochemicals from crude extracts of *Cucumis* africanus fruit for improving growth of tomato plant and suppressing *Meloidogyne incognita* numbers［J］．Journal of Agricultural Science，2012，4（12）：8.

［55］ CHITWOOD D J. Phytochemical based strategies for nematode control［J］．Annual review of phytopathology，2002，40（1）：221-249.

［56］ 刘勇鹏，赵群法，张涛，等．生物杀线虫剂对日光温室番茄根结线虫病防效研究［J］．河南农业大学学报，2017，51（6）：815-821.

［57］ 漆永红，胡冠芳，曹素芳，等．莨菪生物碱类对南方根结线虫卵囊、卵孵化及其2龄幼虫存活的影响［J］．华北农学报，2015，30（S1）：272-277.

［58］ 王宏宝，赵桂东，刘伟中，等．不同药剂对黄瓜根结线虫病防治效果研究［J］．福建农业学报，2012，27（11）：1242-1245.

［59］ 刘晓宇，陈立杰，邢志富，等．4种生物源杀线剂对番茄根结线虫的田间防效［J］．植物保护，2020，46（6）：228-232，253.

［60］ 尼秀媚，李光聚，高珏晓，等．5种闷棚处理防治根结线虫药剂的大田药效试验［J］．农药，2017，56（12）：919-921.

［61］ 刘陈晨，任士伟，王娜，等．氨基寡糖素复配苦参碱对黄瓜根结线虫的药效试验［J］．黑龙江农业科学，2017（12）：47-48.

［62］ TAN Q G，LUO X D. Meliaceous limonoids：chemistry and biological activities［J］．Chemical reviews, 2011, 111 (11): 7437-7522.

[63] KUMAR L, SINGH B, SINGH U. Effect of sorghum allelochemicals on the mortality and egg hatching of root-knot nematode, *Meloidogyne javanica* [J]. International Journal of Bio-resource and Stress Management, 2015, 6 (2): 182-191.

[64] KUMAR L, DEVI U, SINGH B, *et al*. Isolation of root exuded allelochemicals of marigold (*Tagetes erecta*) and their effect on the mortality and egg hatching of root-knot nematode (*Meloidogyne javanica*) [J]. Journal of Food Legumes, 2014, 27 (2): 166-169.

中国线虫学研究（第九卷）Nematology Research in China（Vol. 9）：194-248

矛线目线虫的多样性研究概述*

李红梅**，傅韦棋，薛　清
（南京农业大学植物保护学院/农作物生物灾害综合治理教育部重点实验室，南京　210095）

摘　要：矛线目线虫的种类数量位居线虫门各目之首，是土壤与淡水生境中的主要线虫类群之一，而我国关于自由生活的矛线目线虫研究极其匮乏。本文对矛线目的分类系统演变历史、发育系统学、各形态特征进行了简要描述，对矛线目 2 个亚目和 18 个科的主要形态鉴别特征进行了描述，并提供了 18 个科的检索表和 285 个有效属的名录。此外，对矛线目线虫的生物学和生态学习性、捕食性矛线目线虫的生防潜力以及国内的研究情况进行了概述，以期为今后我国矛线目线虫种类鉴定和土壤线虫生态学研究提供参考。
关键词：矛线目；生物多样性；分类系统；形态鉴定；土壤生态学

Advances in Diversity of Nematodes in Order Dorylaimida*

Li Hongmei**，Fu Weiqi，Xue Qing
(*Key Laboratory of Integrated Management of Crop Diseases and Pests*, *Ministry of Education/ College of Plant Protection*, *Nanjing Agriculture University*, *Nanjing*　210095, *China*)

Abstract: The number of species in order Dorylaimida is ranked the first among all orders in phylum Nematoda. They are one of the major nematode groups in soil and freshwater habitats. The understanding of dorylaims in China is rather limited. In this review, the historical outlines of taxonomic system, the systematic phylogeny, and the morphological characters of Dorylaimida were briefly introduced. The diagnostic characters of two suborders and 18 families were described, and a dichotomous key for the families and a list of 285 valid genera were provided. In addition, the biological and ecological characters of dorylaims, the biocontrol potential of predatory dorylaims, and the domestic research status were summerised. It provided comprehensive imformation for future identification of dorylaims and ecological studies of soil nematodes in China.
Key words: Dorylaimida; biodiversity; taxonomic system; morphological identification; soil ecology

1　前言

线虫的生物多样性仅次于昆虫，广泛分布于海水、淡水、土壤以及寄生于动物和植物，估计描述约有 25 000 种线虫（Hodda，2011），其中植物寄生线虫大约有 4 000 种（Decraemer & Hunt，2013）。绝大多数线虫种类体型小，呈圆柱形，又称圆虫（roundworms）。它们在淡水、海水、陆地上随处可见，不论是个体数还是物种数都超越其他

* 基金项目：国家自然科学基金（32001876）
** 第一作者：李红梅，教授，博士生导师，从事植物线虫学研究。E-mail：lihm@njau.edu.cn

动物，并且在极端的环境如南极和海沟都可发现。除了寄生于植物、动物和人类的线虫种类外，自由生活的线虫种类占大多数，广泛分布于全球各类生境，与环境变化密切相关，因而越来越受到重视。

矛线目（Dorylaimida）线虫，亦称为 dorylaims，隶属于线虫门（Nematoda）嘴刺纲（Enoplea）矛线亚纲（Dorylaimia）。矛线目的种类和数量，是线虫门各目中最多的，也是土壤与淡水线虫中最多样化与生态学上最重要的线虫类群。矛线目线虫的高多样性和高丰度、高度变化的摄食习性，以及虫体形态和大小的巨大变化，使它们成为评估土壤健康状况的良好指示生物（bioindicators），许多矛线目种类对重金属和各种土壤污染物表现出高度的敏感性，更加凸显了它们在土壤生态学领域的重要性（Peña-Santiago，2021）。

矛线目线虫主要是陆生的自由生活线虫，在世界各地的土壤和淡水沉积物中非常丰富和频繁，而在海洋栖息地完全没有，只有少数物种偶尔在河口环境中有所记录。矛线目具有各种各样的生活方式，有些是食真菌（fungivorous）、食藻类（algivorous）、杂食（omnivorous）或捕食（predatory）的线虫，或者是专性（obligate）的植物寄生线虫，如长针总科（Longidoroidea）线虫是一类植物外寄生线虫，一些种类能传播植物病毒，对许多经济作物造成严重的危害。从生态学的角度来看，矛线目由于其世代时间长、繁殖率低、新陈代谢活动低以及运动缓慢，被认为是持久策略（K-strategists）（Peña-Santiago，2021）。

矛线目是形态上很容易识别的一个单系类群，所有成员具有一些基本鉴别特征，如口孔有一个能伸出的壁齿（mural tooth）或轴向齿针（axial odontostyle），食道为瓶状的矛线食道，独有的前直肠（prerectum）和一对雄虫泄殖腔前的交配乳突，尾腺（caudal glands）缺失。然而，矛线目显示出复杂的形态多样性，主要体现在其唇区结构及相关结构如齿针（odontostyle）和齿托（odontophore）、食道、两性生殖管形态以及尾形等。许多矛线目线虫缺乏分子数据，形态学仍然是一个关键的诊断依据，了解并观察矛线目线虫的各种形态特征，是研究其分类学和系统发育的基础。尽管矛线目的许多“经典属”和更高阶的分类群可能是多系（polyphyletic）的，但是越来越多的分子测序将揭示出它们形态学概念和系统发育之间经常存在的复杂而混乱的关系（Peña-Santiago，2021）。

2 矛线目分类系统的演变历史

人类对矛线目线虫了解始于近 2 个世纪前，对该类群线虫的研究历史，特别是对其多样性的研究，可以分为 5 个主要阶段。

2.1 起源（1845—1920 年）

Dujardin（1845）描述了第一个矛线类线虫，即池塘矛线虫（*Dorylaimus stagnalis*），矛线目的名称由此而来。20 年后，Bastian（1865）补充了矛线属（*Dorylaimus*）的诊断特征，描述了一些新种并将另外两个物种转移到该属中。Bütschli（1873）重新描述了 Bastian 的物种，de Man（1876）提出建立矛线科（Dorylamidae），而矛线属（*Dorylaimus*）是其唯一的属。此时的矛线科形态诊断特征，已经基本与目前接受的矛线目（Dorylaimida）特征非常相似，即具有明显口孔的轴向齿针和前直肠，有弓形交合刺（spicules）、侧导片（lateral guiding pieces）以及腹中位乳突（ventromedian supplements）等。几位先驱作者（de Man，1880；Örley，1880；Cobb，1913a，1920）随后将许多属归类在矛线科（Dorylaimidae），并且对该科进行了重大修改。到 19 世纪末，已经描述了 2 属 84 种，在接下来的 30 年里，它们的数量显著增加到 16 属和 208 种。

2.2 辉煌的十年（20 世纪 30 年代）

Filipjev（1927）在矛线科建立了一个矛线亚科（Dorylaiminae），到 1934 年该科已经有 4 个亚科，包括无咽亚科（Alaiminae）、矛线亚科（Dorylaiminae）、伊龙亚科（Ironinae）和垫咽亚科（Tylencholaiminae）。Thorne（1934）将矛线科提升为矛线总科（Dorylaimoidea），下分矛线科（Dorylaimidae）和无咽科（Alaimidae）两个科。Thorne（1935）提出建立新的细齿科（Leptonchidae）和膜皮科（Diphtherophoridae），以及在矛线科下建立长针亚科（Longidorinae）和穿咽亚科（Nygolaiminae）。Thorne 和 Swanger（1936）建立了孔咽属（*Aporcelaimus*）、类矛线属（*Dorylaimoides*）和螯属（*Pungentus*）3 个新属并描述了大量种类，接着 Thorne（1939）对自己前期所提出的矛线总科分类系统进行了调整，新增了颚针科（Belondiridae）并描述了大量的新种，至此，经过详尽的修订和整合后，矛线类线虫的现代分类系统初具雏形，数量已增加至 38 属和 449 种。

2.3 矛线目的建立

Pearse（1936）将矛线总科（Dorylaimoidea）提升为矛线亚目（Dorylaimina），并将其归到嘴刺目（Enoplida）下。Pearse（1942）针对这些矛线类线虫，提出建立矛线目（Dorylaimida）。随着不同类群的矛线类线虫不断发现和丰富，矛线目的分类系统存在着较多的争议和变动。Andrássy（1959，1960a）将矛线属（*Dorylaimus*）进行了修订，提出将矛线属重新划分为 9 个属，这一调整已被广泛接受。随后，Clark（1961）在修订嘴刺目时不承认矛线目的分类地位，认为应作为一个总科。Goodey（1963）对土壤和淡水线虫进行了修订，矛线目的分类地位被广泛接受；此外，他还提出矛线目包括矛线亚目（Dorylaimina）和无咽亚目（Alaimina），矛线亚目仅包括矛线总科（Dorylaimoidea）和单齿总科（Mononchioidea）两个总科（表 1）。Jairajpuri（1969）将单齿线虫类群建立为单齿目（Mononchida）。Coomans 和 Loof（1970）建立膜皮亚目（Diphtherophorina），Siddiqi（1983）将其归入三矛目（Triplonchida Cobb，1920），并将无咽亚目（Alaimina）提升为无咽目（Alaimida）。

2.4 矛线目多样性探索的黄金时代（1960—1990 年）

第二次世界大战后陆续涌现出一批杰出的线虫分类学家聚焦于矛线目多样性研究，包括匈牙利的 I. Andrássy、比利时的 A. Coomans、南非的 J. Heyns、印度的 M. S. Jairajpuri、荷兰的 P. A. A. Loof、英国的 M. R. Siddiqi、美国的 V. Ferris 和 G. Thorne、意大利的 M. T. Vinciguerra 和 A. Zullini 以及他们的合作者，在他们的努力下，矛线目线虫属和种的数量经历了令人难以置信的扩张，但是对矛线目的系统发育和系统学研究关注较少。Andrássy（1976）试图阐明矛线目主要线虫类群之间的进化关系，尽管他的观点没有获得普遍接受（表 1）。此外，Coomans（1985）和 Vinciguerra（1987）也试图用支系学（cladistic）方法来了解矛线目不同分类群之间的进化关系。Jairajpuri 和 Ahmad（1992）对矛线目的分类系统进行了全面的修订，整理了矛线目大约有 200 个属和 1 700 个种，但是该分类系统过于细化，许多亚属后来并没有获得认可。

表 1 过去半个世纪分类学家提出的矛线目（科）分类系统

Goodey（1963） 2 亚目 4 总科 10 科	Andrássy（1976） 4 亚目 6 总科 28 科	Jairajpuri & Ahmad（1992） 3 亚目 7 总科 18 科	de Ley & Blaxter（2004） 2 亚目 4 总科 15 科	Andrássy（2009） 2 亚目 5 总科 21 科	Peña-Santiago（2014） 2 亚目 18 科
Dorylaimina	**Dorylaimina**	**Nygolaimina**	**Nygolaimina**	**Nygolaimina**	**Nygolaimina**
DORYLAIMOIDEA	ENCHOLAIMOIDEA	NYGOLAIMOIDEA	NYGOLAIMOIDEA	NYGOLAIMOIDEA	Aetholaimidae
Dorylaimidae	Encholaimidae	Aetholaimidae	Aetholaimidae	Aetholaimidae	Nygellidae
Opailaimidae	NYGOLAIMOIDEA	Nygellidae	Nygellidae	Nygellidae	Nygolaimellidae
Belondiridae	Nygolaimidae	Nygolaimellidae	Nygolaimellidae	Nygolaimidae	Nygolaimidae
Nygolaimidae	Aetholaimidae	Nygolaimidae	Nygolaimidae	**Dorylaimina**	**Dorylaimina**
[Campydoridae]	DORYLAIMOIDEA	**Dorylaimina**	**Dorylaimina**	DORYLAIMOIDEA	Actinolaimidae
[MONONCHOIDEA]	Prodorylaimidae	DORYLAIMOIDEA	DORYLAIMOIDEA	Actinolaimidae	Aporcelaimidae
[Mononchidae]	Dorylaimidae	Aporcelaimidae	Actinolaimidae	Aporcelaimidae	Aulolaimoididae
[Bathyodontidae]	Thornenematidae	Dorylaimidae	Aporcelaimidae	Dorylaimidae	Belondiridae
[Alaimina]	Qudsianematidae	Nordiidae	Dorylaimidae	Nordiidae	Dorylaimidae
[ALAIMOIDEA]	Aporcelaimidae	Qudsianematidae	Longidoridae	Paraxonchiidae	Leptonchidae
[Alaimidae]	Nordiidae	ACTINOLAIMOIDEA	Nordiidae	Qudsianematidae	Longidoridae
[DIPHTHEROPHOROIDEA]	Longidoridae	Actinolaimidae	Qudsianematidae	Thornenematidae	Mydonomidae
[Diphtherophoridae]	Dorylaimoididae	Carcharolaimidae	BELONDIROIDEA	Thorniidae	Nordiidae
[Trichodoridae]	Crateronematidae	LONGIDOROIDEA	Belondiridae	LONGIDOROIDEA	Qudsianematidae
	Thorniidae	Longidoridae	TYLENCHOLAIMOIDEA	Longidoridae	Thornenematidae
	BELONDIROIDEA	Xiphinematidae	Aulolaimoididae	BELONDIROIDEA	Thorniidae
	Oxydiridae	BELONDIROIDEA	Leptonchidae	Belondiridae	Tylencholaimellidae

（续表）

Goodey（1963） 2 亚目 4 总科 10 科	Andrássy（1976） 4 亚目 6 总科 28 科	Jairajpuri & Ahmad（1992） 3 亚目 7 总科 18 科	de Ley & Blaxter（2004） 2 亚目 4 总科 15 科	Andrássy（2009） 2 亚目 5 总科 21 科	Peña-Santiago（2014） 2 亚目 18 科
	Swangeriidae	Belondiridae	Mydonomidae	Dorylaimellidae	Tylencholaimidae
	Roqueidae	TYLENCHOLAIMOIDEA	Tylencholaimidae	Swangeriidae	
	Belondiridae	Aulolaimoididae		TYLENCHOLAIMOIDEA	
	Dorylaimellidae	Leptonchidae		Aulolaimoididae	
	ACTINOLAIMOIDEA	Mydonomidae		Encholaimidae	
	Trachypleurosidae	Tylencholaimidae		Leptonchidae	
	Actinolaimidae	[**Campydorina**]		Mydonomidae	
	Brittonematidae	[CAMPYDROIDEA]		Tylencholaimellidae	
	Carcharolaimidae	[Campydoridae]		Tylencholaimidae	
	LEPTONCHOIDEA				
	Tylencholaimidae				
	Leptonchidae				
	Tylencholaimellidae				
	Belonenchidae				
	Aulolaimoididae				
	[Campydoridae]				
	[**Mononchina**]				
	[**Diphtherophorina**]				
	[**Mermithina**]				

注：亚目 Suborder 用粗体表示，总科用大写表示。后来被视为非矛线目的分类群用内方括号表示。

2.5 综合性方法的新时代

1990 年至今，矛线目线虫多样性继续以传统的形态学观点进行研究，新属和新种的数量持续增长。Blaxter 等（1998）为线虫门的系统学开辟了一个新时代，而基于形态学和分子特征的方法（de Ley & Blaxter，2002，2004）让矛线目线虫的分类受益颇多。Mullin 等（2005）通过 18S 和 28S rRNA 基因序列分析发现，矛线目及其下的矛线亚目（Dorylaimin）和穿咽亚目（Nygolaimina）是自然的单系（monophyletic）类群，Sun 等（2023）的研究同样支持此观点。Holterman 等（2008）揭示矛线目下的总科和大多数科不是单系的，这意味着基于形态特征的传统分类法并不能完全令人满意。今后可以通过重新分析形态学特征，以及通过对新分类群和/或新基因进行测序来收集更多的分子信息，以解析各类群的非单系观点。

Andrássy（2009）发表了矛线目新的分类系统，整理了 263 个有效属和 2 637 个有效种。该分类系统已被广泛接受，根据齿针的类型将矛线目划分为穿咽亚目和矛线亚目，其中穿咽亚目仅有 1 个总科，即穿咽总科（Nygolaimoidea）（Thorne，1935）；矛线亚目有 4 个总科：矛线总科（Dorylaimoidea）（de Man，1876）、长针总科（Longidoroidea）（Thorne，1935）、颚针总科（Belondiroidea）（Thorne，1939）和垫咽总科（Tylencholaimoidea）（Filipjev，1934）。Peña-Santiago（2014，2021）提出了一个不考虑总科的简化的矛线目分类系统，并整理了矛线目 18 个科的 285 个有效属和 3 088 个有效种。

3 矛线目的分类系统学研究

目前形态学和分子学的证据都支持矛线目（Dorylaimida）是一个单系分类群，与单齿目（Mononchida）和寄生动物的其他各目密切相关，都归入矛线亚纲（Dorylaimia）。从形态学的角度来看，矛线目线虫具有专属的形态特征包括可伸出的口部结构（齿针）、二分式的瓶状食道、前直肠以及泄殖腔前的一对交配乳突，将它们与其他线虫分开。此外，分子研究也证实了矛线目的单系性（Blaxter *et al.*，1998；Mullin *et al.*，2004，2005；Holterman *et al.*，2006，2008；van Megen *et al.*，2009）（图 1）。

形态学特征支持将矛线目（Dorylaimida）分为两个亚目，即穿咽亚目（Nygolaimina）和矛线亚目（Dorylaimina）（Jairajpuri & Ahmad，1992；Peña-Santiago，2006；Andrássy，2009），口腔中突出结构的性质是两个亚目之间的主要区别，穿咽亚目成员的口腔有壁齿，食道-肠道连接有围绕的三个大的贲门腺体（cardiac glands），而矛线亚目成员的口腔有一个轴向齿针，没有类似的大贲门腺体，而分子研究也证实这两个亚目代表了自然群体（Mullin *et al.*，2005；Holterman *et al.*，2006，2008；Sun *et al.*，2023）。

然而，矛线目（Dorylaimida）进化树的二级分支具有极大争议性。一方面，传统的形态学分类法将两个亚目分成数量不等的总科和科（Jairajpuri & Ahmad，1992；Andrássy，2009），穿咽亚目由 3 个或 4 个科组成，只归入一个总科（Nygolaimoidea），而矛线亚目的总科（3~5 个）和科（13~18 个）数量变化更大。矛线目低分类阶元之间的进化关系一直是讨论的问题，原因有几个，首先是对该类群的多样性认识不足，每年都有新种被描述，其中一些新种具有显著的形态特征组合，使它们的分类变得困难，并使人们对矛线目的一般分类系统产生了怀疑。其次是描述线虫的旧资料零散和过时，缺乏详细的信息，难以进行分析。

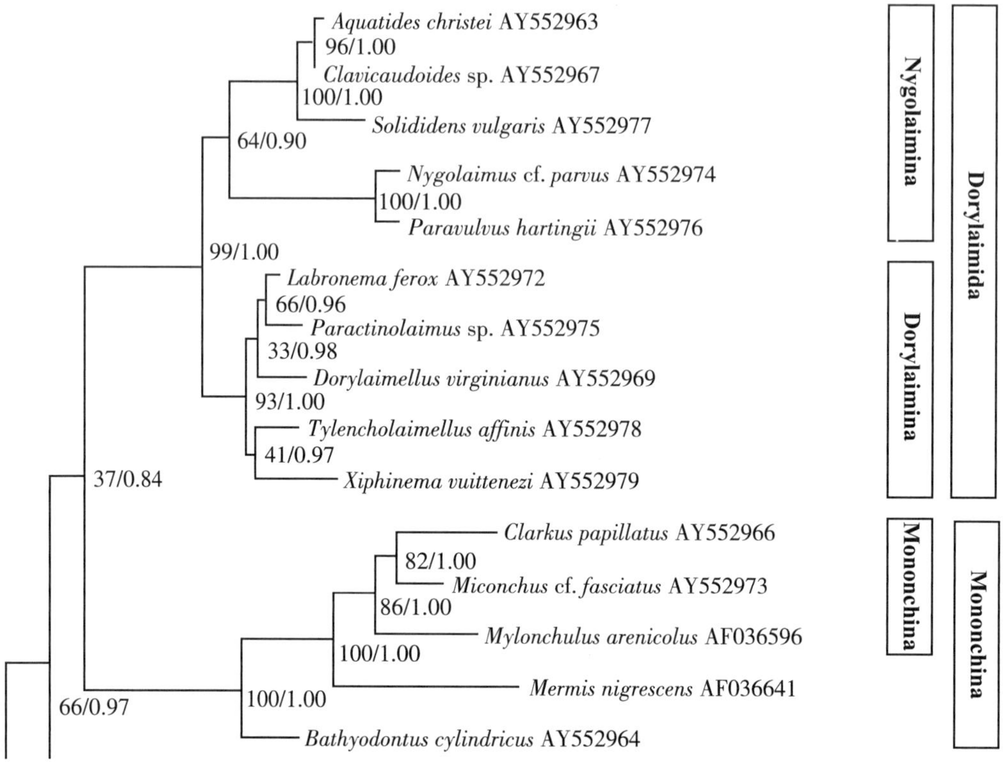

图 1　矛线目的系统进化树（改自 Mullin *et al.*，2005），
显示有穿咽亚目和矛线亚目两个分支，单齿目为外群

最后是迄今为止，只有少数矛线目线虫获得分子信息，而且分子系统进化分析结果与传统（形态学）方法得出的结果有时不相符，例如，Holterman 等（2008）发现矛线亚目中，在属和种的数量上是最重要的总科和几个科，不是自然群体。

矛线亚目是迄今为止最大和最多的亚目，包括 4 个总科 20 个科 2 500 个种以上，占据了矛线目 95.5%的有效种和 95.1%的有效属（Andrássy，2009）。近年来，矛线亚目的分类系统有较大的修订，但是形态学分类与分子系统发育树构建的结果仍然常存在不一致，各科常呈多系分布，各类群之间的系统发育关系仍不明确（Holterman *et al.*，2008；Peña-Santiago & Álvarez-ortega，2014）。Peña-Santiago（2014）提出不考虑总科，建立了一个更简化的矛线目分类系统（表 1），本文有关矛线目各科和属的介绍即采用该新分类系统。

4　矛线目线虫形态特征

矛线目成员体型相对较大，体长从小于 0.5mm 至近 1cm 不等，但大多数种类的长度从 1~3mm 不等（图 2A），杀死固定后，它们通常呈开放的“C”形，但有时体后部比前部更弯曲，呈“G”形，在特殊情况下，虫体几乎是笔直的。

矛线目线虫区别于其他线虫的主要形态特征包括：①口腔内有一个可突出的结构，轴向齿针（图 2B）或壁齿（图 2D）；②食道通常呈瓶状（图 2C），前部细长，后部膨大为一个肌肉发达的基部；③有分化明显的前直肠（图 2E）；④雄虫泄殖腔口前有一对交配乳突（图 2F）和腹中位一系列或一排单乳突。前直肠和一对泄殖腔前乳突是矛线目线虫的 2 个独有特征（Peña-Santiago，2014）。

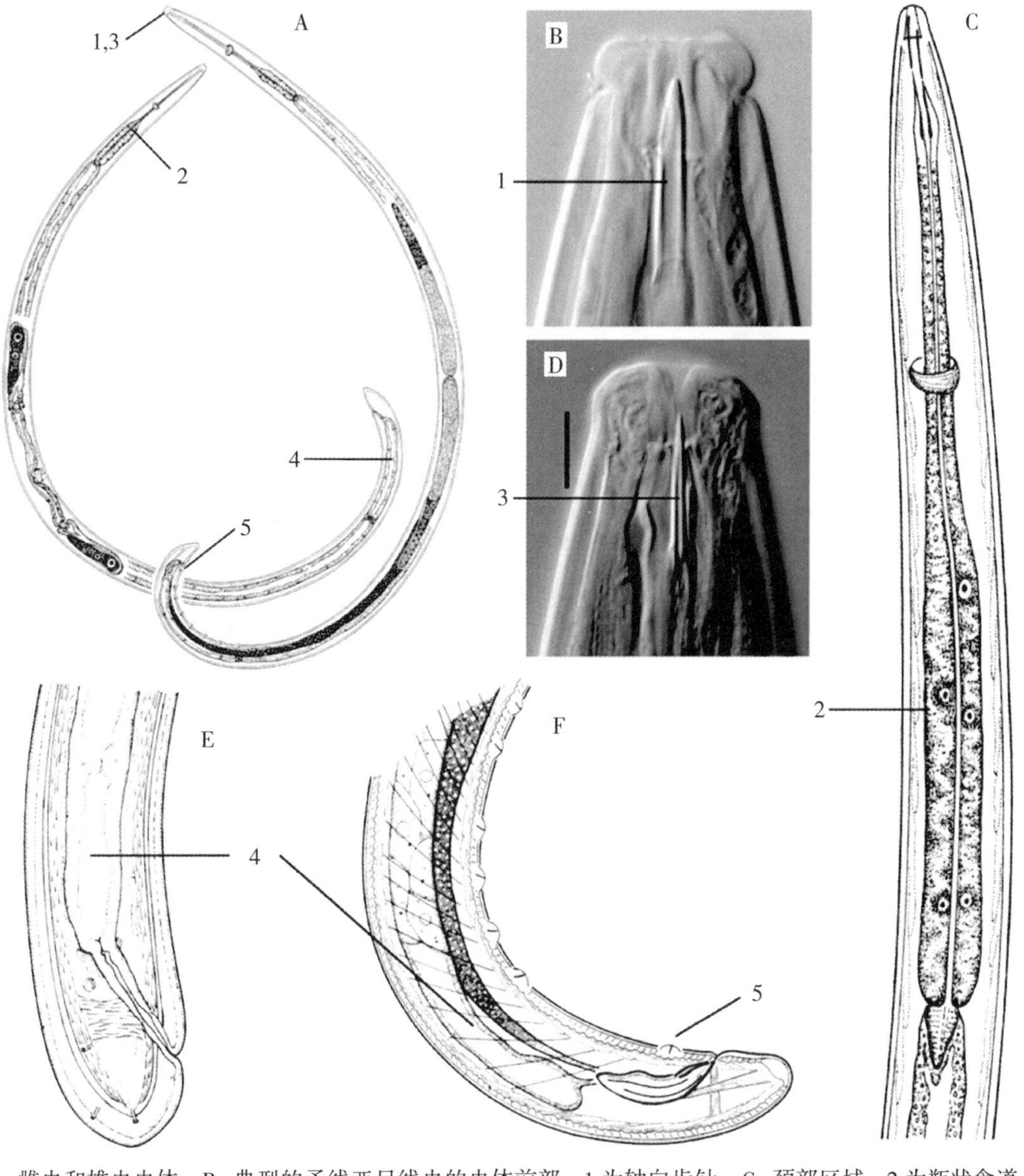

A. 雌虫和雄虫虫体；B. 典型的矛线亚目线虫的虫体前部，1 为轴向齿针；C. 颈部区域，2 为瓶状食道；D. 典型的穿咽亚目线虫的虫体前部，3 为壁齿；E. 雌虫虫体后部；F. 雄虫虫体后部，4 为直肠，5 为泄殖腔口前一对交配乳突

图 2　矛线目的诊断特征

（引自 Peña-Santiago，2014）

4.1 体壁（Cuticle）

矛线目线虫的角质层由内外两层组成，常有略微明显的横纹，偶有环纹，或有粗的纵脊。角质层上体孔（body pore），其连接着神经，感应外界环境的变化，发挥着躯体感应器的作用。体孔沿虫体侧面排成一列或两列，或存在一些背侧和/或腹侧的体孔。体孔在食道、肠和尾等不同部位的数量及其分布方式是一个重要的鉴定特征。

4.2 唇区（Lip region）

矛线目线虫的唇区形态多变，呈圆形、角形、截形等，唇区与相邻身体完全连续，或凹陷，或缢缩。唇区六角对称性，6 枚唇片通常明显分离，偶有融合；口周的唇片可分化为小唇（liplets）或盘状结构等；感觉器通常是 6+6+4 模式（6 个内唇乳突、6 个外唇乳突和 4 个头乳突）。

4.3 侧器（Amphid）

矛线目的侧器由侧器囊（fovea）和侧器口（amphid aperture）组成，侧器口一般位于唇区与身体的交界处，通常是裂缝状或孔状，其大小与相应虫体体宽的比例是重要的形态学鉴定特征；侧器囊位于唇区后，呈杯状、袋状或者高脚杯形等，形态和大小多变，有些种类的侧器囊角质层壁较厚甚至骨化。

4.4 取食器官（Feeding apparatus）

矛线目线虫的口（stoma）通常是一个管状结构，连接口孔和食道衬里。它由 4 个区域组成。①口腔前庭（cheilo-stom 或 vestibulum），从口腔开口到固定导环的这一段区域，前庭的角质层壁通常薄到中等发达，或前庭具有特殊骨化结构，可能呈篮状、圆锥形、长瓶形或管状，或是具小骨化片。②引导器，由一个固定导环和一个薄膜的管状导鞘（guiding sheath）组成，导鞘柔韧可动，其基部与齿针基部相连接，当齿针完全收缩或伸出时，导鞘可形成“单”或“双”导环，导环到虫体前端的距离通常是固定不变的，是鉴定种和属的一个重要形态学特征。③一个突出的壁齿（mural tooth）或轴向齿针（odontostyle）结构，壁齿常缺开口，齿针是一个管状结构，开口于背侧，其长度、开口大小及其长宽比例等是极其重要的分类鉴定依据。④口的最后部分为齿托，其起源和组织学与齿针不同，也是一个管状结构，前端连接齿针，后端连接食道衬里。

4.5 食道（Oesophagus）

矛线目线虫的食道由一个较细长的前部和膨大的基部组成，食道前部和后部的交界处，一般是略微逐渐扩大，或出现一个收缩，或出现一个短的峡部。膨大基部通常占总长度的 2/5～1/2（图 3 C1－C3），它可能更长或非常短，呈球状，占总长度 1/5～1/3（图 3 C4－C5）。在少数属中，食道基部包被螺旋状的肌肉鞘（图 3 C3）。食道基部具有 5 个食道腺细胞（偶尔是 3 个），1 个背食道腺、2 对亚腹食道腺，腺细胞核和开口的位置具有分类学意义。食道与肠的交界处有一个类似于阀门结构的贲门，控制着食物从食道向肠的流动，其前端通常被肠道组织包围。穿咽亚目线虫的贲门周围有三个大的贲门腺细胞（图 3 C2）。矛线目的肠道分为两个部分，一个是前部较长的肠道本体，另一个是后部较短的前直肠。消化道的最后部分是直肠，雌虫的直肠结构简单，而雄虫的直肠与射精管开口相连接，形成泄殖腔，并具有交配结构。

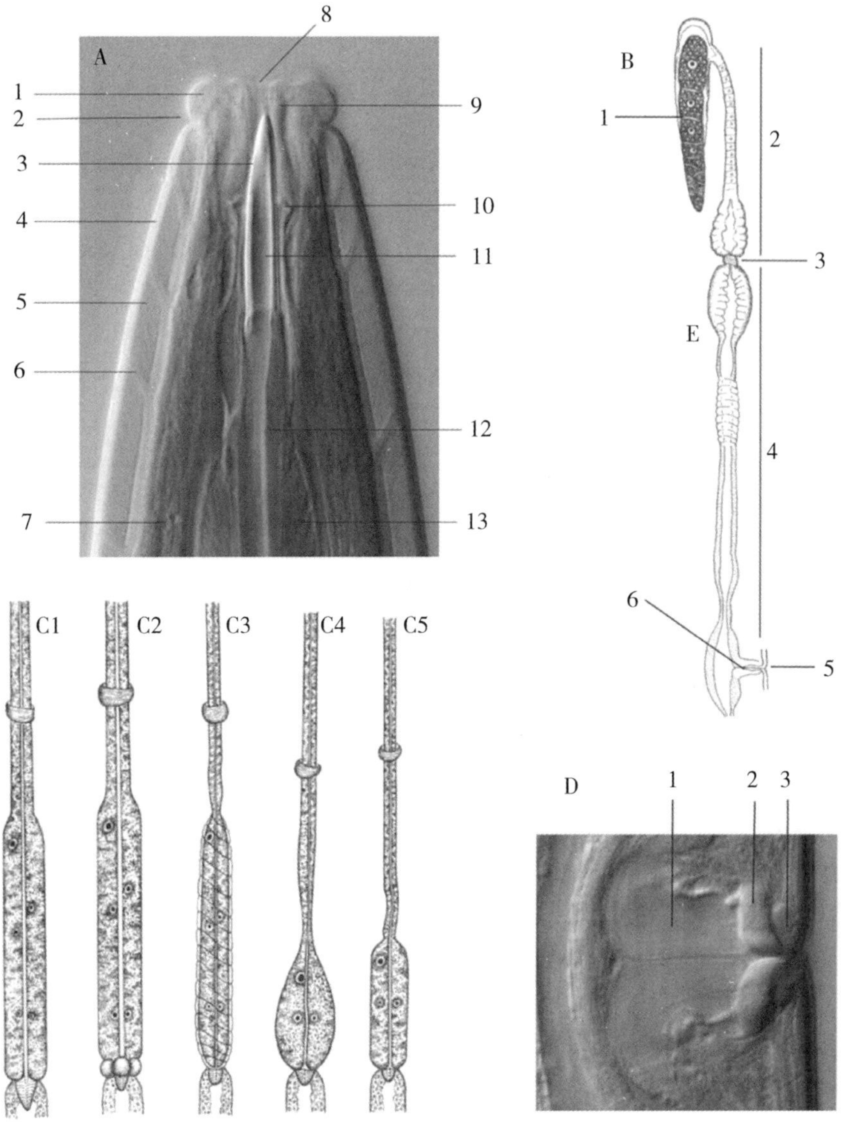

A. 虫体前部区域，显示 1 唇区、2 唇区收缩、3 齿针口孔、4 角质膜外层、5 角质膜内层、6 背侧体孔、7 体壁肌肉组织、8 口腔孔、9 唇口壁、10 导环、11 齿针、12 齿托和 13 食道前端；B. 雌虫前生殖器管，显示 1 卵巢、2 输卵管、3 括约肌、4 复合子宫、5 阴门和 6 阴道；C. 食道类型，C1 为矛线型，C2 为穿咽型，具贲门腺，C3 为颚针型，具螺旋肌肉鞘，C4 和 C5 为细齿型，具食道基球；D. 阴道，显示其通常的 3 个部分，即 1 阴道近端部、2 阴道曲折部和 3 阴道远端部

图 3　矛线目的一般形态特征

（引自 Peña-Santiago，2014）

4.6　雌虫生殖系统（Female reproductive system）

矛线目线虫雌虫通常包括前后 2 个生殖管（didelphic），或者其中一个生殖管退化或完全缺失（monodelphic）。完整的生殖管（genital tract）由卵巢、输卵管和子宫组成，卵巢与

输卵管的交界处总是回折的，其远端指向阴门，而输卵管和子宫由一个括约肌分开。子宫连接阴道，阴道开口于体表的阴门（图 3B）。阴道通常由近端部（pars proximalis）、曲折部（pars reringens）和远端部（pars distalis）3 部分组成（图 3D）。阴门通常横裂，偶尔纵裂，或孔口状。雌性生殖系统的变化具有分类学上的意义，特别是生殖管的数量以及输卵管、子宫和阴道的形态。

4.7 雄虫生殖系统（Male reproductive system）

矛线目雄虫生殖系统结构较为统一，总是由对生的两个精巢组成（diorchic），精巢共同连接一个输精管（vas deferens），然后连接射精管（ejaculatory duct）和直肠，构成泄殖腔（cloaca）。交配器官包括一对交合刺（spicules）、侧导片（lateral guiding pieces）、引带（gubernaculum）、交配乳突（supplements）以及交配肌（copulatory muscles）等。矛线目雄虫交合刺可分为背线（dorsal contour）、腹线（ventral contour）、中段（median piece）、头（capitulum）和薄板（lamina）5 个部分。矛线目雄虫具有泄殖腔口前亚腹位的一对乳突和数量不等的腹中位单乳突，乳突的数量和排列方式具有分类学意义。

4.8 尾部（Caudal region）

矛线目线虫的尾形和长度多变，都具有分类学重要性。从长丝状到短圆状，以及各种中间态，长圆锥状、圆锥形、圆柱形、棍状、圆形、半球形等。雌雄虫尾形相似，或雌雄虫尾形不同，雌虫长丝状，雄虫短圆。

5 矛线目分类系统

自矛线目建立以来，大量的新属和新种被描述，矛线目的分类系统也发生了重大的变化。目前，分子证据已经明确矛线目的两个亚目即穿咽亚目（Nygolaimina）和矛线亚目（Dorylaimina）均为单系类群，而庞大的矛线亚目下的各科大多是多系类群，各类群之间的系统发育关系仍不明确，因此，Peña-Santiago（2014）提出了一个不考虑总科的简化的矛线目分类系统，矛线目包括 2 个亚目和 18 个科。Peña-Santiago（2021）整理了矛线目 285 个有效属和 3 088 个有效种。

5.1 矛线目各科检索表

Peña-Santiago（2014）整理了矛线目 2 亚目 18 科的检索表，如下。

检索表

01a 口腔内有一个无孔口的壁齿，食道与肠交界处有 3 个大的贲门细胞
……［穿咽亚目 Nygolaimina］…02

01b 口腔内有一个有明显孔口的轴向齿针，食道与肠交界处没有大贲门细胞
……［矛线亚目 Dorylaimina］…05

02a 口腔前庭强烈骨化，有唇和后唇两个不同的部分……异咽科 Aetholaimidae

02b 口腔前庭简单，非骨化……03

03a 食道膨大部包被有明显的螺旋状肌肉鞘；雌虫生殖管单后生；雌虫尾长棍棒状
……小穿科 Nygellidae

03b 食道膨大部没有包被螺旋状肌肉鞘，偶尔可能有弱的肌肉鞘；雌虫生殖管双生；雌虫

尾通常较短………………………………………………………………………………… 04

04a 角质层较厚；食道膨大部约为头端至贲门处长度的 2/3；没有贲门细胞，但可能有贲门盘；雄虫腹中位交配乳突众多，且发育良好 ……………… 小穿咽科 Nygolaimellidae

04b 角质层较薄；食道膨大部约为头端至贲门处长度的 1/2（少数超过 1/2）；贲门细胞明显；雄虫腹中位交配乳突很少（1~2 个，偶尔 6~8 个）且发育不良 ……………………………………………………………………………… 穿咽科 Nygolaimidae

05a 食道膨大部包被有发达的螺旋状肌肉鞘 ……………………… 颚针科 Belondiridae

05b 食道膨大部没有包被明显的螺旋状肌肉鞘………………………………………………… 06

06a 食道膨大部较短，球状，约为 1/3 的头端至贲门处长度………………………………… 07

06b 食道膨大部较长，非球状，大于 1/3 的头端至贲门处长度……………………………… 11

07a 食道由三部分组成：细长的前部逐渐向后膨大，中间部，更细长的部分有腺体组织，以及基部膨大呈梨形球体，有发达的瓣门腔 ………………… 类管咽科 Aulolaimoididae

07b 食道由两部分组成：细长的前部和膨大的基部………………………………………… 08

08a 典型的垫咽类（tylencholaimoid）角质层：内层轮廓不规则，与外层分开，有射线状折射或固定褶皱；齿针较细，针腔窄，开口小 ……………………… 细齿科 Leptonchidae

08b 典型的矛线类（dorylaimoid）角质层：内层轮廓规则，靠近外层，没有射线状折射或固定褶皱；齿针较粗壮，有明显的针腔和开口……………………………………………… 09

09a 齿针很长，是唇区直径的 4 倍多，针腔很窄，开口很小 ………… 长针科 Longidoridae

09b 齿针不很长，小于唇区直径的 2 倍，有可见的针腔和开口…………………………… 10

10a 齿针不对称，腹臂短于背臂，开口大；食道球约为头端至贲门处长度的 1/4~1/3 ……………………………………………………………………………… 湿生科 Mydonomidae

10b 齿针对称，管状，通常有小开口；食道球约为头端至贲门处长度的 1/5~1/4 ……… …………………………………………………………………… 小垫咽科 Tylencholaimellidae

11a 典型的垫咽类（tylencholaimoid）角质层：内层轮廓不规则，与外层分开，有射线状折射或固定褶皱；唇区呈帽状，口周区域高或突起 …………… 垫咽科 Tylencholaimidae

11b 典型的矛线类（dorylaimoid）角质层：内层轮廓规则，靠近外层，没有射线状折射或固定褶皱；唇区不呈帽状，口周区域不突起，甚至有些缢缩……………………………… 12

12a 前口腔宽，骨化严重，有 4 个结实的咽齿环绕齿针 …………… 角咽科 Actinolaimidae

12b 前口腔通常狭窄，很少骨化，没有 4 个咽齿……………………………………………… 13

13a 齿针纤细，长于 7 倍针宽），开口小于 1/5 齿针长，针腔狭窄 …… 诺尔迪科 Nordiidae

13b 齿针不纤细（小于 7 倍针宽长），开口明显（大于 1/4 齿针长），针腔宽 ………… 14

14a 齿针相对较短（少于唇区宽）且粗壮（小于 5 倍针宽），开口较长（大于 1/2 齿针长）……………………………………………………………………… 孔咽科 Aporcelaimidae

14b 齿针与唇区宽一样长或更长，更细长（5~7 倍针宽长），开口更短（小于 1/2 齿针长）……………………………………………………………………………………………… 15

15a 唇部和后唇部骨化，且发育良好；S_2N 位置较前，远离食道基部；雌虫生殖管单后生，偶尔双生……………………………………………………… 索恩线虫科 Thornenematidae

15b 唇部非骨化，后唇部偶尔骨化；S_2N 位置靠近食道基部；雌虫生殖管双生 ………… 16

16a 雌虫尾较长（大于 3 个肛门处体宽），很少呈圆锥形、细长和较短；雄虫尾，或长而与雌虫相似，或圆形而与雌虫不同 ………………………………… 矛线科 Dorylaimidae

16b 雌虫尾较短（少于 3 个肛门处体宽），圆锥形至圆形；雄虫尾与雌虫相似 ………… 17

17a 阴道膨胀，缺乏发达的曲折部和远端部；腹中位交配乳突 0~4 个；引带通常存在，且很小 …………………………………………………………………… 索氏科 Thorniidae

17b 阴道有发达的曲折部和/或远端部；腹中位交配乳突很多；缺引带 ……………………………………………………………… 库西亚线虫科 Qudsianematidae

5.2 矛线目各科和属分类

Peña-Santiago（2014）对矛线目 2 个亚目 18 个科的形态鉴别特征分别进行了简要介绍，Peña-Santiago（2021）汇编了矛线目各科各亚科共计 285 个有效属。本文给出了所有线虫物种的中文名，已有中译名的，保留或据原始描述文献资料进行更正，没有中译名的，则根据文献资料以及《生物名称和生物学术语的词源》（耶格，1965）词典进行翻译。各科中各属的种类数量说明，依据 Andrássy（2009）的统计数据。

5.2.1 矛线目 Dorylaimida

鉴别特征：体型通常较大，角质膜光滑，偶尔有脊。唇区形态多变，唇乳突排列模式 6+6+4 个。口腔内有轴向齿或壁齿。瓶状食道，膨大基部有 5 个（或 3 个）腺体，有前直肠。雌虫生殖管双生或单生，卵巢回折，阴道通常有 3 个部分（近端部、曲折部和远端部）；阴门多为横裂，偶有纵裂或孔状。雄虫有两个精巢，有一对泄殖腔口前交配乳突和一系列的腹中位乳突。尾形多变，长丝状到短圆状，雌雄虫尾形相似或不相似。尾腺缺乏。

5.2.2 穿咽亚目 Nygolaimina

鉴别特征：口腔内有位于亚腹口腔壁的壁齿，食道-肠交界处有 3 个大的贲门腺细胞。穿咽亚目是一个自然群（Mullin *et al.*，2005），只包含一个穿咽总科（Nygolaimoidea）（Thorne，1935），有 4 个科，种类多样性较差。Andrássy（2009）列出 13 个有效属和 119 个有效种，但是 Peña-Santiago（2021）列出了 4 科 14 个有效属。

5.2.2.1 异咽科 Aetholaimidae

鉴别特征：体长小于 2mm。唇区略似盘状，略有缢缩，口腔前庭强烈骨化，壁齿呈正三角形至线形。食道膨大部被一个精致的鞘所包被，有时形成基部袋。有 3 个发达的贲门细胞。雌虫双生殖管，阴道有 3 个部分，阴门横裂。雌虫尾圆形、圆锥形至半球形。雄虫未知。该科线虫比较罕见，只有一个有效属即异咽属（*Aetholaimus*）（Williams，1962），有效种 5 个。

5.2.2.2 小穿科 Nygellidae

鉴别特征：体长小于 2mm。角质层薄，有精细的横向条纹。唇区连续，唇片融合。壁齿呈正三角形至线形。口腔前庭简单。食道膨大部被明显的螺旋状鞘所包被。有 3 个贲门细胞。雌虫单生殖管；阴道有两个部分，缺乏曲折部；阴门横裂。雌虫尾棍棒状。雄虫未知。该科线虫比较罕见，只有一个有效属即小穿属（*Nygellus*）（Thorne，1939），有 6 个有效种。

5.2.2.3 小穿咽科 Nygolaimellidae

鉴别特征：体长大于 2mm。角质层厚度中，有明显的侧体孔。唇区连续或缢缩。口腔前庭简单，壁齿三角状或矛状。食道膨大基部的长度大于颈部总长度的一半。食道-肠交界处有或没有贲门盘（cardiac disc），无贲门细胞。雌虫双生殖管；阴道有两段，缺乏曲折部，阴门横裂。雄虫有矛线型交合刺和发达的交配乳突，无引带。雌雄虫尾相似，短圆锥状。该科线虫比较罕见，只有一个有效属即小穿咽属（*Nygolaimellus*）（Loos，1949），有 6 个有效种。

5.2.2.4　穿咽科 Nygolaimidae（图 4）

鉴别特征：体长通常大于 1mm。角质层薄，有细横纹。口腔前庭简单。壁齿的形状和大小不一。食道膨大部占总颈部长度的 1/2 或以上，有时包被一个薄鞘。有 3 个大的贲门细胞。雌虫双生殖管，阴道通常缺乏曲折部，阴门横裂，偶纵裂。雄虫交合刺矛线型，有侧导片，腹中位乳突通常较少且发育不良。尾形和大小多变，雌雄虫尾相似。

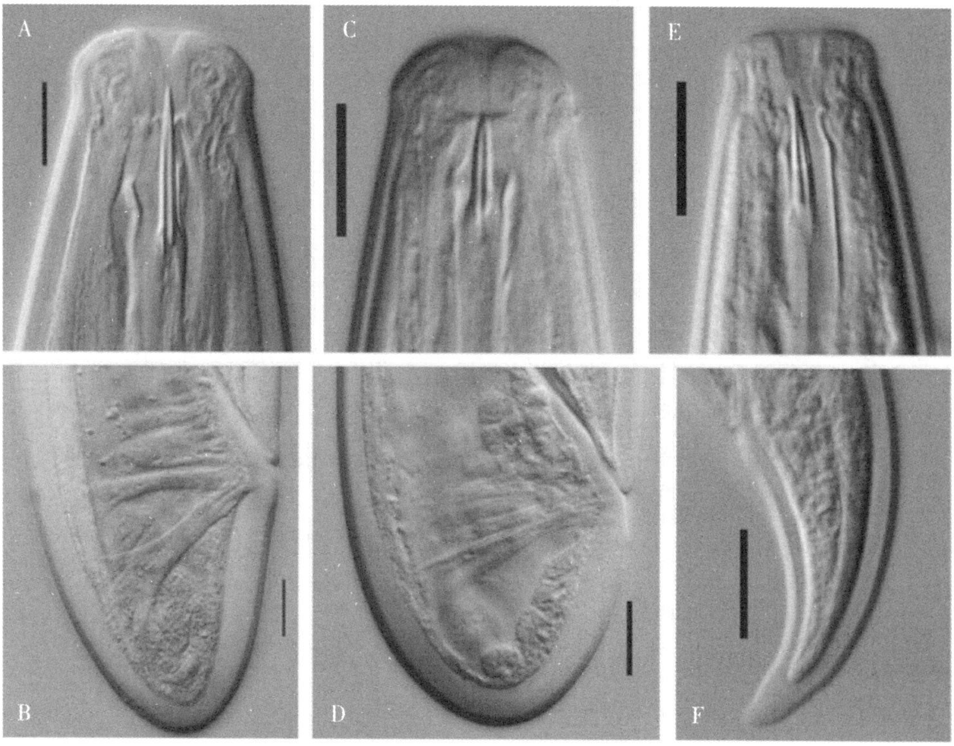

A、B. 水线属；C、D. 滑线属；E、F. 拟阴门属。A、C、E. 虫体前部，示壁齿；B、D、E. 尾部（标尺：10μm）

图 4　穿咽亚目穿咽科线虫的多样性

（引自 Peña-Santiago，2014，以下各图出处同此）

该科是一个重要的线虫类群，有 11 个有效属。在土壤和/或淡水栖息地最为常见的、多样性最高的属有 5 个，分别是水线属（*Aquatides*）（Heyns，1968），13 个有效种（图 4A，B）；类棒尾属（*Clavicaudoides*）（Heyns，1968），11 个有效种；滑线属（*Laevides*）（Heyns，1968），13 个有效种（图 4C，D）；穿咽属（*Nygolaimus*）（Cobb，1913b），34 个有效种，以及拟阴门属（*Paravulvus*）（Heyns，1968），17 个有效种（图 4E，F）。其他 6 个属比较罕见，包括非洲穿咽属（*Afronygus*）（Heyns，1968）、棒尾属（*Clavicauda*）（Heyns，1968）、

猛尖齿属（*Feroxides*）（Heyns，1968）、拟穿咽属（*Paranygolaimus*）（Heyns，1968）、前穿咽属（*Pronygolaimus*）（Sharma & Baqri，2006）、硬齿属（*Solididens*）（Heyns，1968）。

5.2.3 矛线亚目 Dorylaimina

鉴别特征：矛线亚目与穿咽亚目的主要区别是口腔内有轴向齿针，食道-肠道交界处没有大的贲门腺细胞。矛线亚目是一个自然群体，包含 250 个有效属和 2 518 个有效种（Andrássy，2009）。Andrássy（2009）将该亚目分为 4 个总科，分别是矛线总科（Dorylaimoidea）（de Man，1876）、颚针总科（Belondiroidea）（Thorne，1939）、长针总科（Longidoroidea Thorne，1935）以及垫咽总科（Tylencholaimoidea） （Filipjev，1934）。Holterman 等（2008）认为矛线亚目的总科分类系统没有坚实的形态学依据来支持，而分子进化分析揭示这些总科不是自然群体，是并系（paraphyletic）或多系（polyphyletic）类群。基于这些观点，Peña-Santiago（2014）提出矛线亚目的分类不考虑总科，而只考虑科，Peña-Santiago（2021）整理了矛线亚目 14 科的各属名录。

5.2.3.1 角咽科 Actinolaimidae（图 5A）

鉴别特征：体长 1~10mm。角质层光滑或有纵脊或沟纹。唇片融合；唇区有一个前部的角质化环，通常看起来有波纹。口腔前庭大，有骨化的环，偶尔有小齿，口腔后部包围有 4 个巨大的齿。齿针坚固，形态多变化。食道前部有肌肉或无肌肉；膨大基部占颈部总长度的一半。雌虫双生殖管；阴道曲折部通常发育良好；阴门横裂、纵裂或孔状。雄虫交配乳突在腹位排一列，或以 2 个或 3 个簇排列。雌雄虫的尾长丝状，或雌虫尾长丝状，雄虫尾短。

角咽科是一个自然分类群，强烈发达的口腔前庭很容易将其区别于其他科。Vinciguerra（1987）将角咽科分为 3 个亚科，Andrássy（2009）整理该科包含 3 个亚科 18 个有效属和 132 个有效种，其中最具代表性的属有拟角咽属（*Paractinolaimus*，27 个有效种）、埃格属（*Egtitus*，22 个有效种）以及新角咽属（*Neoactinolaimus*，17 个有效种），均分布在淡水栖息地，偶尔也会在土壤中发现。Peña-Santiago（2021）整理角咽科有 3 个亚科 20 个有效属，如下。

角咽科 Actinolaimidae Thorne，1939

角咽亚科 Actinolaiminae Thorne，1939

角咽属 *Actinolaimus* Cobb，1913a

非洲角咽属 *Afractinolaimus* Andrássy，1970

埃格属 *Egtitus* Thorne，1967

亚角咽属 *Mactinolaimus* Andrássy，1970

后角咽属 *Metactinolaimus* Meyl，1957

新角咽属 *Neoactinolaimus* Thorne，1967

伪角咽属 *Nothactinolaimus* Loof，1973

拟类角咽属 *Paractinolaimoides* Khan，Ahmad & Jairajpuri，1994

拟角咽属 *Paractinolaimus* Meyl，1957

硬角咽属 *Scleroactinolaimus* Ahmad，Khan & Ahmad，1992

口拟角咽属 *Stopractinca* Khan，Ahmad & Jairajpuri，1994

布里托线虫亚科 Brittonematinae Thorne，1967

角属 *Actinca* Andrássy, 1964

非洲角属 *Afractinca* Vinciguerra & Clausi, 2000

巴西角咽属 *Brasilaimus* Lordello & Zamith, 1957

布里托线虫属 *Brittonema* Thorne, 1967

拟口舌属 *Parastomachoglossa* Coomans & Loof, 1986

柔角头属 *Practinocephalus* Andrássy, 1974

西印度属 *Westindicus* Thorne, 1967

粗肋亚科 Trachypleurosinae Thorne, 1967

粗角咽属 *Trachactinolaimus* Andrássy, 1963

粗肋属 *Trachypleurosum* Andrássy, 1959

5.2.3.2 孔咽科 Aporcelaimidae（图 5B，图 5C）

鉴别特征：中等到非常大的线虫，长度可达 10mm。角质层厚，光滑或有细横纹，通常有十字线或点状物，体孔多。唇区通常缢缩，唇片分离。齿针粗短，齿针开口大于其长度的一半。雌虫双生殖管；阴道曲折部通常存在；阴门横裂、纵裂或孔状。雄虫交合刺矛线型，交配乳突间隔排列，偶尔连续排列。雌雄虫尾形相似，短圆锥状到圆形。

孔咽科是土壤、淡水沉积物中最常见、最丰富的线虫类群，其中，钝尾小孔咽线虫（*Aporcelaimellus obtusicaudatus*）是全世界分布最广的自由生活线虫。没有充分证据支持孔咽科是一个单系群体。Andrássy（2009）认为有 3 个亚科 13 个有效属和 144 个有效种，其中，最重要的属有小孔咽属（*Aporcelaimellus*，57 个有效种）、裂线虫属（*Sectonema*，24 个有效种）、孔咽属（*Aporcelaimus*，21 个有效种）以及后孔咽属（*Metaporcelaimus*，14 个有效种）。Peña-Santiago（2021）整理孔咽科有 3 个亚科 14 个有效属，如下。

孔咽科 Aporcelaimidae Heyns, 1965

孔咽亚科 Aporcelaiminae Heyns, 1965

强肌属 *Akrotonus* Thorne, 1974

小孔咽属 *Aporcelaimellus* Heyns, 1965

孔咽属 *Aporcelaimus* Thorne & Swanger, 1936

孔属 *Aporcelinus* Andrássy, 2009

小孔属 *Aporcella* Andrássy, 2002

尖咽属 *Epacrolaimus* Andrássy, 2000

马卡廷属 *Makatinus* Heyns, 1965

后孔咽属 *Metaporcelaimus* Lordello, 1965

席尔瓦勒属 *Silvallis* Ahmad & Jairajpuri, 1986

图比沙布属 *Tubixaba* Monteiro & Lordello, 1980

孔矛亚科 Aporcedorinae Andrássy, 2009

孔矛属 *Aporcedorus* Jairajpuri & Ahmad, 1983

裂线虫亚科 Sectonematinae Siddiqi, 1969

类孔咽属 *Aporcelaimoides* Heyns, 1965

竿齿属 *Scapidens* Heyns, 1965

裂线虫属 *Sectonema* Thorne，1930

5.2.3.3　瘤咽科 Aulolaimoididae（图 5D，图 5E）

鉴别特征：体长小于 2mm。角质层为矛线型。唇区连续，唇片合并。口腔前庭为平截的圆锥体，被一个有细小肋骨状成分的篮状结构所包围。齿针渐狭，有狭窄的针腔和开口；齿托基部明显增厚。食道由三部分组成：细长的部分逐渐向后膨大，更细长的中间部分由腺体组织包围，基部膨大为梨形球体，有发达的瓣室。雌虫生殖管双生或单生；阴道膨胀，缺乏曲折部。雄虫交合刺为细长的矛线型，交配乳突少且间隔。雌雄虫尾形相似，近圆筒形至丝状。

瘤咽科线虫非常少见，只有 4 个有效属 15 个有效种。三部分组成的食道是该科的一个独特的特征，并支持其单系。最具代表性的属是腺咽属（*Adenolaimus* Andrássy，1973），有 6 个有效种，瘤咽属（*Aulolaimoides* Micoletzky，1915）也有 6 个有效种。其他两个属分别是枝头属（*Cladocephalus* Swart & Heyns，1991）和奥斯藤布林克属（*Oostenbrinkia* Ali，Suryawanshi & Ahmad，1973）。

5.2.3.4　颚针科 Belondiridae（图 5F-图 5H）

鉴别特征：虫体体长常约 1mm，或更长。唇区狭窄，连续或缢缩。齿针短，很少长于唇区宽度；齿托棒状，偶尔有基部凸缘。食道的细长前部略有肌肉，逐渐延伸，但有时通过一个类似峡部的部分与基部膨大分开；基部膨大部变化很大，总包被一个发达的螺旋状肌肉鞘。贲门常被肠道组织包裹。雌虫生殖管双生或单生；阴道形态多变；阴门横裂、纵裂或孔状。雄虫交合刺大小和形状各异，交配乳突少数到许多，呈间隔或连续。雌雄虫尾形相似或不同，短圆到长丝状。

颚针科成员常习居土壤中，偶尔也出现在淡水沉积物中。食道膨大部包被一个螺旋状的肌肉鞘是颚针科唯一的共同鉴别特征，分子进化树强烈支持其不是单系类群。Andrássy（2009）认为颚针科包括 3 亚科 36 个有效属 242 个有效种，其中多样性最高的属是牙咽属（*Dorylaimellus*，63 个有效种）、缢咽属（*Axonchium*，52 个有效种）以及颚针属（*Belondira*，42 个有效种）。Peña-Santiago（2021）整理颚针科有 3 个亚科 32 个有效属，如下。

颚针科 Belondiridae Thorne，1939

颚针亚科 Belondirinae Thorne，1939

双颚针属 *Amphibelondira* Rahman，Jairajpuri，Ahmad & Ahmad，1987

近颚针属 *Anchobelondira* Coomans & Nair，1971

缢咽属 *Axonchium* Cobb，1920

类缢咽属 *Axonchoides* Thorne，1967

小刺缢咽属 *Belaxellus* Thorne，1974

颚针属 *Belondira* Thorne，1939

小颚针属 *Belondirella* Thorne，1964

泡线虫属 *Bullaenema* Sauer，1968

指缢咽属 *Dactyluraxonchium* Coomans & Nair，1975

卷颚针属 *Helicobelondira* Yeates，1973

海恩斯缢咽属 *Heynsaxonchium* Coomans & Nair，1975

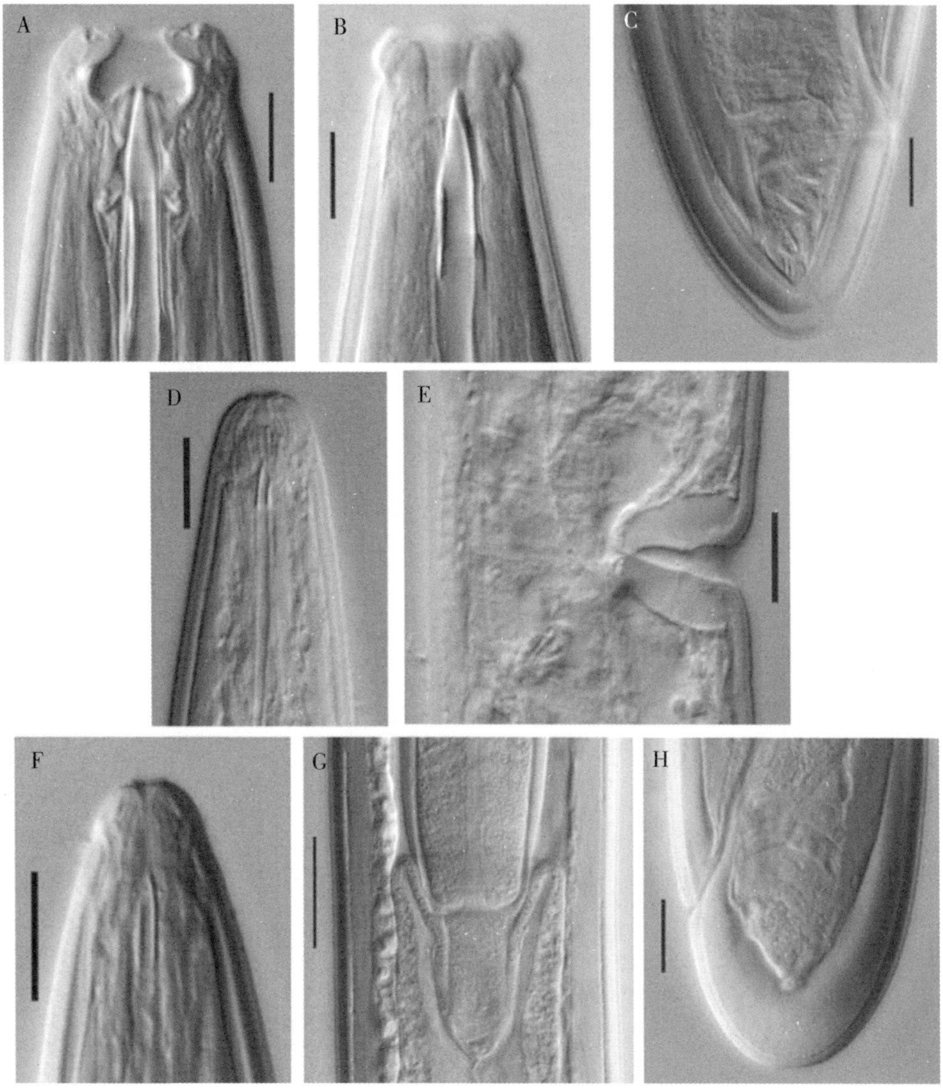

A. 角咽科，角咽属，虫体前部，示强烈骨化的口腔前庭；B、C. 孔咽科，小孔咽属，B 示虫体前部，有大开孔的齿针；C 示尾部有三层角质层；D、E. 瘤咽科，瘤咽属，D 示虫体前部，E 示阴门；F-H. 颚针科，并唇缢咽属，F 示虫体前部，G 示食道与肠交界处，食道膨大部包被明显的肌肉鞘，H 示尾部（标尺：10μm）

图 5 矛线亚目的多样性（一）

巨咽属 *Immanigula* Andrássy，1991

后缢咽属 *Metaxonchium* Coomans & Nair，1975

过咽属 *Nimigula* Andrássy，1985

阴茎缢咽属 *Phallaxonchium* Jairajpuri & Dhanachand，1979

波特线虫属 *Porternema* Suryawanshi，1972

前颚针属 *Probelondira* Andrássy，2009

并唇缢咽属 *Syncheilaxonchium* Coomans & Nair，1975

独缢咽属 *Uniqaxonchium* Dhanam & Jairajpuri，1988

牙咽亚科 Dorylaimellinae Jairajpuri，1964

牙咽属 *Dorylaimellus* Cobb，1913a

斯万格亚科 Swangeriinae Jairajpuri，1964

小硬线虫属 *Durinemella* Andrássy，2009

镰矛属 *Falcihasta* Clark，1964

胡尔克属 *Hulqus* Siddiqi，1982

林德赛属 *Lindseyus* Ferris & Ferris，1973

尖颚针属 *Oxybelondira* Ahmad & Jairajpuri，1979

尖针属 *Oxydirus* Thorne，1939

拟尖颚针属 *Paraoxybelondira* Dhanam & Jairajpuri，1999

拟尖针属 *Paraoxydirus* Jairajpuri & Ahmad，1979

拟库西亚属 *Paraqudsiella* Siddiqi，1982

库西亚属 *Qudsiella* Jairajpuri，1967

罗克属 *Roqueus* Thorne，1964

斯万格属 *Swangeria* Thorne，1939

5.2.3.5　矛线科 Dorylaimidae（图 6A–图 6D）

鉴别特征：虫体体长 0.8~9.0mm。角质层光滑或有细横纹，偶有纵脊。齿针直，针腔较宽，孔口大约是其长度的 1/3；导环单环或双环；齿托棒状。食道肌肉发达，膨大基部约占其长度的一半。雌虫双生殖管；有阴道曲折部；阴门横裂或纵裂。交合刺一般为矛线型；交配乳突在数量和排列上变化大。雌雄虫尾通常异形，雌虫尾丝状，雄虫尾短圆，有些种的雌雄虫尾都是长丝状。

矛线科是一个重要的科，特点是长尾，雌虫总是双生殖管，目前是一个并系类群。Andrássy（2009）认为矛线科包括 6 亚科 24 个有效属 336 个有效种，其中，最多样化的属是中矛线属（*Mesodorylaimus*，145 个有效种）、咽矛属（*Laimydorus*，43 个有效种）、矛线属（*Dorylaimus*，29 个有效种）和前矛线属（*Prodorylaimus*，20 个有效种），在水生或半水生生境中很常见，特别是中矛线属和前矛线属线虫在土壤中常见。Peña-Santiago（2021）整理矛线科有 4 个亚科 40 个有效属，如下。

矛线科 Dorylaimidae de Man，1876

非洲矛线亚科 Afrodorylaiminae Andrássy，1969

非洲矛线属 *Afrodorylaimus* Andrássy，1964

矛线亚科 Dorylaiminae de Man，1876

巴拉矛线属 *Baladorylaimus* Andrássy，2001

距矛线属 *Calcaridorylaimus* Andrássy，1986

丽矛线属 *Calodorylaimus* Andrássy，1969

金矛属 *Chrysodorus* Jiménez-Guirado & Cadenas，1985

黄矛线属 *Crocodorylaimus* Andrássy，1988

矛线属 *Dorylaimus* Dujardin, 1845
镰矛线属 *Drepanodorylaimus* Jairajpuri, 1966
硬唇杆属 *Fuscheila* Siddiqi, 1982
海矛线属 *Halodorylaimus* Andrássy, 1988
殊矛线属 *Idiodorylaimus* Andrássy, 1969
髋矛线属 *Ischiodorylaimus* Andrássy, 1969
顶矛线属 *Karadorylaimus* Andrássy, 2011
凯蒂矛线属 *Kittydorylaimus* Andrássy, 1998
咽矛属 *Laimydorus* Siddiqi, 1969
中矛线属 *Mesodorylaimus* Andrássy, 1959
中新矛线属 *Miodorylaimus* Andrássy, 1986
米拉线虫属 *Miranema* Thorne, 1939
纳马线虫属 *Namaquanema* Heyns & Swart, 1993
拟矛线属 *Paradorylaimus* Andrássy, 1969
犀牛矛线属 *Rhinodorylaimus* Ahmad, Baniyamuddin & Tauheed, 2010
硬矛线属 *Sclerodorylaimus* Ahmad, Tauheed & Baniyamuddin, 2010
唇线虫亚科 Labronematinae Peña-Santiago & Álvarez-Ortega, 2014
厚咽属 *Crassogula* Andrássy, 1991
厚唇属 *Crassolabium* Yeates 1967
海恩斯线虫属 *Heynsnema* Peña-Santiago, Guerrero & Ciobanu, 2008
唇线虫属 *Labronema* Thorne, 1939
小唇线虫属 *Labronemella* Andrássy 1985
内华达线虫属 *Nevadanema* Álvarez-Ortega & Peña-Santiago 2012
刀针属 *Scalpelus* Ahmad 2004
斯基贝线虫属 *Skibbenema* Van Reenen & Heyns 1986
托鲁马纳瓦属 *Torumanawa* Yeates 1967
前矛线亚科 Prodorylaiminae Andrássy, 1969
双矛线属 *Amphidorylaimus* Andrássy, 1960a
离矛线属 *Apodorylaimus* Andrássy, 1988
坤巨矛线属 *Kunjudorylaimus* Dhanam & Jairajpuri, 2000
线缢咽属 *Mitoaxonchium* Yeates, 1973
类尖针属 *Oxydiroides* Altherr, 1972
小前矛线属 *Prodorylaimium* Andrássy, 1969
前矛线属 *Prodorylaimus* Andrássy, 1959
原矛线属 *Protodorylaimus* Andrássy, 1988
莱佛士属 *Rafflesius* Ahmad, 2007

5.2.3.6 细齿科 Leptonchidae（图 6E，图 6F）

鉴别特征：虫体体长可达 3mm。角质层是典型的垫咽型，有松散的内层和丰富的放射状折射。唇区呈帽状，通常缢缩，或有口周盘。齿针非常多变，常渐狭或针状，短于唇区的宽度；齿托杆状，基部分化有或无。食道前部细长，肌肉弱；基部膨大呈球状，占颈部总长度的1/4~1/3，偶有一个瓣室；3个食道腺核。雌虫生殖管双生或单生，阴道没有曲折部，阴门横裂，偶尔纵裂或孔状。雄虫交合刺矛线型，交配乳突少且间隔。雌雄虫尾形相似，短圆，偶尔圆锥形或长丝状。

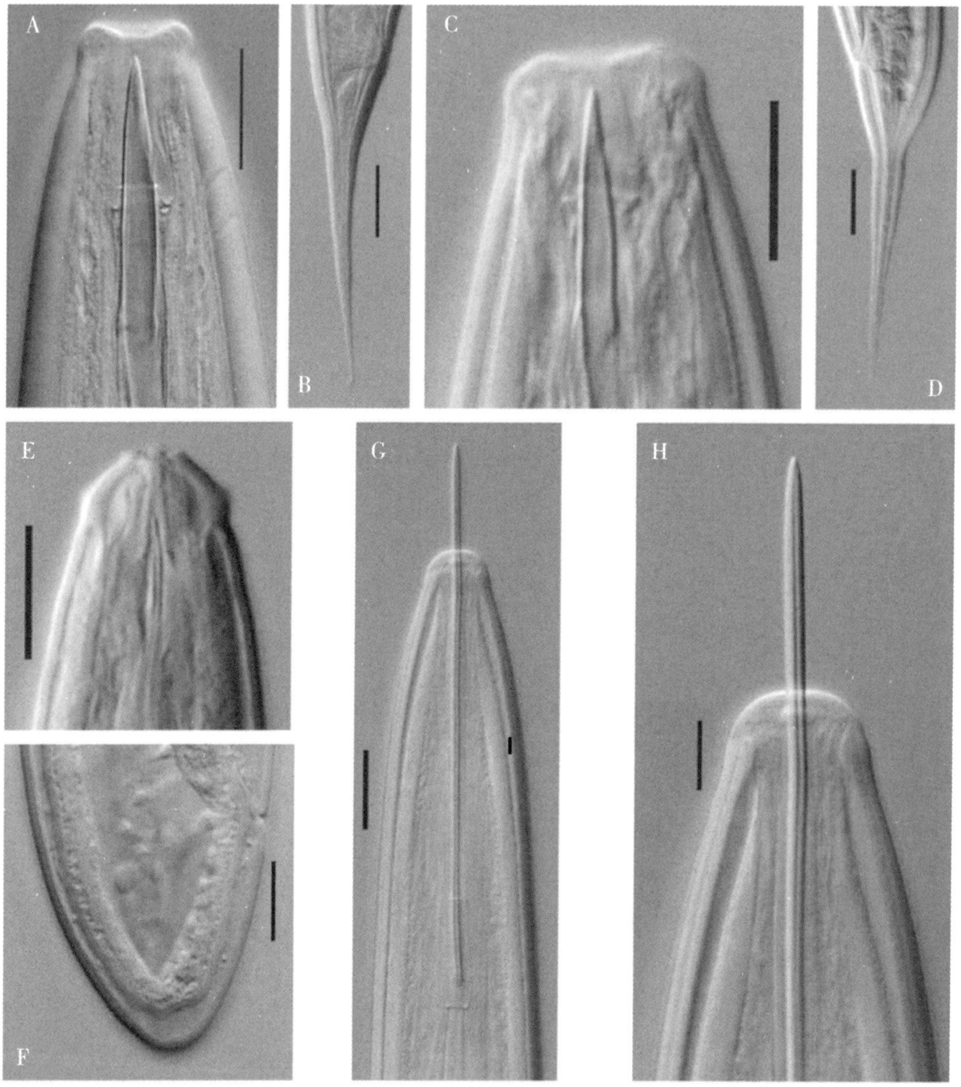

A-D. 矛线科，A、B为矛线属，A示虫体前部、B示尾部，C 、D为中矛属，C示虫体前部、D示尾部；E、F. 细齿科，富纳尔属，E示虫体前部，帽状唇区和细长的齿针，F示尾部，垫咽型角质层；G、H. 长针科，剑线虫属，G示虫体前部，非常细长的齿针，H示虫体前部的细节

（标尺：A、D、G、H为20μm；B为50μm；C、E、F为10μm）

图6　矛线亚目的多样性（二）

细齿科的特点是具有垫咽型角质层，齿针非常细长，食道膨大部呈球状，推测可能是一

个自然群体，但尚无分子数据支持。细齿科线虫在土壤或淡水中都不是很频繁或丰富，其摄食行为也欠了解。Andrássy（2009）认为该科有 6 个亚科 25 个属 134 个有效种，其中最重要的属是底垫裙属（*Basirotyleptus*，26 个有效种）、前细齿属（*Proleptonchus*，15 个有效种）、小剑线虫属（*Xiphinemella*，15 个有效种）、富纳尔属（*Funaria*，12 个有效种）以及细齿属（*Leptonchus*，11 个有效种）。Peña-Santiago（2021）整理细齿科有 6 个亚科 25 个有效属，如下。

细齿科 LeptonchidaeThorne，1935
- 阿甘线虫亚科 Arganematinae Siddiqi，2016
  - 阿甘线虫属 *Arganema* Siddiqi，2016
- 尖齿亚科 Belonenchinae Thorne，1964
  - 针齿属 *Aculonchus* Siddiqi，1983
  - 底垫裙属 *Basirotyleptus* Jairajpuri，1964
  - 小钩矛属 *Glochidorella* Siddiqi，1982
  - 硬针属 *Sclerostylus* Goseco，Ferris & Ferris，1981
  - 毛齿属 *Trichonchium* Siddiqi & Khan，1964
  - 寻咽属 *Zetalaimus* Siddiqi，1983
- 细齿亚科 Leptonchinae Thorne，1935
  - 后细齿属 *Apoleptonchus* Siddiqi，1982
  - 伯特祖克曼属 *Bertzuckermania* Khera，1970
  - 空齿属 *Caveonchus* Siddiqi，1982
  - 棒咽属 *Clavigula* Siddiqi，1995
  - 富纳尔属 *Funaria* Van der Linde，1938
  - 印加线虫属 *Incanema* Andrássy，1997
  - 劳顿线虫属 *Lawtonema* Siddiqi，1999
  - 细齿属 *Leptonchus* Cobb，1920
  - 梅尔属 *Meylis* Goseco，Ferris & Ferris，1974
  - 拟细齿属 *Paraleptonchus* Dhanam & Jairajpuri，1999
  - 类前细齿属 *Proleptonchoides* Ferris，Goseco & Kumar，1979
  - 前细齿属 *Proleptonchus* Lordello，1955
- 刀齿亚科 Scalpenchinae Peña-Santiago，2006
  - 刀齿属 *Scalpenchus* Siddiqi，1995
- 垫裙亚科 Tyleptinae Jairajpuri，1964
  - 裸垫裙属 *Gymnotyleptus* Ahmad & Jairajpuri，1982
  - 垫裙属 *Tyleptus* Thorne，1939
  - 犹他线虫属 *Utahnema* Thorne，1939
- 小剑线虫亚科 Xiphinemellinae Jairajpuri，1964
  - 坎特巴拉属 *Kantbhala* Siddiqi，1982
  - 小剑线虫属 *Xiphinemella* Loos，1950

5.2.3.7 长针科 Longidoridae（图 6G，图 6H）

鉴别特征：虫体细长，中至大型，体长可达 12mm。侧器形态各异，偶尔为两叶状；侧器口为裂缝状或孔状。齿针非常细长，有细小的针腔和针孔；导环单环或双环，位置可变；齿托棒状或有基部凸缘。食道前半部分是非肌肉的管状，膨大基部占颈部长度的 1/4，只有 3 个食道腺细胞核。雌虫生殖管双生或单生，阴道没有曲折部。交合刺为矛线型，有间隔的交配乳突。雌雄虫尾形相似，短圆锥形到圆形，偶尔长丝状。

长针科的一个高度多样化的类群，形态学和分子证据支持其为单系类群，其主要特征是齿针细长，食道膨大部短小呈球状，食道腺核只有 3 个。它们经常习居在土壤中，是具有农业重要性的植物寄生线虫，有些种类是植物病毒的载体。Andrássy（2009）将其列为总科，有 8 个有效属和 490 个有效种，其中最重要的属是剑线虫属（*Xiphinema*，248 个有效种），是迄今为止物种数量最丰富的矛线目属，以及长针属（*Longidorus*，150 个有效种）。Peña-Santiago（2021）整理长针科有 3 个亚科 8 个有效属，如下。

长针科 Longidoridae Thorne，1935

长针亚科 Longidorinae Thorne，1935

类长针属 *Longidoroides* Khan，Chawla & Saha，1978

长针属 *Longidorus* Micoletzky，1922

拟长针属 *Paralongidorus* Siddiqi，Hooper & Khan，1963

西迪克属 *Siddiqia* Khan，Chawla & Saha，1978

剑针亚科 Xiphidorinae Khan，Chawla & Saha，1978

澳洲剑针属 *Australodorus* Coomans，Olmos，Casella & Chaves，2004

拟剑针属 *Paraxiphidorus* Coomans & Chaves，1995

剑针属 *Xiphidorus* Monteiro，1976

剑线虫亚科 Xiphinematinae Dalmasso，1969

剑线虫属 *Xiphinema* Cobb，1913b

5.2.3.8 湿生科 Mydonomidae（图 7A）

鉴别特征：虫体体长超过 1mm。角质层矛线型，没有放射状折射。齿针相对粗壮，不对称，两臂长度不同；引导环薄或厚且有曲折；齿托直，简单或弧形，略微有骨化。食道基部膨大呈圆筒状，偶尔梨形，占总颈部长度的 1/3 以下。雌虫生殖管双生或单后生，阴道缺曲折部。雄虫交合刺矛线型，交配乳突间隔。尾形多变，从短圆到长丝状，雌雄虫相似或不相似。

湿生科线虫不很常见，主要在土壤和石灰质的栖息地收集到。该科的进化关系尚需澄清，基于食道膨大部较短和阴道没有曲折部，传统上认为其与细齿科（Leptonchidae）和小垫咽科（Tylencholaimellidae）相近。Andrássy（2009）认为该科有 2 个亚科 5 个有效属和 86 个有效种，最重要的世界性的属是类矛线属（*Dorylaimoides*，69 个有效种）。Peña-Santiago（2021）整理湿生科有 2 个亚科 6 个有效属，如下。

湿生科 Mydonomidae Thorne，1964

美咽亚科 Calolaiminae Goseco，Ferris & Ferris，1976

美咽属 *Calolaimus* Timm，1964

拟蒂姆属 *Paratimmus* Baniyamuddin & Ahmad, 2009
蒂姆属 *Timmus* Goseco, Ferris & Ferris, 1976
湿生亚科 Mydonominae Thorne, 1964
类矛线属 *Dorylaimoides* Thorne & Swanger, 1936
泽地属 *Morasia* Baqri & Jairajpuri, 1969
湿生属 *Mydonomus* Thorne, 1964

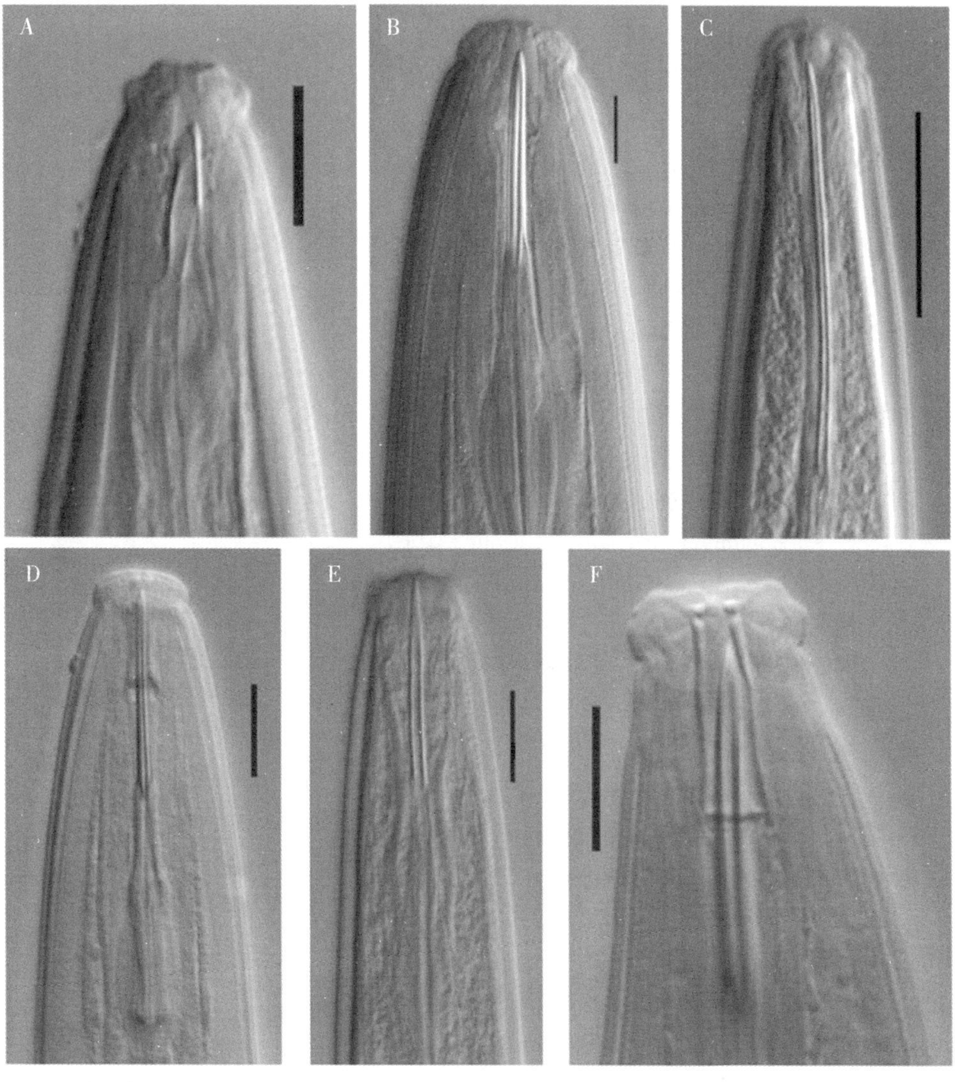

A. 湿生科，类矛线属；B-F. 北印度科，B 为异矛属，C 为小长针属，D 为大矛属，E 为口矛属；F 为鳌属（标尺：A、B、D-F 为 10μm；C 为 20μm）

图 7　矛线亚目的多样性（三）

5.2.3.9　北印度科 Nordiidae（图 7B-图 7E）

鉴别特征：虫体体长 0.5～8.0mm。角质层光滑或有非常细的横纹。齿针细长，渐狭，有狭窄的针腔和针孔；导环单环或双环；齿托杆状或有凸缘。食道肌肉发达，后部膨大。雌虫生殖管双生，偶尔单生；阴道通常有曲折部；阴门横裂。交配乳突数量变化大，几乎总是间隔的。雌雄虫尾形相似，短圆，偶尔长丝状。

北印度科线虫是土壤微动物群落中非常常见的成员，也出现在石灰岩中，分布广泛，成员唯一的共同特征是渐狭的细长齿针，是一个并系类群。Andrássy（2009）认为该科有 3 个亚科 18 个有效属 200 个有效种，其中最多样化的属是小长针属（*Longidorella*，39 个有效种）、口矛属（*Oriverutus*，27 个有效种）、异矛属（*Heterodorus*，25 个有效种）、大矛属（*Enchodelus*，23 个有效种）和鳌属（*Pungentus*，21 个有效种）。Peña-Santiago（2021）整理北印度科有 5 个亚科 30 个有效属，如下。

北印度科 Nordiiae Jairajpuri & Siddiqi，1964

类角咽亚科 Actinolaimoidinae Jairajpuri & Ahmad，1992

类角咽属 *Actinolaimoides* Meyl，1957

口矛亚科 Oriverutinae Andrássy，2009

马雷克属 *Malekus* Thorne，1974

秃头属 *Oonaguntus* Thorne，1974

类口矛属 *Oriverutoides* Ahmad & Sturhan，2002

口矛属 *Oriverutus* Siddiqi，1971

拟口矛属 *Paroriverutus* Carbonell & Coomans，1982

胃咽属 *Stomacholaimus* Andrássy，2011

矛咽亚科 Encholaiminae Golden & Murphy，1967

无头矛线属 *Acephalodorylaimus* Ahmad & Jairajpuri，1983

头矛线属 *Cephalodorylaimus* Jairajpuri，1967

刺矛属 *Echinodorus* Siddiqi，1995

矛咽属 *Encholaimus* Golden & Murphy，1967

蠕生属 *Helmabia* Siddiqi，1971

线生属 *Nemabia* Siddiqi，1995

北印度亚科 Nordiinae Jairajpuri & Siddiqi，1964

小针线虫属 *Acunemella* Andrássy，2002

小长针属 *Longidorella* Thorne，1939

全索氏属 *Thornedia* Husain & Khan，1965

鳌亚科 Pungentinae Siddiqi，1969

加利福尼亚矛属 *Californidorus* Robbins & Weiner，1978

类显矛属 *Enchodeloides* Elshishka，Lazarova，Radoslavov，Hristov & Peneva，2017

大矛属 *Enchodelus* Thorne，1939

齿矛属 *Enchodorus* Vinciguerra，1976

异矛属 *Heterodorus* Altherr，1952

柯赫线虫属 *Kochinema* Heyns，1963

兰扎维奇属 *Lanzavecchia* Zullini，1988

软齿属 *Lenonchium* Siddiqi, 1965
巴布亚矛属 *Papuadorus* Andrássy, 2009
拟皱槽属 *Pararhyssocolpus* Elshishka, Lazarova, Radoslavov, Hristov & Peneva, 2015
小鳌属 *Pungentella* Andrássy, 2009
鳌属 *Pungentus* Thorne & Swanger, 1936
皱槽属 *Rhyssocolpus* Andrássy, 1971
狭矛线属 *Stenodorylaimus* Álvarez-Ortega & Peña-Santiago, 2011

5.2.3.10 库西亚线虫科 Qudsianematidae（图 8A-图 8E）

鉴别特征：虫体体长 0.3~6.0mm。角质层光滑或有细横纹，偶有纵脊。唇区缢缩，唇片通常明显。齿针直，有明显的针腔和开口，开孔小于齿针长的一半；齿托杆状。食道基部膨大约占其长度的一半。雌虫生殖管通常双生，偶尔单生，阴道通常有曲折部。交合刺矛线型，腹中位交配乳突在数量和排列上变化大，多数间隔。雌雄虫尾形相似，短圆锥形至圆形，有时长圆锥形。

库西亚线虫科在全世界的土壤和淡水沉积物中都非常频繁和丰富，是一个多系或并系的群体，没有任何坚实的形态特征和分子数据来支持。Andrássy（2009）认为该科有 5 个亚科 31 属 402 个有效种，其中最多样化的世界性的属是真矛线属（*Eudorylaimus*，95 个有效种）、盘咽属（*Discolaimus*，41 个有效种）、唇线虫属（*Labronema*，41 个有效种）、厚唇属（*Crassolabium*，34 个有效种）、小盘咽属（*Discolaimium*，30 个有效种）和异矛线属（*Allodorylaimus*，28 个有效种）。Peña-Santiago（2021）认为库西亚线虫科有 8 个亚科 40 个属，如下。

库西亚线虫科 Qudsianematidae Jairajpuri, 1965
窄矛线亚科 Arctidorylaiminae Mulvey & Anderson, 1979
窄矛线属 *Arctidorylaimus* Mulvey & Anderson, 1979
锐咽亚科 Carcharolaiminae Thorne, 1967
明咽属 *Antholaimus* Cobb, 1913a
锐咽属 *Carcharolaimus* Thorne, 1939
加勒比线虫属 *Caribenema* Thorne, 1967
混血属 *Caryboca* Lordello, 1967
杯线虫亚科 Crateronematinae Siddiqi, 1969
金线虫属 *Chrysonema* Thorne, 1929
杯线虫属 *Crateronema* Siddiqi, 1969
盘咽亚科 Discolaiminae Siddiqi, 1969
小盘咽属 *Discolaimium* Thorne, 1939
类盘咽属 *Discolaimoides* Heyns, 1963
盘咽属 *Discolaimus* Cobb, 1913a
丝盘咽属 *Filidiscolaimus* Siddiqi, 1995
偏头属 *Latocephalus* Patil & Khan, 1982
类磨盘属 *Mylodiscoides* Lordello, 1963

磨盘属 *Mylodiscus* Thorne，1939

萨利姆属 *Salimella* Siddiqi，2005

后弯线虫亚科 Lordellonematinae Siddiqi，1969

后弯线虫属 *Lordellonema* Andrássy，1960a

摩萨基属 *Moshajia* Siddiqi，1982

小孔线虫属 *Poronemella* Siddiqi，1969

希科线虫属 *Sicorinema* Siddiqi，1982

小希科线虫属 *Sicorinemella* Andrássy，2009

厚矛线亚科 Pachydorylaiminae Andrássy，2009

厚矛线属 *Pachydorylaimus* Siddiqi，1983

拟缢咽亚科 Paraxonchiinae Dhanachand & Jairajpuri，1981

颈线虫属 *Cerviconema* Andrássy，2009

戈帕尔属 *Gopalus* Khan，Jairajpuri & Ahmad，1988

拟戈帕尔属 *Parapalus* Loof & Zullini，2000

拟缢咽属 *Paraxonchium* Krall，1958

钩矛线属 *Ramphidorylaimus* Baniyamuddin，Ahmad & Jairajpuri，2010

延展线虫属 *Tendinema* Siddiqi，1995

库西亚线虫亚科 Qudsianematinae Jairajpuri，1965

异矛线属 *Allodorylaimus* Andrássy，1986

钝矛线属 *Amblydorylaimus* Andrássy，1998

巴克瑞属 *Baqriella* Ahmad & Jairajpuri，1989

北咽属 *Boreolaimus* Andrássy，1988

环矛线属 *Cricodorylaimus* Ahmad & Sturhan，2001

矛针属 *Dorydorella* Andrássy，1987

通俗属 *Ecumenicus* Thorne，1974

上矛线属 *Epidorylaimus* Andrássy，1986

真矛线属 *Eudorylaimus* Andrássy，1959

美矛线属 *Kallidorylaimus* Andrássy，1989

鞘矛线属 *Kolodorylaimus* Andrássy，1998

小矛线属 *Microdorylaimus* Andrássy，1986

塔拉线虫属 *Talanema* Andrássy，1991

5.2.3.11　索恩线虫科 Thornenematidae

鉴别特征：虫体体长 0.3～2.3mm。唇区连续，唇片合并，唇架骨化。唇后亚角质层骨化。齿针矛线型，引导环薄。食道膨大部小于总颈部长度的一半；后方的一对腺核位于食道底部的远端。雌虫生殖管单后生，极少为双生。雌虫交配乳突有间隔，数量很少（1～11个）。雌虫尾长丝状到短圆形；雄虫尾形与雌虫尾形相似或不同。

索恩线虫科一个多样性差的类群，是陆地栖息的稀有物种。它的特点是有唇和/或唇后的骨化，雌虫的生殖管大多数是单后生，尾长。这个科似乎是一个单系，但是其进化关系，特别是与矛线科的关系，需通过研究加以澄清。Andrássy（2009）将该科分为两个亚科 11 个属 70 个有效种，其中最重要的属是索恩线虫属（*Thornenema*，25 个有效种）和后矛线属（*Opisthodorylaimus*，11 个有效种）。Peña-Santiago（2021）认为索恩线虫科有 2 个亚科 14 个属，如下。

索恩线虫科 Thornenematidae Siddiqi，1969

索恩线虫亚科 Thornenematinae Siddiqi，1969

库曼斯线虫属 *Coomansinema* Ahmad & Jairajpuri，1969

类库曼斯线虫属 *Coomansinemoides* Sen，Chatterjee & Manna，2012

印度矛线属 *Indodorylaimus* Ali & Prabha，1974；*Lagenonema* Andrássy，1987

瓶线虫属 *Laurophragus* Nesterov，1976

前索恩线虫属 *Prothornenema* Baqri & Bohra，2003

后矛线属 *Opisthodorylaimus* Ahmad & Jairajpuri，1982

剑咽属 *Sicaguttur* Siddiqi，1971

索恩线虫属 *Thornenema* Andrássy，1959

威利线虫亚科 Willinematinae Andrássy，1987

高小矛属 *Anadorella* Siddiqi，2005

拟蒂姆线虫属 *Paratimminema* Rahaman，Ahmad & Khan，1993

硬唇属 *Sclerolabia* Carbonell & Coomans，1986

蒂姆线虫属 *Timminema* Khan，1978

威利线虫属 *Willinema* Baqri & Jairajpuri，1967

5.2.3.12 索氏科 Thorniidae（图 8F）

鉴别特征：虫体体长 0.5~2.0mm。唇区连续或缢缩，两圈唇乳突互相靠近或不靠近。齿针形态多变，或者弱且有点不对称，或者是典型的矛线型。食道膨大部占颈部长度的一半；后部腺体的细胞核相当前置。雌虫双生殖管，阴道没有曲折部或不良。交合刺为矛线型或无咽型（alaimoid）。常有引带，腹中位乳突 0~4 个。雌雄虫尾形相似，圆筒状，可达肛门处体宽的 3 倍。

索氏科一般来说是罕见的类群，习居在石灰质和泥质的生境，可能不是一个单系群。Andrássy（2009）将该科分为两个亚科 6 个有效属和 20 个有效种，其中最重要的属是索氏属（*Thornia*，11 个有效种）和类穿咽属（*Nygolaimoides*，5 个有效种），Peña-Santiago（2021）认同索氏科有 2 个亚科 6 个属，如下。

索氏科 Thorniidae De Coninck，1965

索氏亚科 Thorniinae De Coninck，1965

类穿咽属 *Nygolaimoides* Meyl in Andrássy，1960a

索氏属 *Thornia* Meyl，1954

多索氏属 *Thorniosa* Andrássy，1996

小索氏亚科 Thorneellinae Andrássy，1987，

卢夫矛线属 *Loofilaimus* Jairajpuri，Ahmad & Sturhan，1998，

球侧器属 *Sphaeroamphis* Ahmad & Sturhan，2000

小索氏属 *Thorneella* Andrássy，1960a

5.2.3.13　小垫咽科 Tylencholaimellidae（图 8G，图 8H）

鉴别特征：虫体体长小于 1.8mm。角质层为矛线型，缺乏放射状折射。唇区呈帽状，唇合并。齿针直管状，短于唇区宽度；齿托分化有或无。食道膨大为一个略微的梨形球体，占总颈部长度的 1/4。雌虫生殖管双生或单生，阴道缺乏曲折部，阴门横裂。雄虫交合刺矛线型，腹中位交配乳突少且有间隔。雌雄虫尾形相似，长丝状至短圆形。

小垫咽科似乎是一个单系类群，除了几个属外，它们是罕见的线虫。Andrássy（2009）认为该科有 3 个亚科 9 个有效属和 66 个有效种，其中多样性最高的属是小垫咽属（*Tylencholaimellus*，36 个有效种）和短矛属（*Doryllium*，14 个有效种），而 Peña-Santiago（2021）认为该科内有 2 个亚科 9 个属，如下。

小垫咽科 Tylenchollaimellidae Jairajpuri，1964

伊瑟线虫亚科 Athernematinae Ahmad & Jairajpuri，1978

伊瑟线虫属 *Athernema* Ahmad & Jairajpuri，1978

小垫咽亚科 Tylenchollaimellinae Jairajpuri，1964

艾根矛属 *Agmodorus* Thorne，1964

小矛属 *Dorella* Jairajpuri，1964

短矛属 *Doryllium* Cobb，1920

高弗属 *Goferus* Jairajpuri & Ahmad，1992

边垫咽属 *Margollus* Peña-Santiago，Peralta & Siddiqi，1993

奥斯藤布林克属 *Oostenbrinkella* Jairajpuri，1965

石质土线虫属 *Phellonema* Thorne，1964

小垫咽属 *Tylencholaimellus* Cobb in Cobb，1915

5.2.3.14　垫咽科 Tylencholaimidae（图 8I，图 8J）

鉴别特征：虫体体长从小于 0.5mm 到 3mm 以上，通常小于 1mm。角质层垫咽型，有松散的内层和丰富的放射状皱褶。唇区呈帽状，多变化；唇片合并，口周区分化为盘状结构。齿针变化大，有明显的针腔和开口，相对较短；齿托呈棒状，常有基部球，偶有凸缘。食道前部略有肌肉；基部膨大部占颈部总长度的 1/3 以上，通常有 5 个食道腺核。雌虫生殖管双生或单生；阴道曲折部通常无；阴门横裂、纵裂或孔状。雄虫交合刺矛线型，腹中位交配乳突数量不等，通常间隔，偶有连续。雌雄虫尾形相似，长丝状到短圆。

垫咽科是一个多系类群，常习居陆地土壤中，偶尔在淡水栖息地，有些种类是食菌性的。Andrássy（2009）认为该科有 5 个亚科 24 属和 122 个有效种，其中最重要的属是垫咽属（*Tylencholaimus*，52 个有效种）和盘喙属（*Discomyctus*，10 个有效种）。Peña-Santiago（2021）认为该科有 3 个亚科 21 个属，如下。

垫咽科 Tylencholaimidae Filipjev，1934

次矛线亚科 Metadorylaiminae Andrássy，1976

次矛线属 *Metadorylaimus* Jairajpuri & Goodey，1966

新次矛线属 *Neometadorylaimus* Jairajpuri & Ahmad，1992

垫咽亚科 Tylencholaiminae Filipjev，1934

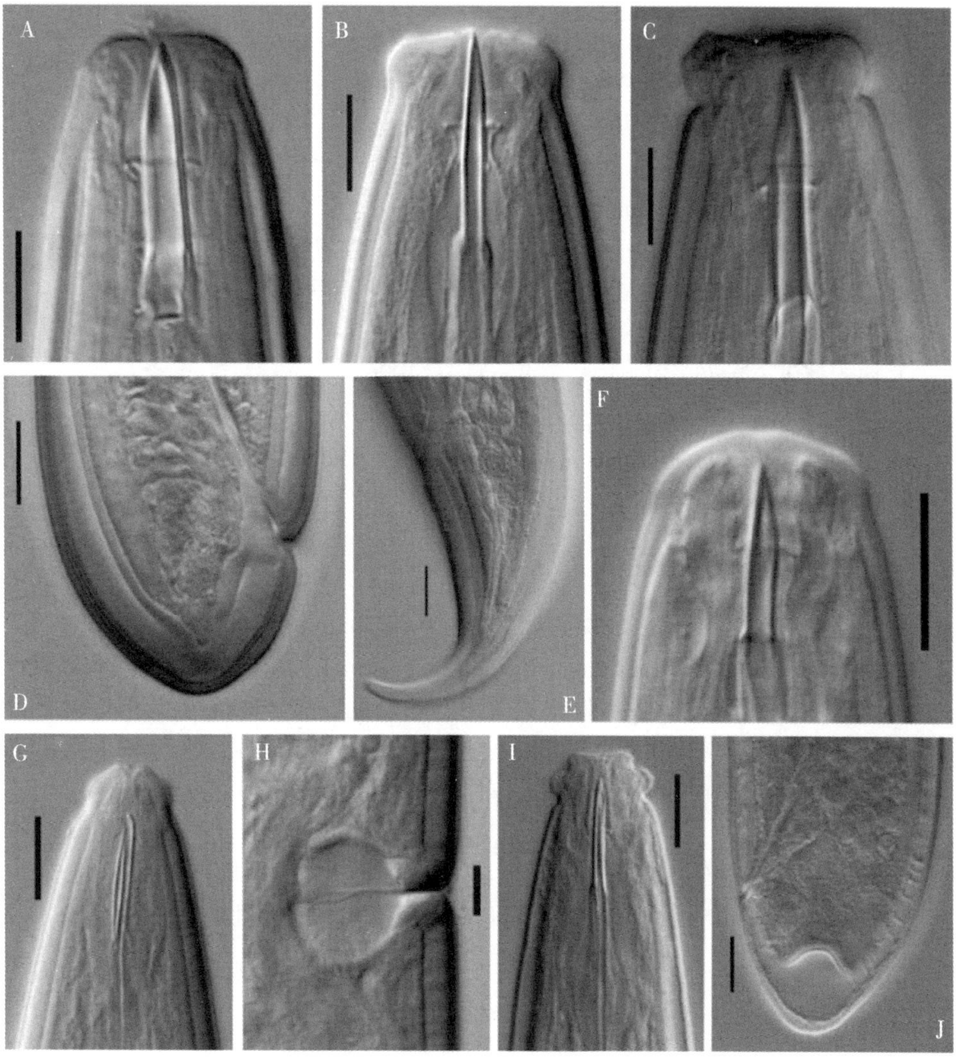

A-E. 库西亚线虫科，A、D 为厚唇属，A 示虫体前部，D 示雌虫尾部；B、E. 真矛线属，B 示虫体前部，E 示雌虫尾部；C. 唇线虫属，虫体前部；F. 索氏科，索氏属，示虫体前部；G，H. 小垫咽科，小垫咽属，G 示虫体前部，H 示阴门区；I，J. 垫咽科，奇伍德属，I 示虫体前部，J 示雌虫尾部（标尺：A-G、I、J 为 10μm；H 为 5μm）

图 8　矛线亚目的多样性（四）

细矛属 *Capilonchus* Siddiqi，1982

类小奇伍德属 *Chitwoodielloides* Ahmad & Araki，2003

小奇伍德属 *Chitwoodiellus* Jiménez-Guirado & Peña-Santiago，1992

奇伍德属 *Chitwoodius* Furstenberg & Heyns，1966

盘喙属 *Discomyctus* Thorne，1939

美矛线虫属 *Loncharionema* Goseco，Ferris & Ferris，1974

梅尔线虫属 *Meylonema* Andrássy，1960a

穆塔兹属 *Mumtazium* Siddiqi，1969

前穆塔兹属 *Promumtazium* Siddiqi，1982

假垫咽属 *Pseudotylencholaimus* Jairajpuri & Ahmad，1992

小喙属 *Rostrulium* Siddiqi，1995

硬咽属 *Sclerolaimus* Jairajpuri & Ahmad，1992

小线属 *Tantunema* Siddiqi，1982

类垫矛属 *Tylenchodoroides* Gagarin，2001

垫矛属 *Tylenchodorus* Siddiqi，1983

垫咽属 *Tylencholaimus* de Man，1876

瓦西姆属 *Wasimellus* Bloemers & Wanless，1996

范德林亚科 Vanderlindiinae Siddiqi，1969

弯矛线属 *Curvidorylaimus* Jairajpuri & Rahman，1983

范德林属 *Vanderlindia* Heyns，1964

6 矛线目线虫的生物学和生态学

矛线目大多数种类是自由生活的杂食性类群（Yeates *et al.*，1993，2009），少数种类是其他线虫的捕食者（Small & Grootaert，1983；Bilgrami *et al.*，1985），或是食真菌者（Okada *et al.*，2005），而长针科线虫是众所周知的植物外寄生虫，部分种类可传播植物病毒。矛线目是陆地土壤和淡水沉积物中线虫群落的主要组成部分，多样性、丰度和生物量都很高。Liébanas 等（2002）发现在伊比利亚半岛东南部每 100cm^{3}的土壤样品中，矛线目线虫平均有 11.6 种，最多达 20 种。在自然（非栽培）生境中，矛线目的多样性总体上比任何其他线虫种类要高，在恢复的生态系统中矛线目一些属的丰度明显增加（Todd *et al.*，2006；Briar *et al.*，2012）。施用有机施肥会使矛线目代表种群的数量增加（McSorley & Frederick，1999），灌溉或耕作技术也会改变矛线目种群的丰度（McSorley *et al.*，2012，2013；Postma-Blaauw *et al.*，2012）。

许多环境变量可能会影响矛线目种类的分布、频率和丰度等，Liébanas 等（2002，2004）分析了一个自然区的 131 种矛线目线虫的空间分布，发现海拔、土壤特征和相关植物群落是影响物种分布的最重要因素。土壤特性如土壤颗粒的大小可能会影响线虫的运动，土壤的化学成分可能影响许多线虫种类的生物学。某些土壤元素的水平可能会影响矛线目物种的分布，一些土壤成分（阳离子、盐类、重金属）的存在可能对矛线目种群产生负面影响（McSorley *et al.*，2012）。植物群落性质（多年生/落叶，封闭/开放）和树龄会影响矛线目的分布和数量（Cesarz *et al.*，2013），矛线目在捷克落叶林土壤中的丰富度高于针叶林土壤（Hánel，2008）。

由于矛线目的摄食习性和生物学信息了解并不充分，它们在土壤和沉积物中的生态作用还没有得到很好的阐释（McSorley *et al.*，2012）。然而，它们完成世代的时间长、繁殖率低、代谢活动低、运动缓慢，所有这些生物特性支持它们是 K-战略家（K-strategists）。矛线目与其他土壤线虫相比，体积（生物量）相对较大，导致对能量流和代谢有更大的影响（Ferris，2010；McSorley *et al.*，2012）。矛线目线虫的角质层渗透性很特别，对污染物和其他环境干扰和压力非常敏感（Tenuta & Ferris，2004），因此成为监测和评估土壤质量（健康）的优秀生物指标（Wasilewska，1997；Todd *et al.*，2006）。当栖息地发生重金属污染、杀虫剂等干扰时，矛线目线虫会发生物种和数量的急剧下降或完全消失（Ekschmitt & Korthals，2006）。

矛线目的世界地理分布范围，除欧洲、北美和少数国家（印度、南非、新西兰）有较多研究外，其他国家和地区的矛线目类线虫分布几乎知之甚少。少数种类如钝尾小孔咽线虫（*Aporcelaimellus obtusicaudatus*）和单宫通俗线虫（*Ecumenicus monohystera*）可能是世界性分布的，而许多种类的分布可能仅局限在特定地区（Andrássy，2009）。Decraemer 和 Coomans（1994）汇编了古老湖泊包括贝加尔湖（Baikal）、坦噶尼喀湖（Tanganyika）、奥赫里德湖（Ohrid）、金奈特湖（Kinneret）记录的线虫动物群，其中有矛线目 13 个属和 19 个种显示出高度的地方性，每个湖中有 1~3 个种是特有的。Abebe 等（2008）分析了淡水线虫在动物地理区域的分布情况，发现共有矛线目 610 种线虫栖息在全世界的淡水水体中。Du Preez 等（2017）汇编了世界各地洞穴中出现的线虫动物群，矛线目有 22 个属是穴居类群。

矛线目线虫在土壤中能以自主运动的方式移动，取决于其栖息地的物理和化学性质，也可以通过灌溉水而扩散，人类的活动包括栽培作物、森林和观赏植物的交换或贸易都可造成矛线目物种的扩散。此外，Chizov（1996）发现嗜蝇类穿咽线虫（*Nygolaimoides borborophilus*）可通过介体苍蝇传播到不同地区。矛线目线虫如真矛线虫（*Eudorylaimus*）也可以通过脱水（anhydrobiosis）机制作为一种生存策略以对抗不利的环境条件（Lee，1961；Wall & Virginia，1999），虫体能形盘绕成螺旋状，极大减少身体表面暴露，一旦环境条件改善，它们就可恢复活力（Freckman *et al.*，1977）。

7 捕食性矛线目线虫的生物防治潜力

植物寄生线虫侵染植物造成巨大的损失，利用捕食性线虫防治植物寄生线虫具有重要意义。线虫捕食有双重功能，可以减少土壤中植物寄生线虫的数量，同时以植物可用的形式释放营养物质，使植物能够更好地承受线虫对根部的损害（Yeates & Wardle，1996），作为一种生防因子，捕食性线虫提供了一种潜在的生态安全的化学杀线虫剂的替代品。Cobb（1917）首次提出可利用单齿线虫防治植物寄生线虫，此后陆续有利用捕食线虫的生防事例（Cassidy，1931；Christei，1960；Banage，1963；Esser & Sobers，1964；Esser，1987）。矛线目捕食者可以控制土壤中的线虫数量，然而关于矛线目捕食者的研究相对较少（Boosalis & Mankau，1965；Wyss & Grootaert，1977；Small & Evans，1981；Bilgrami，1993；Khan *et al.*，1991），表 2 列出了捕食性矛线目线虫与它们的植物寄生线虫猎物（Khan & Kim，2007）。

矛线目用齿针来刺穿猎物的角质层，并通过它吸食虫体内容物，然后进入食道。有些矛线目线虫有小齿等，辅助于捕捉和刺伤猎物，并将其切成碎片（Jairajpuri & Bilgrami，1990）。穿咽亚目线虫，如水线虫（*Aquatides*）有一个大的细长壁齿，用来刺穿或切割它们的猎物。矛线目线虫捕食能力的研究大部分是基于体外实验，Wyss 和 Grootaert（1977）和 Bilgrami 等（1985）分别详细描述了阴门乳突唇线虫（*Labronema vulvapapillatum*）和索氏水线虫（*Aquatides thornei*）的捕食能力。矛线目捕食猎物有偏好性，偏拟角咽线虫（*Paractinolaimus elongatus*）偏好捕食南方根结线虫（*Meloidogyne incognita*）和小麦粒线虫

表 2　捕食性矛线目线虫与其植物寄生线虫猎物

捕食线虫	植物寄生线虫猎物	参考文献
美洲异矛线虫 *Allodorylaimus americanus*	小麦粒线虫 *Anguina tritici*，滑刃线虫属 *Aphelenchoides*，巴兹尔属 *Basiria*，印度螺旋线虫 *Helicotylenchus indicus*，狄氏鞘线虫 *Hemicycliophora dhirenderi*，香附子孢囊线虫 *Heterrodera mothi*，水稻潜根线虫 *Hirschmanniella oryzae*，长针属 *Longidorus*，南方根结线虫 *Meloidodera incognita*，毛刺属 *Trichodorus*，马氏矮化线虫 *Tylenchorhynchus mashhoodi*，半穿刺线虫 *Tylenchulus semipenetrans*，巴氏剑线虫 *Xiphinema basiri*	Khan *et al.*，1995a
雪小孔咽线虫 *Aporcelaimellus nivalis*	小麦粒线虫 *A. tritici*，滑刃线虫属 *Aphelenchoides*，巴兹尔属 *Basiria*，印度螺旋线虫 *H. indicus*，杧果半轮线虫 *Hemicriconemoides mangiferae*，狄氏鞘线虫 *H. dhirenderi*，香附子孢囊线虫 *H. mothi*，水稻潜根线虫 *H. oryzae*，印度纽带线虫 *Hoplolaimus indicus*，长针属 *Longidorus*，南方根结线虫 *M. incognita*，柑橘拟长针线虫 *Paralongidorus citri*，盾线虫属 *Scutellonema*，马氏矮化线虫 *T. mashhoodi*，半穿刺线虫 *T. semipenetrans*，毛刺属 *Trichodorus*，美洲剑线虫 *X. americanum*，标明剑线虫 *Xiphinema insigne*	Khan *et al.*，1991；Bilggrami，1993
暗色小孔咽线虫 *Aporcelaimellus obscurus*	甜菜孢囊线虫（卵）*H. schachtii*（eggs）	Thorne & Swanger，1936
孔咽线虫 *Aporcelaimus* spp.	车轴草孢囊线虫 *Heterodera trifolii*，根结线虫属 *Meloidogyne*	Doncaster，1962；Esser，1987
索恩水线虫 *Aquatides thornei*	小麦粒线虫 *A. tritici*，印度螺旋线虫 *H. indicus*，香附子孢囊线虫 *H. mothi*，水稻潜根线虫 *H. oryzae*，长针属 *Longidorus*，南方根结线虫 *M. incognita*，柑橘拟长针线虫 *P. citri*，马氏矮化线虫 *T. mashhoodi*，拟毛刺属 *Paratrichodorus*，美洲剑线虫 *X. americanum*	Bilgrami *et al.*，1985；Bilgrami，1992
沙地盘咽线虫 *Discolaimus arenicolus*	南方根结线虫 *M. incognita*	Esser，1963，1987
森林盘咽线虫 *Discolaimus silvicolus*	小麦粒线虫 *A. tritici*，滑刃线虫属 *Aphelenchoides*，巴兹尔属 *Basiria*，印度螺旋线虫 *H. indicus*，狄氏鞘线虫 *H. dhirenderi*，香附子孢囊线虫 *H. mothi*，水稻潜根线虫 *H. oryzae*，长针属 *Longidorus*，南方根结线虫 *M. incognita*，毛刺属 *Trichodorus*，马氏矮化线虫 *T. mashhoodi*，半穿刺线虫 *T. semipenetrans*，巴氏剑线虫 *X. basiri*	Khan *et al.*，1995a
暗色矛线虫 *Dorylaimus obscurus*	甜菜孢囊线虫（卵）*H. schachtii*（eggs）	Thorne & Swanger，1936
池塘矛线虫 *Dorylaimus stagnalis*	小麦粒线虫 *A. tritici*，印度螺旋线虫 *H. indicus*，香附子孢囊线虫 *H. mothi*，水稻潜根线虫 *H. oryzae*，长针属 *Longidorus*，南方根结线虫 *M. incognita*，柑橘拟长针线虫 *P. citri*，拟毛刺属 *Paratrichodorus*，马氏矮化线虫 *T. mashhoodi*，美洲剑线虫 *X. americanum*	Shafqat *et al.*，1987；Bilgrami，1992

（续表）

捕食线虫	植物寄生线虫猎物	参考文献
钝尾真矛线虫 *Eudorylaimus obtusicaudatus*	甜菜孢囊线虫（卵）*H. schachtii*（eggs）	Thorne，1928
真矛线虫 *Eudorylaimus* sp.	肾形肾状线虫 *R. reniformis*	Wang *et al.*，2015
阴门乳突唇线虫 *Labronema vulvapapillatus*	小麦粒线虫 *A. tritici*，马铃薯金线虫 *Globodera rostochiensis*，纳西根结线虫 *Meloidogyne naasi*，标准剑线虫 *Xiphinema index*	Wyss & Grootaert，1977；Grootaert & Small，1982；Small & Grootaert，1983
巴氏中矛线虫 *Mesodorylaimus bastiani*	小麦粒线虫 *A. tritici*，滑刃属 *Aphelenchoides*，巴兹尔属 *Basiri*，印度螺旋线虫 *H. indicus*，狄氏鞘线虫 *H. dhirenderi*，香附子孢囊线虫 *H. mothi*，水稻潜根线虫 *H. oryzae*，长针属 *Longidorus*，南方根结线虫 *M. incognita*，柑橘拟长针线虫 *P. citri*，拟毛刺属 *Paratrichodorus*，马氏矮化线虫 *T. mashhoodi*，美洲剑线虫 *X. americanum*，巴氏剑线虫 *X. basiri*	Bilgrami，1992
敏捷新角咽线虫 *Neoactinolaimus agilis*	小麦粒线虫 *A. tritici*，滑刃属 *Aphelenchoides*，巴兹尔属 *Basiria*，印度螺旋线虫 *H. indicus*，水稻潜根线虫 *H. oryzae*，香附子孢囊线虫 *H. mothi*，南方根结线虫 *M. incognita*，强壮盘旋线虫 *Rotylenchus robustus*，柑橘拟长针线虫 *P. citri*，拟毛刺线虫属 *Paratrichodorus*，马氏矮化线虫 *T. mashhoodi*，半穿刺线虫 *T. semipenetrans*，美洲剑线虫 *X. americanum*	Khan *et al.*，1995b
双齿新角咽线虫 *Neoactinolaimus duplicidentatus*	伪强壮盘旋线虫 *Rotylenchus fallorobustus*	Small & Grootaert，1983
新角咽线虫 *Neoactinolaimus* sp.	肾形肾状线虫 *Rotylenchus reniformis*	Wang *et al.*，2015
偏拟角咽线虫 *Paractinolaimus elongatus*	小麦粒线虫 *A. tritici*，印度螺旋线虫 *H. indicus*，水稻潜根线虫 *H. oryzae*，长针属 *Longidorus*，南方根结线虫 *M. incognita*，马氏矮化线虫 *T. mashhoodi*，巴氏剑线虫 *X. basiri*	Khan & Jairajpuri，1997
索氏线虫 *Thornia*	轮线虫属 *Criconemoides*，佛罗里达根结线虫 *Meloidodera floridensis*，弯曲针线虫 *Paratylenchus curvitatus*，穿刺短体线虫 *Pratylenchus penetrans*，伤残短体线虫 *Pratylenchus vulnus*，半穿刺线虫 *T. semipenetrans*	Boosalis & Mankau，1965；Esser，1987
贪婪西印度线虫 *Westindicus rapax*	剑线虫属 *Xiphinema*	Hunt，1978

（*Anguina tritic*）的 2 龄幼虫（Khan & Jairajpuri，1997）；新角咽线虫（*Neoactinolaimus*）偏好取食肾形肾状线虫（*Rotylenchulus reniformis*）（Wang *et al.*，2015）。大部分的矛线目捕食者是无情的杀手，例如阴门乳突唇线虫每天可以杀死大约 100 条全齿线虫（*Panagrellus*）（Jairajpuri & Bilgrami，1990）。猎物密度会影响捕食率，在较高的猎物密度下，捕食概率会增加。此外，钝尾真矛线虫（*Eudorylaimus obtusicaudatus*）还可取食甜菜孢囊线虫（*Heterodera schachtii*）孢囊内的卵，表明具有生防潜力。

矛线目捕食者的繁殖力通常很低，可能取决于捕食者的大小，例如阴门乳突唇线虫（*L. vulvipapillatum*）每天产 1~10 粒卵。温度对线虫的生命周期、发育、繁殖和生存至关重要，根据温度的不同，阴门乳突唇线虫完成一代需 27~126d，凶猛唇线虫（*L. ferox*）则需 90~120d（Grootaert & Small，1982）。小孔咽线虫（*Aporcelaimellus*）在 18~20℃时完成一代需 95~130d（Wood，1974）。雄虫在许多矛线目种类中不常见，在这种情况下营孤雌生殖。矛线目线虫不是强制性的捕食者，除了猎物线虫之外，还可取食细菌、藻类和真菌，具杂食性（Russell，1986；Yeates *et al.*，1993）。为了评估矛线目捕食者的生防应用潜力，必须大量人工培养它们，可以在含有一种或多种猎物的单异培养基上饲养它们（Khan *et al.*，1991，1994；Khan *et al.*，1995a，1995b；Khan & Jairajpuri，1997）。由于杂食的习惯，可以通过在农业土壤中添加有机物来提高它们的种群密度（Webster，1972；Akhtar，1998），这为有机耕作下的植物寄生线虫防治开辟了新领域。

捕食性线虫的生物防治潜力取决于其猎物搜索能力、猎物特异性、捕食效率、生命周期的持续时间和寿命、繁殖潜力、生存以及对生态条件的适应等，人工大量培养以便能在田间条件下应用，是捕食性线虫作为成功的生防因子的另一个重要特征（Kanwar *et al.*，2021）。矛线目在田间广泛而丰富地分布，也是更好的生物防治候选者，可以通过添加有机营养物提高它们在田间的数量。然而，它们的生命周期长和繁殖率低是令人困扰的原因。在推荐广泛使用之前，需要进行大量的研究来评估矛线目捕食者在不同农业气候区的多样性、它们的生物学、生态学、行为、繁殖潜力和生命周期，以及开发它们的大规模培养技术、配方、时间和应用模式。

8 我国矛线目线虫研究现状

矛线目线虫是土壤线虫中的最大类群，我国土壤线虫类群的研究大多数集中于植物寄生线虫类群，它们主要分布在小杆目（Rhabditida）中的垫刃次目（Tylenchomorpha）、矛线目（Dorylaimida）的长针科（Longidoridae）以及三矛目（Triplonchida）的毛刺科（Trichodoridae）。相比国外，我国关于非植物寄生性的土壤线虫研究起步较晚，虽然在一些土壤生态学和土壤动物多样性调查研究中有部分线虫的简单记录，但缺少具体的属和种类鉴定，也无相应的分子序列信息（李红梅等，2021）。

我国土壤线虫的研究最早可追溯到 1929 年，Reinhard Hoeppli 博士（Morley，2021）和同事伍献文院士（中国科学院水生生物研究所，1985）首先报道了福建和浙江省的 14 种线虫（Wu & Hoeppli，1929；Hoeppli，1932），接着描述了福建、江苏、台湾和北京温泉中的 9 种线虫（Hoeppli & Chu，1932）。Kreis（1929，1930）报道了来自北京的 15 种线虫。Rahm（11937，1938）报道了北京的 20 种线虫和海南省的 35 种线虫。Li（1951）报道了北京的 2 种线虫。Andrássy（1960b）描述了北京和江西省的 14 种线虫，此外还列出了在我国首次报道的 38 种淡水和土壤线虫。

20 世纪 80 年代后期，土壤线虫研究开始逐渐受到关注，孙希达等（1987）对浙江省西天目山土壤线虫种类发生情况进行了调查，发现 5 目 23 属 40 种线虫，其中单齿目线虫有 5 种，乳突单齿线虫（*Mononchus papillatus*）和苔藓锯齿线虫（*Prionchulus muscorum*）是西天目山土壤线虫中的优势种群。姜德全（1988）记述了四川成都、内江、乐山等地区发现的 7 种单齿目线虫，其中在污水沟、浅水池等水生环境发现的平截单齿线虫（*Mononchus truncatus*）分布很广，为我国首次报道。伍惠生和孙希达（1992）报道在浙江、湖南等地发现有乳突克拉克线虫（*Clarkus papillatus*），该种为广泛分布种。林秀敏等（1999）记述了福建省南部地区菜地、甘蔗地、沼泽地的单齿目线虫 4 属 7 种。

吴纪华（1999）历时 5 年对我国长江流域土壤和底栖生境的自由生活线虫进行了系统调查研究，从采自湖北、安徽、浙江、江西、四川等 17 个省份样品中发现 154 种线虫，其中线虫新种 11 个（Wu & Ahmad，1998；Wu & Liang，1999；Ahmad & Wu，1999；Ahmad *et al.*，2002），国内新记录种 79 个（吴纪华和梁彦龄，1997；吴纪华等，1997）。吴纪华（1999）不仅对自己发现的大多数种类都作了较为详细的描述和图示，还对前人报道的 162 种线虫进行了复核，对其中大量存在的同物异名或业已重新整合的物种名进行了逐一校订，总共记述了我国淡水和土壤线虫 255 种，隶属于 2 纲 7 目 61 科 142 属。

此后，Li 等（2008）对来自我国云南省哀牢山和玉龙山常绿阔叶混交林，以及西藏色季拉山森林、灌木和草甸土壤的矛线目垫咽总科（Tylencholaimoidea）的 4 个新种和 4 个已知种进行了描述。王旭（2008）在辽宁省发现平截单齿线虫、贝尔单齿线虫（*Mononchus bellus*）和薄片单齿线虫（*Mononchus laminatus*），其中后 2 种线虫为我国新纪录种。赵春丹和赵洪海（2010）对来自山东省胶南市蓝莓根部的济南小孔咽线虫（*Aporcelaimellus jiaonanensis*）新种进行了描述，李建立和赵洪海（2012）对来自山东省沂水县烟草根围土的钝尾小孔咽线虫（*Aporcelaimellus obtusicaudatus*）进行了记述。Zhang 等（2012）报道了来自中国长白山的小垫咽属（*Tylencholaimellus*）一个新种和 3 个已知种。潘玉雯等（2012）对江苏省杨树根际几种垫刃目和矛线目线虫种类进行了记述。李志辉和赵洪海（2016）对来自山东省青岛市牡丹花根际土壤的大盘咽线虫（*Discolaimus major*）进行了记述。

吴文佳（2018）对我国多个省份不同生境土壤中的矛线目线虫进行了鉴定，共鉴定和描述了 1 个新属、15 个新种、1 个中国新纪录属、1 个中国新纪录种和 2 个四川省新纪录种，其中正式发表的有 9 种（Wu *et al.*，2016a，2016b，2017a，2017b，2018a，2018b，2019；Peña-santiago，2020）。于焦（2019）对海南省土壤线虫种类多样性进行了调查和鉴定，对采自海南省各县市及群岛的不同生境的 150 份土壤样品进行线虫的分离与鉴定，共鉴定出土壤线虫 8 目 23 科 41 属 29 种，其中矛线目线虫有 6 种。刘姝含（2021）对来自我国 7 省份植被土壤样品的单齿目 8 个线虫种群和矛线目 4 个种群进行了形态和分子鉴定。Zhang 等（2023）描述了从南京市阳山碑材公园的苔藓上分离到的新种南京粗角咽线虫（*Trachactinolaimus nanjingenesis*）。Sun 等（2023）描述了从南京市钟山风景区草地分离到的新种钟山拟阴门线虫（*Paravulvus zhongshanensis*）。

截至 2023 年 4 月，我国报道的矛线目线虫种类共有 11 科 49 属 101 种，各种类的发生分布情况整理见表 3。

表 3　我国矛线目线虫种类发生分布情况

目、科、属	种名	采集地	参考文献
穿咽亚目　Nygolaimina Ahmad & Jairajpuri，1979			
穿咽科　Nygolaimidae Thorne，1935			
水线属 *Aquatides* Heyns，1968	正水线虫 *A. aquaticus*（Thorne，1930）Thorne，1974	湖北保安湖的底泥	吴纪华，1999
类棒尾属 *Clavicaudoides* Heyns，1968	类棒尾属未知种 *Clavicaudoides* sp.	南京农业大学校园苔藓土，黑龙江齐齐哈尔森林土壤	Sun *et al.*，2023
滑线属 *Laevides* Heyns，1968	正滑线虫 *L. laevis*（Thorne，1939）Thorne，1974	湖北武汉的草地和保安湖的底泥，安徽太平湖岸边的土壤，浙江金华仙华山的土壤	吴纪华，1999
	迅猛滑线虫 *L. rapax*（Thorne，1939）Ahmad & Jairajpuri，1982	江西鄱阳湖和湖北保安湖的底泥	Wu & Liang，1997
拟阴门属 *Paravulvus* Heyns，1968	哈廷吉拟阴门线虫 *P. hartingii*（de Man，1880）Heyns，1968	南京农业大学校园苔藓土，西藏林芝草地	Sun *et al.*，2023
	钟山拟阴门线虫 *P. zhongshanensis* Sun，Zeng，Qing，Li & Álvarez-Ortega，2023	南京钟山风景区草地	Sun *et al.*，2023
硬齿属 *Solididens* Heyns，1968	硬齿属未知种 *Solididens* sp.	西藏林芝草地	Sun *et al.*，2023
矛线亚目 Dorylaimina Pearse，1936			
角咽科 Actinolaimidae Thorne，1939			
角咽属 *Actinolaimus* Cobb，1913a	涩角咽线虫 *A. perplexus* Heyns & Argo，1969	湖北保安湖的底泥	吴纪华，1999
非洲角咽线虫 *Afractinolaimus* Andrássy，1970	诺氏非洲角咽线虫 *A. noblei* Andrássy，1970	湖北保安湖的水草，安徽太平的溪流，湖南洞庭湖的底泥，浙江仙华山的山泉	吴纪华，1999

（续表）

目、科、属	种名	采集地	参考文献
埃格属 *Egtitus* Thorne，1967	裸埃格线虫 *E. nudus*（Wu & Hoeppli，1929）Thorne，1967	浙江温州一溪流	Wu & Hoeppli，1929
	中华埃格线虫 *E. sinensis* Wu & Liang，1999	湖北省咸宁的枯枝落叶层	吴纪华，1999
亚角咽属 *Mactinolaimus* Andrássy，1970	奇氏亚角咽线虫 *M. chitwoodi*（Moorthy，1937）Andrássy，1970	湖北东湖、保安湖和牛山湖的水草与底泥，安徽太平湖沿岸带	吴纪华，1999
拟角咽属 *Paractinolaimus* Meyl，1957	大咽拟角咽线虫 *P. macrolaimus*（de Man，1880）Meyl，1957	安徽黄山光明顶和立马桥的苔藓和枯枝落叶层	吴纪华，1999
	小齿拟角咽线虫 *P. microdentatus*（Thorne，1939）Meyl，1957	浙江天目山	伍惠生和孙希达，1992
粗角咽属 *Trachactinolaimus* Andrássy，1963	短尾粗角咽线虫 *T. brevicaudatus* Wu & Liang，1999	四川小三峡岩石壁上的苔藓	吴纪华，1999
	南京粗角咽线虫 *T. nanjingenesis* Zhang，Ji，Guo，Qing & Li，2023	南京阳山碑材公园苔藓	Zhang *et al.*，2023
孔咽科 Aporcelaimidae Heyns，1965			
小孔咽属 *Aporcelaimellus* Heyns，1965	嗜粉小孔咽线虫 *A. amylovorus*（Thorne & Swanger，1936）Heyns，1965	浙江杭州的芋头地和苔藓，浙江金华仙华山的地钱，湖北武汉的草地和菜地、东湖的底泥和岸边土壤，湖南洞庭湖岸边的土壤	吴纪华，1999
	胶南小孔咽线虫 *A. jiaonanensis* Zhao & Zhao，2010	山东胶南市蓝莓根部	赵春丹和赵洪海，2010
	钝尾小孔咽线虫 *A. obtusicaudatus*（Bastian，1865）Altherr，1968	湖北武汉的土壤，东湖、保安湖的底泥，安徽贵池、黟县和黄山的草地、苔藓和枯枝落叶层，四川成都、云南和吉林长春的土壤	吴纪华，1999
		山东沂水县烟草根围土	李建立和赵洪海，2012
		四川崇州市崇阳镇金鸡乡的柑橘根际	吴文佳，2018

（续表）

目、科、属	种名	采集地	参考文献
孔咽属 *Aporcelaimus* Thorne & Swanger，1936	亚唇孔咽线虫 *A. sublabiatus*（Thorne & Swanger，1936） Brzeski，1962	吉林长白山的土壤	吴纪华，1999
小孔属 *Aporcella* Andrássy，2002	透明小孔线虫 *A. vitrinus*（Thorne & Swanger，1936）Álvarez-Ortega，Subbotin & Peña-Santiago，2013）	内蒙古阿拉善左旗敖登格日勒嘎查的四合木根际土壤	吴文佳，2018
颚针科 Belondiridae Thorne，1939			
缢咽属 *Axonchium* Cobb，1920	后钝尾缢咽线虫 *A. metobtusicaudatum*（Sch. Stekhoven & Teunissen，1938）Nair & Coomans，1973	安徽黄山北海的枯枝落叶层和太平的茶叶地，湖北武汉的土壤	吴纪华，1999
	石门台缢咽线虫 *A. shimentai*	广东英德市石门台自然保护区	吴文佳，2018
颚针属 *Belondira* Thorne，1939	八公山颚针线虫 *B. bagongshanensis* Wu，Huang，Xie，Wang & Xu，2017	安徽淮南市八公山森林公园草地	Wu *et al.*，2017a
	广州颚针线虫 *B. guangzhouensis*	华南农业大学校内树木园竹子根际	吴文佳，2018
	尾颚针线虫 *B. caudata* Thorne，1939	浙江天目山	伍惠生和孙希达，1992
	棒状颚针线虫 *B. clara* Thorne，1939	浙江天目山	伍惠生和孙希达，1992
牙咽属 *Dorylaimellus* Cobb，1913a	黑山牙咽线虫 *D. montenegricus* Andrássy，1960a	北京和江西庐山	Andrássy，1960b
	高山牙咽线虫 D. *monticolus* Clark，1963	甘肃武威市草地	刘姝含，2021

（续表）

目、科、属	种名	采集地	参考文献
尖颚针属 *Oxybelondira* Ahmad & Jairajpuri，1979	拟涩尖颚针线虫 *O. paraperplexa* Ahmad & Jairajpuri，1979	南京农业大学校园苔藓土	刘姝含，2021
矛线科 Dorylaimidae de Man，1876			
非洲矛线属 *Afrodorylaimus* Andrássy，1964	比氏非洲矛线虫 *A. beaumonti*（Altherr，1952）Andrássy，1969	安徽黄山立马桥苔藓，湖北保安湖的底泥	吴纪华，1999
距矛线属 *Calcaridorylaimus* Andrássy，1986	腾格里距矛线虫 *C. tengeri*	内蒙古自治区阿拉善左旗腾格里沙漠的通湖草原	吴文佳，2018
库曼斯线虫属 *Coomansinema* Ahmad & Jairajpuri，1989	异尾库曼斯线虫 *C. dimorphicauda* Ahmad & Jairajpuri，1989	湖北咸宁的苔藓	吴纪华，1999
矛线属 *Dorylaimus* Dujardin，1845	翼状矛线虫 *D. alaeus* Thorne，1939	浙江天目山，湖南衡山、岳麓山	伍惠生和孙希达，1992
	池塘矛线虫 *D. stagnalis* Dujardin，1845	浙江杭州的芋头地和茭白地，湖北东湖和保安湖的底泥，安徽仙源的溪流，山东青岛的池塘 Rahm（1938）在海南发现本种	吴纪华，1999
镰矛线属 *Drepanodorylaimus* Jairajpuri，1966	捷氏镰矛线虫 *D. szekessyi*（Andrássy，1960b）Andrássy，1969	中国江西庐山	Andrássy，1960b
咽矛属 *Laimydorus* Siddiqi，1969	顶尖咽矛线虫 *L. acris*（Thorne，1939）Andrássy，1969	浙江天目山	伍惠生和孙希达，1992
	粗咽矛线虫 *L. crassus*（de Man，1884）Thorne，1974	海南	Rahm，1938
	伪池塘咽矛线虫 *L. pseudostagnalis*（Micoletzky，1927）Siddiqi，1969	安徽太平的溪流	吴纪华，1999

（续表）

目、科、属	种名	采集地	参考文献
中矛线属 *Mesodorylaimus* Andrássy，1959	高山中矛线虫 *M. alpestris*（Thorne，1939）Andrássy，1959	浙江天目山	伍惠生和孙希达，1992
	巴氏中矛线虫 *M. bastiani*（Bütschli，1873）Andrássy，1959	我国	张云美等，1998
	博罗中矛线虫 *M. boluoensis*	广东省惠州市博罗县的柑橘根际	吴文佳，2018
	浙江中矛线虫 *M. chekiangensis*（Wu & Hoeppli，1929）Andrássy，1986	浙江温州	Wu & Hoeppli，1929
	中华中矛线虫 *M. chinensis* Wu & Ahmad，1998	安徽黄山玉屏楼（海拔 1 680m）的苔藓	Wu & Ahmad，1998
	黄斑中矛线虫 *M. flavomaculatus*（Linstow，1876）Goodey，1963	湖北东湖和保安湖的底泥，江西鄱阳湖和湖南洞庭湖的底泥	吴纪华，1999
	纤巧中矛线虫 *M. graciosus* Andrássy，1986	安徽黄山光明顶枯枝落叶层和半山寺的泉流	吴纪华，1999
	午夜中矛线虫 *M. mesonyctius*（Kreis，1930）Andrássy，1959	湖北宜昌的草地和武汉的土壤，湖北东湖和保安湖的底泥和水草，四川成都的草地和小三峡岩壁的苔藓，浙江肖山的草地和苔藓，河南新乡的土壤	吴纪华，1999
	主要中矛线虫 *M. major*	青海果洛州玛沁县格姆滩	吴文佳，2018
	裸中矛线虫 *M. nudus*（Thorne，1939）Andrássy，1959	浙江天目山	伍惠生和孙希达，1992
	斯氏中矛线虫 *M. spengelii*（de Man，1912）Andrássy，1959	安徽黄山莲花沟的土壤	吴纪华，1999

（续表）

目、科、属	种名	采集地	参考文献
	细微中矛线虫 *M. subtilis*（Thorne & Swanger，1936）Andrássy，1959	浙江金华仙华山的土壤，安徽太平菊科植物根部的土壤	吴纪华，1999
		广东深圳南山区白掌种植基质土	吴文佳，2018
原矛线属 *Prodorylaimium* Andrássy，1969	高山原矛线虫 *P. alpinum* Andrássy，1969	湖北咸宁的枯枝落叶层	吴纪华，1999
	布里赫原矛线虫 *P. brigdammensis*（de Man，1876）Andrássy，1969	海南	Rham，1938
前矛线属 *Prodorylaimus* Andrássy，1959	似线前矛线虫 *P. filiarum* Andrássy，1992	山东青岛牡丹花根际土壤	李志辉和赵洪海，2016
索恩线虫属 *Thornenema* Andrássy，1959	巴耳多索恩线虫 *T. baldum*（Thorne，1939）Andrássy，1959	浙江天目山	伍惠生和孙希达，1992
	平滑索恩线虫 *T. lissum*（Thorne，1939）Andrássy，1959	浙江天目山	伍惠生和孙希达，1992
细齿科 Leptonchidae Thorne，1935			
底垫裙属 *Basirotyleptus* Jairajpuri，1964	松底垫裙线虫 *B. pini* Siddiqi & Khan，1965	云南哀牢山草原，常绿阔叶混交林中树木根部周围的土壤	Li *et al.*，2008
细齿属 *Leptonchus* Cobb，1920	颗粒细齿线虫 *L. granulosus* Cobb，1920	云南	张云美等，1998
前细齿属 *Proleptonchus* Lordello，1955	中华前细齿线虫 *P. sinensis* Li，Baniyamuddin，Ahmad & Wu，2008	云南哀牢山常绿阔叶混交林中树木根部周围的土壤	Li *et al.*，2008
垫裙属 *Tyleptus* Thorne，1939	耸突垫裙线虫 *T. projectus* Thorne，1939	安徽太平的枯枝落叶层	吴纪华，1999
湿生科 Mydonomidae Thorne，1964			
类矛线属 *Dorylaimoides* Thorne & Swanger，1936	高山类矛线虫 *D. alpinus* Li，Baniyamuddin，Ahmad & Wu，2008	云南哀牢山和玉龙山常绿阔叶混交林中树木根部周围的土壤	Li *et al.*，2008

（续表）

目、科、属	种名	采集地	参考文献
	精致类矛线虫 *D. elaboratus* Siddiqi，1965	广东深圳的土壤	吴纪华，1999
	迈克类矛线虫 *D. micoletzkyi*（de Man，1921）Thorne & Swanger，1936	北京	Andrássy，1960b
	雷耶斯类矛线虫 *D. reyesi* Ahmad，Mushtaq & Baniyamuddin，2003	西藏色季拉山灌木和草甸土壤	Li *et al.*，2008
	沙坡头类矛线虫 *D. shapotouensis* Wu，Xu，Xie &Wang，2019	宁夏中卫市沙坡头的苹果树根际	Wu *et al.*，2019
	光滑类矛线虫 *D. teres* Thorne & Swanger，1936	湖南洞庭湖的底泥	吴纪华，1999
泽地属 *Morasia* Basqi & Jairajpuri，1969	杆齿泽地线虫 *M. rhabdontus*（Kreis，1930）Basqi & Jairajpuri，1969	北京和海南	Kreis，1930；Rahm，1938
北印度科 Nordiidae Jairajpuri & Siddiqi，1964			
大矛属 *Enchodelus* Thorne，1939	粗大矛线虫 *E. hopedorus*（Thorne，1929）Thorne，1939	安徽黄山玉屏楼和北海的苔藓	吴纪华，1999
	索氏大矛线虫 *E. southeyi* Jairajpuri & Ahmad，1986	中国北京	Jairajpuri & Ahmad，1986
齿矛属 *Enchodorus* Vinciguerra，1976	新锥齿矛线虫 *E. neodolichurus* Ahmad & Wu，1999	安徽黄山玉屏楼（海拔 1 680m）的苔藓	吴纪华，1999
异矛属 *Heterodorus* Altherr，1952	青海异矛线虫 *Heterodorus qinghaiensis* Wu，Yan，Xu，Yu，Wang，Jin & Xie，2016	青海省果洛州玛沁县格姆滩的草地	Wu *et al.*，2016b
库西亚线虫科 Qudsianematidae Jairajpuri，1965			
异矛线属 *Allodorylaimus* Andrássy，1986	高山异矛线虫 *A. alpinus*（Steiner，1914）Andrássy，1986	安徽黄山光明顶枯枝落叶层	吴纪华，1999

（续表）

目、科、属	种名	采集地	参考文献
	整齐异矛线虫 *A. uniformis*（Thorne，1929）Andrássy，1986	安徽黄山立马桥苔藓	吴纪华，1999
小盘咽属 *Discolaimium* Thorne，1939	草莓小盘咽线虫 *D. fragaria*	河南许昌草莓根际	吴文佳，2018
盘咽属 *Discolaimus* Cobb，1913a	阿尼玛卿盘咽线虫 *D. anemaqen* Wu，Yan，Xu，Wang，Jin & Xie，2016	青海果洛州玛沁县格姆滩的赖草封育地	Wu *et al.*，2016a
	大盘咽线虫 *D. major* Andrássy，2002	山东青岛牡丹花根际土壤	李志辉和赵洪海，2016
通俗属 *Ecumenicus* Thorne，1974	单宫通俗线虫 *E. monohystera*（de Man，1880）Thorne，1974	Kreis（1930）报道在北京土壤中发现两个种，*Dorylaimus monohyslera* 和 *Dorylaimus gibberoaculeatus* 实为同一种。Thorne（1974）建立 *Ecumemcus* 属，该种是这个属的模式种，也是迄今记录唯一的种（Jairjpuri & Ahmad，1992）	
上矛线属 *Epidorylaimus* Andrássy，1986	相关上矛线虫 *E. consobrinus*（de Man，1918）Andrássy，1986	北京、河北、内蒙古	Rahm，1937
	吕格德上矛线虫 *E. lugdunensis*（de Man，1880）Andrássy，1986	湖北武汉和广西桂林的土壤，浙江湖州和杭州的苔藓，安徽黄山北海的枯枝落叶层和苔藓	吴纪华，1999
真矛线属 *Eudorylaimus* Andrássy，1959	卡氏真矛线虫 *E. carteri*（Bastian，1865）Andrássy，1959	湖北武汉、咸宁和神农架等地的土壤和枯枝落叶层，安徽黄山的枯枝落叶层和苔藓，广东中山的土壤，重庆的土壤，四川大足和小三峡的苔藓，吉林长白山的苔藓，浙江湖州的草地，湖北东湖和保安湖的底泥	吴纪华，1999
	王冠真矛线虫 *E. diadematus*（Cobb in Thorne & Swanger，1936）	海南三沙市南沙群岛的行道树、草坪根际土壤	于焦，2019

（续表）

目、科、属	种名	采集地	参考文献
	云杉真矛线虫 *E. piceae* Wu，Yu，Xu，Wang & Xie，2018.	内蒙古阿拉善左旗贺兰山的青海云杉根际	Wu *et al.*，2018a
唇线虫属 *Labronema* Thorne，1939	纤细唇线虫 *L. fimbriatum* Thorne，1939	浙江天目山	伍惠生和孙希达，1992
	透明唇线虫 *L. hyalinum*（Thorne & Swanger，1936）Thorne，1939	四川小三峡岸边湿土，安徽太平和茶叶地和黄山的苔藓，湖北武汉和咸宁的枯枝落叶层，保安湖和梁子湖的岸边土壤，江西鄱阳湖的底泥	吴纪华，1999
	施特歇林唇线虫 *L. stechlinense* Altherr，1968	安徽黄山始信峰和莲花峰的枯枝落叶层	吴纪华，1999
小唇线虫属 *Labronemella* Andrássy，1985	大小唇线虫 *L. major* Wu，Yan，Xie，Xu，Yu，Wang & Jin，2017	青海果洛州玛沁县格姆滩的黑土滩放牧草地	Wu *et al.*，2017b
微矛线虫属 *Microdorylaimus* Andrássy，1986	微矛线虫属未知种 *Microdorylaimus* sp.	海南三沙市南沙群岛	于焦，2019
磨盘属 *Mylodiscus* Thorne，1939	矮小磨盘线虫 *Mylodiscus nanus* Thorne，1939	湖北咸宁的土壤	吴纪华，1999
小垫咽科 Tylencholaimellidae Jairajpuri，1964			
小垫咽属 *Tylencholaimellus* Cobb in Cobb，1915	具环小垫咽线虫 *T. cinctus* Orr & Dickerson，1965	东北长白山针叶阔叶混交林	Zhang *et al.*，2012
	高山小垫咽线虫 *T. montanus* Thorne，1939	东北长白山针叶阔叶混交林	Zhang *et al.*，2012
	中华小垫咽线虫 *T. sinensis* Zhang，Ahad，Baniyamuddin，Liang & Ahmad，2012	东北长白山针叶阔叶混交林	Zhang *et al.*，2012

（续表）

目、科、属	种名	采集地	参考文献
	具纹小垫咽线虫 *T. striatus* Thorne，1939，	东北长白山暗针叶云杉林	Zhang *et al.*，2012
垫咽科 Tylencholaimidae Filipjev，1934			
垫咽属 *Tylencholaimus* de Man，1876	犬齿垫咽线虫 *T. cynodonti* Nasira，Erum & Shahina，2005 同种异名：三沙拟垫咽线虫 *Paratylencholaimus sanshaensis* Wu，Xu，Xie & Wang，2019	海南省三沙市永兴岛一品红根际，广东省广州市花都区绿萝和中山市的蝴蝶兰培养基质，惠州市博罗县的柑橘根际	Wu *et al.*，2019 Peña-santiago，2020
	贺兰垫咽线虫 *T. helanensis* Wu，Yu，Xie，Xu，Yu & Wang，2018	内蒙古自治区阿拉善左旗贺兰山的草地	Wu *et al.*，2018b
	伊比利亚垫咽线虫 *T. ibericus* Peña-Santiago & Coomans，1994 同种异名：中山垫咽线虫 *T. zhongshanensis* Wu，Xu，Xie & Wang，2019	广东中山蝴蝶兰培养基质	Wu *et al.*，2019 Peña-santiago，2020
	日本垫咽线虫 *T. japonicus* Ahmad & Araki，2003	云南哀牢山常绿阔叶混交林中树根周围土壤，中国西藏米拉、德木拉和色季拉山的草甸和灌木丛	Li *et al.*，2008
	矮小垫咽线虫 *T. nanus* Thorne，1939	北京	Andrássy，1960b
	东方垫咽线虫 *T. orientalis* Li，Baniyamuddin，Ahmad & Wu，2008	云南哀牢山常绿阔叶林树木根部周围的土壤	Li *et al.*，2008
	近垫咽线虫 *T. proximus* Thorne，1939	西藏米拉山草甸的土壤	Li *et al.*，2008
	中华垫咽线虫 *T. sinensis* Li，Baniyamuddin，Ahmad & Wu，2008	西藏色季拉山米拉草甸和灌丛根部周围的土壤	Li *et al.*，2008
	史氏垫咽线虫 *T. stecki* Steiner，1914	安徽黄山慈光阁枯枝落叶层，吉林小天池的苔藓	吴纪华，1999

9 研究展望

矛线目的种类和数量，是线虫门各目中最多的，也是陆地土壤与淡水沉积物中最多样化与生态学上最重要的线虫类群之一。矛线目线虫的高多样性、高丰度、高度变化的取食习性，以及虫体形态和大小的巨大变化，使它们成为评估土壤健康状况的良好指示生物，许多矛线目种类对重金属和各种土壤污染物表现出高度的敏感性，更加凸显了它们在土壤生态学领域的重要性（Peña-Santiago，2021）。矛线目分为 2 个亚目 18 个科，全世界已描述的有效属达 285 个，我国记录有 49 个属（植物寄生的长针科线虫除外），占已描述属的 17.2%，前世界已描述 3 000 多个有效种，我国记录的自由生活的矛线目线虫有 101 种，约占世界总描述种的 3.3%。我国国土辽阔，生态环境多样，土壤线虫物种资源非常丰富，然而关于矛线目线虫多样性的研究极为稀缺，矛线目线虫在土壤生态学研究的重要性还没有得到更多关注。

研究矛线目的生物多样性是一项真正具挑战性的研究，解析线虫种类和环境变量之间关系的主要障碍之一是如何准确鉴定线虫类群，这对矛线目线虫尤为困难。尽管一些学者认为种水平的鉴别是研究线虫生态学的基本要求（Yeates，2003；Hánel，2010），但是许多关于矛线目的生态学工作可能只涉及属或科，很少关注具体的种类（Liébanas *et al.*，2002，2004）。鉴定土壤中自由生活的矛线目线虫种群，不仅需要消耗大量的精力和资源，还需要具有坚实的分类学背景。而当今的传统分类学研究常常被低估，全球从事分类学研究的人员面临急剧缩减，且仅有的分类工作者多关注具有经济危害的植物寄生线虫，全球从事矛线目分类进化研究的专业工作者寥寥无几。

培训线虫学分类人才是一项长期的艰巨的系统工程，在比利时政府和欧盟的持续资金支持下，根特大学开设的两年国际线虫研究生课程（The European Master of Science in Nematology，http：//www. eumaine. ugent. be/）得以在过去的 30 多年间，为全球 80 多个国家培养了大量的线虫学研究人才。此外，荷兰的瓦赫宁根大学也会不定期举办土壤和淡水自由生活线虫以及植物寄生线虫的鉴定培训课程，为世界各国相关研究人员提供 1~2 个月的培训。然而，这些努力难以从根本上解决分类鉴定专业人士的缺口，目前矛线目线虫的鉴定仍较为混乱，相关论文中存在大量鉴定错误。针对这些问题，目前已有研究尝试使用图像人工智能识别技术鉴定线虫（Qing *et al.*，2022）。该技术利用专家人工鉴定的图片对系统进行训练，目前能够鉴定包括真矛线属（*Eudorylaimus*）、矛线属（*Dorylaimus*）、孔咽属（*Aporcelaimus*）在内的 19 个种常见土壤线虫，具有广阔的应用前景。

近年来，基于 rRNA 与线粒体 COI 基因的分子条形码（DNA barcoding）技术被广泛应用于线虫的鉴定，其与高通量测序相结合的分子宏条形码（metabarcoding）技术，更是成为土壤线虫生态学研究的重要工具（Porazinska *et al.*，2009；Ahmad *et al.*，2015）。然而，因缺乏相关矛线目分类专家，NCBI 数据库中目前仅有不到 10%的矛线目种类具有分子条形码序列。同时，由于缺乏矛线目线虫参考序列，现有分子鉴定与宏条形码引物大多基于秀丽隐杆线虫（*Caenorhabditis elegans*）或植物寄生线虫，尚无针对矛线目线虫设计的通用引物，导致 PCR 扩增中矛线目线虫的扩增效率和成功率往往较低。这些问题导致矛线目线虫难以进行分子鉴定，极大地阻碍了该类群相关生态学研究。解决这些问题的重要途径之一是对更多种类的矛线目线虫进行分子条形码测序，建立分子条形码数据库。目前虽然已有线虫分子条形码数据库 PPNID（Qing *et al.*，2020），但其包含的种类仅为植物寄生线虫，目前尚无针对矛线目、小杆目等自由生活类线虫数据库，亟须国内外相关分类工作者共同合作，建立更为全面的分子数据库。

参考文献

姜德全，1988. 四川捕食性线虫（单齿科）的记述［J］. 四川动物，7：4-72.

李红梅，刘姝含，薛清，2021. 单齿目线虫的多样性及其生防潜力研究概述［C］//彭德良．中国线虫学研究．北京：中国农业科学技术出版社，8：210-249.

李建立，赵洪，2012. 中国小无环咽属线虫一新纪录种（线虫纲　矛线目　无环咽科）［J］. 动物分类学报，37：425-428.

李志辉，赵洪海，2016. 矛线总科线虫 2 个新纪录种［J］. 青岛农业大学学报（自然科学版），33：174-177.

林秀敏，陈美，陈清泉，等，1999. 闽南地区捕食性线虫（Nematoda，Mononchida）的记述［J］. 厦门大学学报（自然科学版），1：112-116.

刘姝含，2021. 单齿目和矛线目线虫的种类鉴定及食性研究［D］. 南京：南京农业大学.

潘玉雯，谈家金，叶建仁，2012. 江苏省杨树根际几种垫刃目和矛线目线虫种类记述［J］. 林业科学，48（8）：161-165.

孙希达，赵英，胡江琴，等，1989. 西天目山土壤线虫区系与生态的初步探讨［J］. 杭州师范学院学报，3：62-68.

王旭，2008. 捕食性线虫多样性及生防潜力初探［D］. 沈阳：沈阳农业大学.

吴纪华，1999. 中国淡水和土壤线虫的研究［D］. 武汉：中国科学院水生物研究所.

吴纪华，梁彦龄，1997. 中国长江流域自由生活线虫初步研究［J］. 水生生物学报，21（增）：114-122.

吴纪华，梁彦龄，孙希达，1997. 中国自由生活线虫新记录（色矛目、嘴刺目和窄咽目）［J］. 水生生物学报，21（4）：312-321.

吴文佳，2018. 矛线亚目线虫的种类鉴定和系统分类研究［D］. 广州：华南农业大学.

伍惠生，孙希达，1992. 线虫纲．中国亚热带土壤动物（尹文英主编）［M］. 北京：科学出版社，161-189.

耶格，1965. 生物名称和生物学术语的词源［M］. 滕砥平，蒋芝英，译．北京：科学出版社.

于焦，2019. 海南省土壤线虫种类多样性调查和鉴定［D］. 广州：华南农业大学.

张云美，王敏，陈建英，等，1998. 线虫动物门．中国土壤动物检索图鉴（尹文英主编）［M］. 北京：科学出版社：437-475.

赵春丹，赵洪海，2010. 无孔小咽属线虫一新种（线虫纲，矛线目，孔咽科）［J］. 动物分类学报，35：876-879.

中国科学院水生生物研究所，1985. 怀念伍献文所长［J］. 水生生物学报，9：195-202.

ABEBE E，DECRAEMER W，DE LEY P，2008. Global diversity of nematodes（Nematoda）in freshwater［J］. Hidrobiologia，595：67-78.

AHMAD M，SAPP M，PRIOR T，*et al.*，2015. Nematode taxonomy：from morphology to metabarcoding［J］. Soil Discussions，2：1175-1220.

AHMAD W，WU J H，1999. A new species of the rare nematode genus *Enchodorus Vinciguerra*（Nematoda：

Dorylamida) from China [J]. International Journal of Nematology, 9: 181-184.

AHMAD W, WU J H, SHAHEEN A, 2002. Studies on the genus *Enchodelus* Thorne, 1939 (Dorylaimida: Nordiidae) from China [J]. Journal of Nematode Morphology and Systematics, 4: 83-90.

AKHTAR M, 1998. Biological control of plant parasitic nematodes by neem products in agricultural soils [J]. Applied Soil Ecology, 7: 219-223.

ANDRÁSSY I, 1959. Taxonomische Übersicht der Dorylaimen (Nematoda), Ⅰ [J]. Acta Zoologica Academiae Scientiarum Hungaricae, 5: 191-240.

ANDRÁSSY I, 1960a. Taxonomische Übersicht der Dorylaimen (Nematoda), Ⅱ [J]. Acta Zoologica Academiae Scientiarum Hungaricae, 6: 1-28.

ANDRÁSSY I, 1960b. Beiträge zur Kenntnis der freilebenden Nematoden Chinas [C]. Annales Historico Naturales Musei Nationalis Hungarici, 52: 202-216.

ANDRÁSSY I, 1976. Evolution as a basis for the systematization of nematodes [M]. London, Pitman Publishing, 288 pp.

ANDRÁSSY I, 2009. Free-living nematodes of Hungary (Nematoda errantia), Ⅲ [J]. Hungary: Budapest, Hungarian Natural History Museum, 586.

BANAGE W B, 1963. The ecological importance of free-living soil nematodes with special reference to those of moorland soil [J]. Journal of Animal Ecology, 32: 133-140.

BASTIAN H C, 1865. Monograph on the Anguillulidae, free nematoids, marine, land, and freshwater with descriptions of 100 new species [J]. Transactions of the Linnean Society of London, 25: 173-184.

BILGRAMI A L, 1992. Resistance and susceptibility of prey nematodes to predation and strike rate of the predators, *Mononchus aquaticus*, *Dorylaimus stagnalis*, and *Aquatides thornei* [J]. Fundamental and Applied Nematology, 15: 265-270.

BILGRAMI A L, 1993. Analysis of relationships between predation by *Aporcelaimellus nivalis* and different trophic categories [J]. Nematologica, 39: 356-365.

BILGRAMI A L, AHMAD I, JAIRAJPURI M S, 1985. Predatory behaviour of *Aquatides thornei* (Nygolaimina: Nematoda) [J]. Nematologica, 30: 457-462.

BLAXTER M L, DE LEY P, GAREY J R, *et al.*, 1998. A molecular evolutionary framework for the phylum Nematoda [J]. Nature, 392: 71-75.

BOOSALIS M G, MANKAU R, 1965. Parasitism and predation of soil microorganisms. In: Baker K F, Snyder W C (Eds), Ecology of Soil-borne Plant Pathogens [M]. Berkeley: University of California Press, 374-391.

BRIAR S S, BARKER C, TENUTA M, *et al.*, 2012. Soil nematode responses to crop management and conversion to native grasses [J]. Journal of Nematology, 44: 245-254.

CASSIDY G H, 1931. Some mononchs of Hawaii [J]. Hawaiian Planters' Record, 35: 305-339.

CESARZ S, RUESS L, JACOB M, *et al.*, 2013. Tree species diversity versus tree species identity: driving forces in structuring forest food webs as indicated by soil nematodes [J]. Soil Biology and Biochemistry, 62: 36-45.

CHIZHOV I, CHERNAVSKII D S, ENGELHARD M, *et al.*, 1996. Spectrally silent transitions in the bacteriorho-dopsin photocycle [J]. Biophysical Journal, 71: 23-45.

CHRISTEI J R, 1960. Biological control-predaceous nematodes. In: Sasser J M, Jenkins W R (Eds), Nematology: Fundamentals and Recent Advances with Emphasis on Plant Parasitic and Soil Forms [M]. Chapel Hill: University of North Carolina Press: 466-468.

CLARK W C, 1961. A revised classification of the order Enoplida (Nematoda) [J]. New Zealand Journal of Science, 4: 123-150.

COBB N A, 1913a. New nematode genera found inhabiting fresh water and non-brackish soils [J]. Journal of the Washington Academy of Sciences, 3: 432-444.

COBB N A, 1913b. Nematoda: Longidoridae [J]. Annalen Zoologische Wetenschappen, 287: 1-239.

COBB N A, 1917. The *Mononchs* (*Mononchus* Bastian, 1865). A genus of free-living predatory nematode [J]. Soil Science, 3: 431-486.

COBB N A, 1920. Transfer of nematodes (*Mononchs*) from place to place for economic purposes [J]. Science, 51: 640-641.

COOMANS A, 1985. A phylogenetic approach to the classification of the Longidoridae (Nematoda: Dorylaimida) [J]. Agriculture, Ecosystems and Environment, 12: 335-354.

COOMANS A, LOOF P A A, 1970. Morhology and taxonomy of Bathyodontina (Dorylaimida) [J]. Nematologica, 16: 180-196.

DE LEY P, BLAXTER M L, 2002. Systematic position and phylogeny. In: Lee D L (Ed), The Biology of Nematodes [M]. London: Taylor & Francis: 1-30.

DE LEY P, BLAXTER M L, 2004. A new system for Nematoda: combining morphological characters with molecular trees, and translating clades into ranks and taxa [C]. In: Cook R, Hunt D J (Eds), Proceedings of the Fourth International Congress of Nematology, 8-13 June 2002, Tenerife, Spain. Leiden, The Netherlands, Brill: 633-653.

DE MAN J G, 1876. Onderzoekingen over vrij in de aarde levende Nematoden [J]. Tijdschrift der Nederlandische Dierkundige Vereeniging, 2: 78-196.

DE MAN J G, 1880. Die einheimischen, frei in der reinen Erde und im süssen Wasser lebenden Nematoden [J]. Tijdschrift der Nederlandische Dierkundige Vereeniging, 5: 1-104.

DECRAEMER W, COOMANS A, 1994. A compendium of our knowledge of the free-living nematofauna of ancient lakes [J]. Advances in Limnology, 44: 173-181.

DECRAEMER W, HUNT D J, 2013. Structure and classification. In: Perry R N, Moens M (Eds), Plant Nematology [M]. 2nd, Wallingford, CAB International: 3-39.

DONCASTER C C, 1962. Predators of soil nematodes (fifilm) [J]. Parasitology, 52: 19.

DU PREEZ R, BENDIXEN M, ABRATT R, 2017. The behavioral consequences of internal brand management among frontline employees [J]. Journal of Product and Brand Management, 26: 251-261.

DUJARDIN F, 1845. Histoire Naturelle des Helminthes ou vers Intestinaux [M]. Paris, France, Libraire Encyclopedique de Roret.

EKSCHMITT K, KORTHALS G W, 2006. Nematodes as sentinels of heavy metals and organic toxicants in the soil [J]. Journal of Nematology, 38: 13-19.

ESSER R P, 1963. Nematode interactions in plates of no sterile water agar [J]. Soil and Crop Science Society of Florida Proceedings, 23: 121-128.

ESSER R P, 1987. Biological control of nematodes by nematodes. I. *Dorylaims* (Nematoda: Dorylaimida) [J]. Nematology Circulars, Florida Department of Agriculture and Consumer Service, Division of Plant Industry, 144: 1-4.

ESSER R P, SOBERS E K, 1964. Natural enemies of nematodes [J]. Proceedings of Soil and Crop Science Society of Florida, 24: 326-353.

FERRIS H, 2010. Form and function: metabolic footprints of nematodes in the soil food web [J]. European

Journal of Soil Biology, 46: 97-104.

FILIPJEV I N, 1927. Les Némato des libres des mers septentrionales appartenant a la famille des Enoplidae [J]. Archiv für Naturgeschichte, 91: 1-216.

FRECKMAN D W, KAPLAN D T, VAN GUNDY S D, 1977. A comparison of techniques for extraction and study of anhydrobiotic nematodes from dry soils [J]. Journal of Nematology, 9: 176-181.

GOODEY T, 1963. Soil and Freshwater Nematodes. Rewritten by J B. Goodey [M]. London/New York: Science.

GROOTAERT P, SMALL R W, 1982. Aspects of biology of *Labronema vulvapapillatum* (Meyl) (Nematoda: Dorylaimida) in laboratory culture [J]. Biologisch Jaarboek-Dodonaea, 50: 135-148.

HÁNEL L, 2008. Nematode assemblages indicate soil restoration on colliery spoils afforested by planting different tree species and by natural succession [J]. Applied Soil Ecology, 40: 86-99.

HÁNEL L, 2010. An outline of soil nematode succession on abandoned fields in South Bohemia [J]. Applied Soil Ecology, 46: 355-371.

HODDA M, 2011. Phylum nematoda Cobb 1932. In: Zhang Z Q (Ed), Animal Biodiversity: An Outline of Higher-level Classification and Survey of Taxonomic Richness [M]. Zootaxa Monograph 3148. Auckland, New Zealand: Magnolia Press: 63-95.

HOEPPLI R, 1932. Parasitic and free-living nematodes found on the island of Amoy [C]. Marine Biology Association of China First Annual Report, 57-63.

HOLTERMAN M, RYBARCZYK K, VAN DER WURFF A, *et al.*, 2006. Phylum-wide analysis of SSU rDNA reveals deep phylogenetic relationships among nematodes and accelerated evolution towards crown clades [J]. Molecular Biology and Evolution, 23: 1792-1800.

HOLTERMAN M, RYBARCZYK K, VAN DEN ELSEN S, *et al.*, 2008. A ribosomal DNA-based framework for the detection and quantification of stress-sensitive nematode families in terrestrial habitats [J]. Molecular Ecology Notes, 8: 23-34.

HUNT D J, 1978. *Stomachoglossa bryophilum* n. sp. (Brittonematidae), *Westindicus rapax* n. sp. (Paractinolaimidae), *Caribinema siddiqi* n. sp. and *C. longidens* Thorne, 1967 (Charcharolaimidae) from St. Lucia [J]. Nematologica, 24: 175-183.

JAIRAJPURI M S, 1969. Studies on Mononchida of India Ⅰ. The genera *Hadronchus*, *Iotonchus*, and *Miconchus* and a revised classfication of Mononchida, new order [J]. Nematologica, 15: 557-581.

JAIRAJPURI M S, BILGRAMI A L, 1990. Predatory nematodes. In: Jairajpuri M S, Alam M M, Ahmad I (Eds), Nematode Biocontrol [M]. Delhi: CBS Publishers: 95-125.

JAIRAJPURI M S, AHMAD W, 1992. Dorylaimida. Free-living, Predaceous and Plant-Parasitic Nematodes [M]. Netherlands: Leiden University, Brill.

KANWAR R S, PATIL J A, YADAV S, 2021. Prospects of using predatory nematodes in biological control for plant parasitic nematodes-A review [J]. Biological Control, 160: 104668.

KHAN Z, BILGRAMI A L, JAIRAJPURI M S, 1991. Some observations on the predation ability of *Aporcelaimellus nivalis* (Altherr, 1952) Heyns 1966 (Nematoda: Dorylaimida) [J]. Nematologica, 37: 333-342.

KHAN Z, BILGRAMI A L, JAIRAJPURI M S, 1994. Attraction and food preference behaviour of predatory nematodes, *Allodorylaimus americanus* and *Discolaimus silvicolus* (Nematoda: Dorylaimida) [J]. Indian Journal of Nematology, 24: 168-175.

KHAN Z, BILGRAMI A L, JAIRAJPURI M S, 1995a. A comparative study on predation by *Allodorylaimus americanus* and *Discolaimus silvicolus* (Nematoda: Dorylaimida) on different species of plant parasitic nematodes. Indian [J]. Nematology, 25: 94-103.

KHAN Z, BILGRAMI A L, JAIRAJPURI M S, 1995b. Observations on the predation abilities of *Neoactinolaimus agilis* (Dorylaimida: Actinolaimoidea) [J]. Indian Journal of Nematology, 25: 129-135.

KHAN Z, JAIRAJPURI M S, 1997. Predatory behavior of *Paractinolaimus elongatus* on plant parasitic nematodes [J]. Indian Journal of Nematology, 8: 178-181.

KHAN Z, KIM Y H, 2007. A review on the role of predatory soil nematodes in the biological control of plant parasitic nematodes [J]. Applied Soil Ecology, 35: 370-379.

KREIS H A, 1929. Freilebende terrestrische Nematoden aus der Umgebung von Peking (China) Ⅰ [J]. Zoologischer Anzeiger, 84: 283-294.

KREIS H A, 1930. Freilebende terrestrische Nematoden aus der Umgebung von Peking (China) Ⅱ [J]. Zoologischer Anzeiger, 87: 67-87.

LEE D L, 1961. Two new species of cryptobiotic (anabiotic) freshwater nematodes, *Actinolaimus hintoni* sp. nov. and *Dorylaimus keilini* sp. nov. (Dorylaimidae) [J]. Parasitology, 51: 237-240.

LI F, 1951. On two species of free-living nematodes from latrines in Peking which may contaminate improperly collected stool samples [J]. Peking Natural History Bulletin, 19: 363-373.

LI Y, BANIYAMUDDIN M, AHMAD W, *et al.*, 2008. Four new and four known species of Tylencholaimoidea (Dorylaimida: Nematoda) from China [J]. Journal of Natural History, 42: 1991-2010.

LIÉBANAS G, PEÑA-SANTIAGO R, REAL R, *et al.*, 2002. Spatial distribution of dorylaimid and mononchid nematodes from southeast Iberian Peninsula: chorological relationships among species [J]. Journal of Nematology, 34: 390-395.

LIÉBANAS G, GUERRERO P, MARTÍN-GARCíA J M, *et al.*, 2004. Spatial distribution of dorylaimid and mononchid nematodes from southeast Iberian Peninsula: environmental characterization of chorotypes [J]. Journal of Nematology, 36: 114-122.

MCSORLEY H J, O'GORMAN M T, BLAIR N, *et al.*, 2012. Suppression of type 2 immunity and allergic airway inflammation by secreted products of the helminth *Heligmosomoides polygyrus* [J]. European Journal of Immunology, 42: 2667-2682.

MCSORLEY H J, HEWITSON J P, MAIZELS R M, 2013. Immunomodulation by helminth parasites: defining mechanisms and mediators [J]. International Journal for Parasitology, 43: 301-310.

MCSORLEY R, FREDERICK J J, 1999. Nematode population fluctuations during decomposition of specific organic amendments [J]. Journal of Nematology, 31: 37-44.

MORLEY N J, 2021. Reinhard Hoeppli (1893-1973): The life and curious afterlife of a distinguished parasitologist [J]. Journal of Medical Biography, 29: 162-169.

MULLIN P G, HARRIS T S, POWERS T O, 2004. Systematic status of *Campydora* Cobb, 1920 (Nematoda: Campydorina) [J]. Nematology, 5: 699-711.

MULLIN P G, HARRIS T S, POWERS T O, 2005. Phylogenetic relationships of Nygolaimina and Dorylaimina (Nematoda: Dorylaimida) inferred from small subunit ribosomal DNA sequences [J]. Nematology, 7: 59-79.

OKADA T, UMINO N, MATSUZAWA T, *et al.*, 2005. Aftershock distribution and 3D seismic velocity structure in and around the focal area of the 2004 mid Niigata prefecture earthquake obtained by applying double-difference tomography to dense temporary seismic network data [J]. Earth Planets Space, 57: 435-440.

ÖRLEY L, 1880. Monographie der Anguilluliden [M]. Eine von der K. ung. Naturhistorischen Gesellschraft Gekrönte Preisschrift, Budapest, Hungary.

PEARSE A S, 1936. Zoological Names. A List of Phyla, Classes, and Orders Prepared for Section F [M]. American Associtaion for the Advancement of Science, Durhan, USA, Duke University Press.

PEARSE A S, 1942. Introduction to Parasitology [M]. Baltimore, USA, Springfield Springfield, ILL.

PEÑA-SANTIAGO R, 2006. Dorylaimida Part I: Superfamilies Belondiroidea, Nygolaimoidea and Tylencholaimoidea. In: Abebe E, Andrássy I, Traunspurger W (Eds), Freshwater Nematodes: Ecology and Taxonomy [M]. Wallingford, UK, CAB International: 326-391.

PEÑA-SANTIAGO R, 2014. Order Dorylaimida Pearse, 1942. In: Schmidt Rhaesa A (Ed), Handbook of zoology-Gastrotricha, Cycloneuralia and Gnathifera. Volume 2. Nematoda. Berlin, Germany, De Gruyter: 277-297.

PEÑA-SANTIAGO R, 2020. On the identity of some *Dorylaims* (Dorylaimida) recently described from China [J]. Nematology, 22: 957-960.

PEÑA-SANTIAGO R, 2021. Morphology and Bionomics of *Dorylaims* (Nematoda: Dorylaimida) [M]. Leiden: The Netherlands, Brill: 278.

PEÑA-SANTIAGO R, ÁLVAREZ-ORTEGA S, 2014. An integrative approach to assess the phylogeny and the systematics of rounded-tailed genera of the subfamily Qudsianematinae (Nematoda, Dorylaimida) [J]. Zoologica Scripta, 43: 418-428.

PORAZINSKA D L, GIBLIN-DAVIS R M, FALLER L, *et al.* 2009. Evaluating high-throughput sequencing as a method for metagenomic analysis of nematode diversity [J]. Molecular Ecolology Resources, 9: 1439-1450.

POSTMA-BLAAUW M B, DE GOEDE R G M, BLOEM J, *et al.*, 2012. Agricultural intensification and de-intensification differentially affect taxonomic diversity of predatory mites, earthworms, enchytraeids, nematodes and bacteria [J]. Applied Soil Ecology, 57: 32-49.

QING X, WANG M, KARSSEN G, *et al.*, 2020. PPNID: a reference database and molecular identification pipeline for plant-parasitic nematodes [J]. Bioinformatics, 36: 1052-1056.

QING X, WANG Y H, LU X Q, *et al.*, 2022. NemaRec: a deep learning-based web application for nematode image identification and ecological indices calculation [J]. European Journal of Soil Biology, 110: 103408.

RAHM P G, 1937. Freilebende nematoden von Yan-Chia-Ping-Tal (Nordchina) [J]. Zoologischer Anzeiger, 119: 87-97.

RAHM P G, 1938. Freilebende und saprophytische Nematoden der Insel Hainan [J]. Annotations Zoologicae Japonenses, 17: 646-667.

RUSSELL C C, 1986. The feeding habits of a species of *Mesodorylaimus* [J]. Journal of Nematology, 18: 641. (Abstract)

SHAFQAT S, BILGRAMI A L, JAIRAJPURI M S, 1987. Evaluation of the predatory behaviour of *Dorylaimus stagnalis* Dujardin, 1845 (Nematoda: Dorylaimida) [J]. Révue de Nematologie, 10: 455-461.

SIDDIQI M R, 1983. Phylogenetic relationships of the soil nematode orders Dorylaimida, Mononchida, Triplonchida and Alaimida, with a revised classification of the subclass Enoplia [J]. Pakistan Journal of Nematology, 26: 79-110.

SMALL R W, EVANS A A F, 1981. Experiments on the population growth of the predatory nematode *Prionchulus punctatus* in laboratory culture with observations on life history [J]. Révue de Nematologie, 4: 261-270.

SMALL R W, GROOTAERT P, 1983. Observations on the predation abilities of some soil-dwelling predatory nematodes [J]. Nematologica, 29: 109-118.

SUN S, ZENG Q L, QING X, *et al.*, 2023. Description of *Paravulvus zhongshanensis* sp. nov. (Dorylaimida: Nygolaimidae) from Nanjing, China [J]. Journal of Helminthology, 97: e19.

TENUTA M, FERRIS H, 2004. Relationship between nematode life-history classify cation and sensitivity to stressors: ionic and osmotic effects of nitrogenous solutions [J]. Journal of Nematology, 36: 85-94.

THORNE G, 1928. Nematodes inhabiting the cysts of the sugar beet nematode (*Heterodera schachtii*) (Schmidt) [J]. Journal of Agricultural Research, 37: 571-575.

THORNE G, 1934. The classification of the higher groups of Dorylaims [J]. Proceedings of the Helminthological Society of Washington, 1: 19.

THORNE G, 1935. Notes on free-living and plant-parasitic nematodes. II-Higher classification groups of Dorylaimoidea [J]. Proceedings of the Helminthological Society of Washington, 2: 96-98.

THORNE G, 1939. A monograph of the nematodes of the superfamily Dorylaimoidea [J]. Capita Zoologica, 8: 1-261.

THORNE G, SWANGER H H, 1936. A monograph of the nematode genera *Dorylaimus Dujardin*, *Aporcelaimus* n. g., *Dorylaimoides* n. g. and *Pungentus* n. g. [J]. Capita Zooloogica, 6: 1-223.

TODD T C, POWERS T O, MULLIN P G, 2006. Sentinel nematodes of land-use change and restoration in tallgrass prairie [J]. Journal of Nematology, 38: 20-27.

VAN MEGEN H, VAN DEN ELSEN S, HOLTERMAN M, *et al.*, 2009. A phylogenetic tree of nematodes based on about 1200 full-length small subunit ribosomal DNA sequences [J]. Nematology, 11: 927-950.

VINCIGUERRAM T, 1987. A new classification of Actinolaimoidea (Thorne, 1939) using a cladistic approach [J]. Nematologica, 33: 251-277.

WALL D H, VIRGINIA R A, 1999. Controls on soil biodiversity: insights from extreme environments [J]. Applied Soil Ecology, 13: 137-150.

WANG K H, MYERS R Y, SRIVASTAVA A, *et al.*, 2015. Evaluating the predatory potential of carnivorous nematodes against *Rotylenchulus reniformis* and *Meloidogyne incognita* [J]. Biological Control, 88: 54-60.

WASILEWSKA L, 1997. Soil invertebrates as bioindicators, with special reference to soil-inhabiting nematodes [J]. Russian Journal of Nematology, 5: 113-126.

WEBSTER P J, 1972. Response of the tropical atmosphere to local, steady forcing [J]. Monthly Weather Review, 100: 518-541.

WOOD F H, 1974. Biology of *Seinura demani* (Nematoda: Aphelenchoididae) [J]. Nematologica, 20: 347-353.

WU H W, HOEPPLI R J C, 1929. Free-living nematodes from Fookien and Chekiang [J]. Archiv für Schiffs-und Tropenhygiee, 33: 35-43.

WU J H, AHMAD W, 1998. *Mesodorylaimus chinensis* n. sp. (Nematode: Dorylaimida) from China [J]. International Journal of Nematology, 8: 123-125.

WU J H, LIANG Y L, 1999. Two new species of Actinolaimidae Thorne, 1939 (Nemata: Dorylaimida) from China [J]. Journal of Nematology, 31: 475-481.

WU W J, YAN L, XU C L, *et al.*, 2016a. A new species of the genus *Discolaimus* Cobb, 1913 (Nematoda: Dorylaimida: Qudsianematidae) from Qinghai, China [J]. Zootaxa, 4088: 129-138.

WU W J, YAN L, XU C L, *et al.*, 2016b. Morphology and morphometrics of *Heterodorus qinghaiensis* n. sp. (Dorylaimida, Nordiidae) from soil samples in China [J]. Journal of Helminthology, 90: 385-391.

WU W J, HUANG X, XIE H, *et al.*, 2017a. Morphometrics and molecular analysis of the free-living nematode, *Belondira bagongshanensis* n. sp. (Dorylaimida, Belondiridae), from China [J]. Journal of Helminthology, 91: 7-13.

WU W J, YAN L, XIE H, *et al.*, 2017b. Morphology of *Labronemella major* n. sp. (Nematoda: Dorylaimida), a soil-dwelling nematode from China, including a revised key to species of the genus [J]. Journal of Helminthology, 91: 80-86.

WU W J, YU L, XIE H, *et al.*, 2018a. Description and molecular analysis of *Tylencholaimus helanensis* n. sp. from China (Dorylaimida, Tylencholaimidea) [J]. Zookeys, 792: 1-14.

WU W J, YU L, XU C L, *et al.*, 2018b. A new species of the genus *Eudorylaimus* Andrάssy, 1959 (Nematoda: Dorylaimida: Qudsianematidae) associated with *Picea crassifolia* in China [J]. Zootaxa, 4526: 576-588.

WU W J, XU C L, XIE H, *et al.*, 2019. Three new species, one new genus and subfamily of Dorylaimida (de Man, 1876) Pearse, 1942, and revisions on the families Tylencholaimellidae Jairajpuri, 1964 and Mydonomidae Thorne, 1964 (Nematoda: Dorylaimida) [J]. Peer J, 7: e7541.

WYSS U, GROOTAERT P, 1977. Feeding mechanism of *Labronewza vulvapapillatum* [J]. Mededelingen Faculteit Landbouwwoheschool Rijksuniversiteit Gent, 42: 1521-1527.

YEATES G W, 2003. Nematodes as soil indicators: functional and biodiversity aspects [J]. Biology and Fertility of Soils, 37: 199-210.

YEATES G W, WARDLE DA, 1996. Nematodes as predators and prey: relationships to biological control and soil processes [J]. Pedobiologia, 40: 43-50.

YEATES G W, BONGERS T, GOEDE R G D, *et al.* , 1993. Feeding habits in soil nematode families and genera-An outline for soil ecologists [J]. Journal of Nematology, 25: 315-319.

YEATES G W, FERRIS H, MOENS T, *et al.*, 2009. The role of nematodes in ecosystems. In: Wilson M J, Kakouli-Duarte T (Eds), Nematodes as Environmental Indicators [M]. Wallingford, UK, CAB International: 1-44.

ZHANG M, AHAD S, BANIYAMUDDIN M, *et al.*, 2012. A new and three known species of the genus *Tylencholaimellus* Cobb in MV Cobb, 1915 (Nematoda: Dorylaimida) from Changbai Mountain, China [J]. Zootaxa, 3499: 46-62.

ZHANG Q, JI H, GUO F, *et al.*, 2023. Morphological and molecular characterisation of *Trachactinolaimus nanjingensis* n. sp. (Dorylaimida: Actinolaimidae) from Nanjing, China [J]. Nematology. DOI: 10. 1163/15685411-bja10222.